Genomic Imprinting and Uniparental Disomy in Medicine

Genomic Imprinting and Uniparental Disomy in Medicine

Clinical and Molecular Aspects

Eric Engel
Stylianos E. Antonarakis
Division of Medical Genetics, University of Geneva Medical School

With Contributions by

Célia D. Delozier-Blanchet
and Robert Lyle

WILEY-LISS

A JOHN WILEY & SONS, INC., PUBLICATION

This book is printed on acid-free paper. ∞

Published simultaneously in Canada.

For ordering and customer service information please call 1-800-CALL-WILEY.

Library of Congress Cataloging-in-Publication Data:

Engel, Eric, Dr.

Uniparental disomy and genomic imprinting in medicine : clinical and molecular aspects/Eric Engel, Stylianos E. Antonarakis; with contributions by Célia D. Delozier-Blanchet and Robert Lyle.

p. ; cm.

Includes bibliographical references and index.

ISBN 0-471-35126-1 (cloth : alk. paper)

1. Genetic disorders. 2. Genomic imprinting. I. Antonarakis, Stylianos E. II. Title.

[DNLM: 1. Genomic imprinting. 2. Chromosome Abnormalities. QH 450 E57u 2002]

RB155.5 .E54 2002

616′.042–dc21

2001026921

Printed in the United States of America.

10 9 8 7 6 5 4 3 2 1

Contents

Foreword

In 1980, Eric Engel proposed a new concept in clinical genetics—uniparental disomy (*Am J Med Genet* **6**:137). From the known information regarding chromosomal abnormalities and spontaneous abortions, he emphasized that it was "statistically likely and foreseeable" that uniparental disomy (UPD) would occur from the union of a disomic gamete with a gamete nullisomic for the same chromosome. He correctly anticipated the potential for male-to-male transmission of an X-linked trait and the likelihood of homozygosity for recessive mutations by virtue of a common origin for part or all of two chromosomal homologs. His ability to anticipate the occurrence of UPD was remarkable. We now know that trisomy or monosomy rescue are also important mechanisms for the occurrence of UPD.

Also in 1980, there were two reports (*Clin Genet* **17**:418 and 18:456) of the transmission of a translocation (22q;22q) from a mother to a normal daughter. In both cases, it was hypothesized that the normal daughter had either resulted from a trisomic zygote with rescue or from gametic complementation through fertilization of a disomic oocyte by a nullisomic sperm. Clearly these reports represented UPD in the context of an abnormal karyotype. In 1984, there was a report (*Am J Hum Genet* **36**:123) of four cases of mosaic Down syndrome; it was concluded that in at least two and possibly three of these individuals, the diploid cell lines represented UPD for chromosome 21. These could be described as a mosaic trisomy rescue. UPD with a normal chromosome analysis was given little attention until we had the remarkable opportunity to see a patient with cystic fibrosis and maternal UPD for chromosome 7 (*Am J Hum Genet* **42**:217, 1988).

By now it is clear that our report of the first documented case of uniparental disomy with a normal karyotype in 1988, while searching for something entirely different, has proven to be a career-altering event for me. The story begins at a North

American Cystic Fibrosis Conference in what must have been Fall of 1986. The CF gene had been mapped to chromosome 7, and a small group of human geneticists were providing didactic sessions on the new potential to use linkage analysis and linkage disequilibrium data for prenatal diagnosis and risk assessment. I asked the large gathering of clinicians and caregivers for help in identifying any patients with cystic fibrosis and other clinical abnormalities. The plan was to find a patient with a small deletion that would allow our research group to be the first to identify the CF gene. This was not to be, but Dr. Ronald Perciaccante told me of a patient of his who had been reported in 1980, the year of Eric Engel's proposal for UPD, to have short stature and growth hormone deficiency (*Am J Dis Child* **134**:317). I was swift to visit Dr. Perciaccante in upstate New York near the Canadian border where he hosted me in his home. At his office, he introduced me to a young woman with CF and her father; unfortunately her mother was deceased.

I enthusiastically collected blood and rushed to Houston where Ed Spence, a clinical fellow in the lab, performed Southern blotting and found that the patient was homozygous or hemizygous for more than one RFLP while her father was apparently homozygous for the corresponding alternate allele. Given a normal karyotype performed in David Ledbetter's laboratory, two explanations seemed most plausible. Either the patient had a *de novo* or inherited deletion that might help lead us to the CF gene, or her father was not her biological parent. Dr. Perciaccante avowed that nonpaternity was implausible, and DNA markers, other than those on chromosomes 7, proved him correct. The deletion hypothesis was suspect from the beginning because the dosage, as crudely assessed on Southern blots, did not appear consistent with haploidy for the markers. It is difficult to recall the exact circumstances when the correct answer fell into place, but a collaboration with Hunt Willard for analysis of centromeric alphoid repeat polymorphisms was a turning point. The data strongly suggested that the patient had inherited two identical copies of the centromeric region of chromosome 7 from her maternal grandmother.

We encountered many skeptics, but three aspects of the literature at that time were remarkable. First, Eric Engel had wrestled with the possibility of UPD in humans as a theoretical topic some seven or more years earlier, and he had laid down the concepts and nosology for us. Second, there were published cases of presumptive UPD in the context of an abnormal karyotpe as noted above. Third, there was extensive relevant and reassuring information becoming available in mouse genetics. The occurrence of UPD in the context of a normal karyotype gave an inkling of the various forms of unusual or nontraditional inheritance that would grow to include occult gonadal mosaicism for mutations, anticipation in triplet repeat disorders, and others.

Our report was surrounded and followed by a few low points including a rejection letter from *Science*, a commentary suggesting that UPD was unlikely to be anything more than an interesting rarity, and our failure to recognize the potential for UPD to cause Prader-Willi syndrome. Although David Ledbetter and I were regularly discussing UPD and the possible etiologies for the nondeletion cases of PWS, we left it to Rob Nicholls to put two and two together in 1989 (*Nature* **342**:281).

There have been compensatory highs, and the opportunity to comment on this outstanding contribution by Eric Engel and Stylios Antonarakis is one of the most pleasant. As they so thoroughly document, UPD is now recognized as a clinically relevant occurrence that can affect the majority of human chromosomes, and maybe even all the chromosomes if one considers the possibility of UPD causing spontaneous abortions. The authors complement each other well, and they represent the lineage of human genetics in Geneva. *Genomic Imprinting and Uniparental Disomy in Medicine* will stand as a landmark contribution in biology and medicine. The book thoroughly documents the early years of the recognition of the importance of genomic imprinting and uniparental disomy in human biology and medicine. The book focuses on UPD but the ties to various other aspects of genomic imprinting are inevitable.

UPD has served as the clinical entry point into genomic imprinting in humans. Where is the field of human epigenetics today and where is it headed? Having just returned from two Gordon Conferences on human genetics and epigenetics, I am exhilarated by the prospect that we have barely scratched the surface of the importance of genomic imprinting in human variation and in evolution more generally. There are hints that autism might soon merit a chapter in future books reviewing imprinting, since maternal but not paternal duplications of chromosome 15q11-q13 cause the condition (*Am J Hum Genet* **60**:928, 1997). Since genes affecting behavior seem to be imprinted in some cases, perhaps nontraditional inheritance has a major role to play in the etiology of psychiatric disease. There are suggestions that "genomic imprinting may have facilitated a rapid non-linear expansion of the brain, especially the cortex, during development over evolutionary time" (*Brain Res Dev Brain Res* **92**:91, 1996). While UPD is just one part of the story of genomic imprinting and epigenetics in humans, the broader view suggests that there are many important discoveries yet to be made. The comprehensive and authoritative character of this volume lays down the gauntlet for future editions of this or similar texts. I expect that it will soon be inconceivable that one could undertake to cover all of genomic imprinting as it relates to medicine in a single book.

ARTHUR L. BEAUDET
Baylor College of Medicine

Preface

In the field of human development, disomies occasionally arising from a single parent have their ugly side as well as some fascinating aspects. They challenge the traditional path of normal chromosome transmission through various mechanisms, among them: pathologically altering parental gene expression in an offspring, preventing the ill-effects of a trisomy or monosomy by "rescuing" them, and bringing about recessive traits or taking part in an oncogenic process. As such, these peculiar disomies have helped us to understand some new rules in the ever-evolving field of human inheritance. It is the purpose of this book to illustrate the various facets of this major but, until recently, unknown chromosomal aberration.

The emphasis of this book is on the clinical manifestations of known phenotypes related to uniparental disomy or genomic imprinting. It is likely, however, that not every clinical phenotype related to these two conditions is known, and that others will be added as our knowledge increases. The field is in full development and evolution, with a more complete understanding of molecular mechanisms and numerous other unanswered questions still to come. Moreover, molecular mechanisms presented herein may be modified as investigators continue to uncover more mysteries. It would give us pleasure if some readers become themselves contributors in the elucidation of the molecular pathophysiology of disorders related to uniparental disomy and genomic imprinting.

The authors are in different ways indebted to colleagues and friends whose inspiration, help, or participation played a decisive role in their work. It is E.E.'s utmost pleasure to acknowledge the people, who made it possible to establish and render credible the once far-fetched concept of UPD. John Opitz, some 20 years ago, called attention to it by publishing an article in the *Journal of Medical Genetics*, the budding idea of which was then based on purely circumstantial evidence derived from the high frequency of gametal aneuploidy in humans. This article, entitled "A new genetic concept: uniparental disomy and its potential effect, isodisomy," was planted into a risky and unbroken speculative ground. That seed only came to fruition when Arthur Beaudet and his colleagues gave it, in 1988, clinical life, by uncovering and describing the first clinical story of a disease due to uniparental disomy for the maternal chromosome 7. Tribute is also paid to the Wiley Editor of this book for advising its initiator (E.E.) to team with a molecular geneticist to carry out the task of writing it.

It was S.E.A. who, in the midst of so many major endeavors, enthusiastically took on this task and invigorated and endowed our work with his command of the molecular biology and the power of his brilliant intellect. Then, others joined the undertaking when the seas were rough and helped to bring it to shore. Celia Dawn Delozier and Robert Lyle are foremost among them, in contributing one chapter each and giving skilled advice.

One also thinks of other friends and colleagues who, without being directly involved, personally influenced the thinking of the authors or the progress of the field, a few of whom only can be named: Drs. Suzanne B. Cassidy, Patricia Jacobs, Rowena James, I. Karen Temple, Joan Knoll, Susan Malcolm, Wendy Robinson, Lisa Shaeffer, Karen Buiting, G. Gillessen-Kaesbach, Ellen Magenis, Susan L. Christian, Claudine Junien, David L. Ledbetter, Albert Schinzel, Robert Nicholls, Daniel Driscoll, Kurt Hirschorn, Bernard Horsthemke, and closer to us, in Geneva, Siv Fokstuen and Armand Bottani. While many others could be named, those who most deserve E.E.'s final words are all the patients, wherever in the world, their parents and caretakers without whom medical history could not be written by their many humble and compassionate medical observers.

It was S.E.A.'s great pleasure to have the opportunity to work with E.E. on this project. When Eric asked me to share the responsibility for the development of this book, I saw it as an opportunity to learn from him and as an historical necessity. Professor Engel is my predecessor in the Chair of Medical Genetics of the University of Geneva Medical School (he himself took the torch from Professor David Klein). What a better way to establish a long-term link and friendship than by agonizing over the various chapters of a book together? Another attraction for participating in this effort was that it provided me with an opportunity to learn more about a fascinating truth that every toddler knows all too well: the father and the mother are not indeed the same, and that both are needed for a healthy development. As I consider myself an eternal student, the pleasure was substantial and continuous: every paper mentioned in the book was an eye opener and teaching tool. My thanks therefore naturally go to Eric Engel for his generous offer to share his exciting obsession, to all the clinical and basic investigators for all the new information they offer to their colleagues, readers, medical caretakers and patients, to all the members of the Division of Medical Genetics for their help, to Drs. C. DeLozier and R. Lyle for their insightful contributions to two chapters, and to all the patients and their families for their confidence in our ignorance and assistance in our curiosity. We also thank the funding agencies, most notably the Swiss National Science Foundation for supporting our research projects. We would also wish to thank Mrs. Georgette Chapuis, Tu-Mai Besson, and Katia Casada for their secretarial assistance, and Luna Han and Collette Bean of John Wiley & Sons for the expert editorial assistance.

ERIC ENGEL
STYLIANOS E. ANTONARAKIS

Acknowledgments

Eric Engel is most grateful to Oliver P. Engel, Ph.D. electronic engineer, for his consistent help and warm encouragements in preparing the first draft of this manuscript.

Stylianos E. Antonarakis thanks all members of this pedigree for the genes and life sharing.

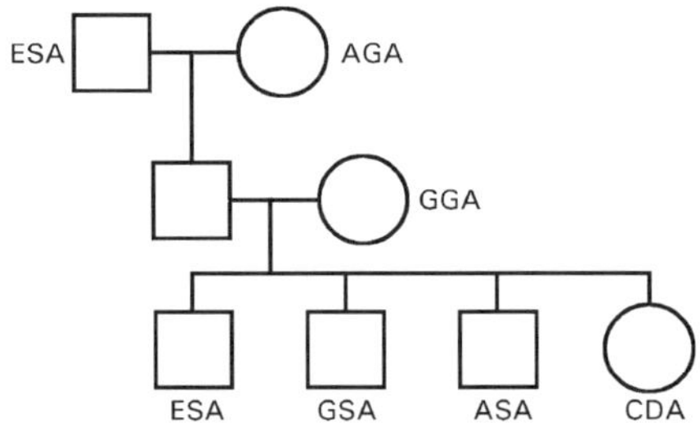

Genomic Imprinting and Uniparental Disomy in Medicine

Chapter 1

Introduction

In humans, after the two meiotic divisions, only one member of each of the 23 chromosomal pairs is normally included in each germ cell. As a result, copies of half of the chromosomes and one copy of each gene from a parent-to-be are included in each reproductive cell. On the basis of random chromosomal assortment alone, millions (2^{23}) of combinations are possible. This enormous diversity of the heritable parental chromosomal material becomes even greater when crossing-over between homologous chromatids at the first meiotic division (Me1) is considered. The latter process further enhances the reshuffling of the chromosomal pieces that are included in the gametes into billions of possible combinations.

The acquisition of such genetic variation is carried out through complex cytological mechanisms, themselves liable to error. Most chromosome anomalies present at conception arise through meiotic errors that modify the number of chromosomes segregating into germ cells, the result being either a supernumerary member (disomy), or a missing one (nullisomy). Upon fertilization, such gametes lead to zygotes that are, respectively, trisomic or monosomic for that particular chromosome.

DEVELOPMENT OF THE CONCEPTS OF UNIPARENTAL DISOMY AND ISODISOMY

The systematic study of large series of conceptuses, resulting in either liveborns or spontaneous abortions, has revealed a high frequency of numerical chromosomal abnormalities, most of them lethal. Trisomies, in particular, have been observed for almost all human chromosomes (Kajii et al., 1973; Hassold et al., 1978; Hassold, 1980). The high rate of human aneuploidy, apparently resulting from the high incidence of abnormal gametes, suggested the theoretical possibility that some apparently normal diploid (2n) genomes occurred through "gamete complementation"; by this process a chromosomal pair would be derived from only one parent, by the chance meeting of two abnormal gametes, coincidentally disomic and nullisomic for a same chromosome. This theoretical possibility was proposed in 1980 and given the name "uniparental disomy" (UPD) (Engel, 1980). UPD is thus a situation in which the chromosome number remains normal and diploid, even though one of the chromosomal pairs derives from only a single parent instead of the usual biparental origin. The original article included sample calculations of the potential frequency of UPD, based on a 20% spontaneous abortion rate and the observed incidence of aneuploidy for chromosomes 15, 16, 21, 22, and the sex chromosomes in first-trimester miscarriages; as many as 3 individuals out of 10,000 might be uniparentally disomic for one of those four autosomes or for the maternal X, paternal X or the XY pair (Engel, 1980).

Considering this possibility, the next step was to foresee, in theory at least, the potential consequences of such an event. Among unlikely outcomes, one was the transmission of the XY pair from father to son, infringing on the time-honored rule of no male-to-male transmission of X-linked genes! But, a more usual consequence of UPD might have been the duplication of recessive alleles, causing recessive diseases, the occurrence of which would depend on the type of nondisjunction and recombination involved prior to the production of UPD. To this theoretical possibility was assigned the name "isodisomy" (Engel, 1980). Isodisomy can be defined as the inheritance of two identical copies of a chromosome (or a chromosomal segment or a gene) from one parent. This is in contrast to "heterodisomy," the inheritance of two different copies of a chromosome (or a chromosome segment or a gene) from a parent (Voss et al., 1989; Knoll et al., 1990).

THE MECHANISMS AND CONSEQUENCES OF UPD

As of December 2000, UPD has been documented for chromosomes 1, 2, 4, 5, 6, 7, 8, 9, 10, 11, 13, 14, 15 ,16, 17, 20, 21, 22, X, and the XY pair (Figure 1 in Chapter 4). In theory, maternal or paternal UPD could exist for all chromosomes; lethality may, however, be the major limiting factor for observing UPD for some chromosomes, as exemplified by complementation studies in mouse experiments in which some uniparental combinations are nonviable (Searle and Beechey, 1978, 1985; Cattanach and Kirk, 1985; Takagi and Sasaki, 1975).

The mechanisms resulting in UPD are multiple. Paradoxically, the best documented one generating UPD arises from the loss of a chromosome from an initial trisomy, the so-called "trisomy rescue" (Spence et al., 1988; Cassidy et al., 1992; Purvis-Smith et al., 1992). Such reduction of a trisomy to a disomy represents the concomittence of two errors, one meiotic, leading to the trisomic state after fertilization by a normal gamete, the other mitotic, removing the extra chromosome by nondisjunction or anaphase lag. In the case of nondisjunction of a trisomic pair, the two resulting cells of a zygote will be, respectively, disomic and tetrasomic, the latter generally lethal, at least for most autosomes. In the case of chromosomal lag, one daughter cell will remain trisomic (unchanged) with emergence of a new disomic type, the source of a potentially viable mosaicism.

The word "rescue" in such cases may be too optimistic, as the salvage may at times be as bad as the initial woe. For example, the loss of one member of a trisomy, most likely to be postzygotic, may result in either a (wrong) uniparental or a normal biparental pair. In either case, a residual trisomic cell line may persist, hindering embryonic or extra-embryonic development, or both. Such a rescue may also circumvent the natural process of lethal selection, thus allowing as the case may be, survival of a liveborn with a severe recessive disorder or a grave imprinting pathology (see below). Trisomy rescue as a cause of UPD should contribute primarily to cases of maternal origin, since most identified segregational errors have occurred in the course of oogenesis (Hassold and Jacobs, 1984; Jacobs and Hassold, 1995; Antonarakis, 1991).

A bias also exists favoring the recognition of trisomy rescue as a cause of UPD, since the clues suggesting such an event, i.e., detection of a trisomic component amidst diploid cells (often at chorionic villus sampling, CVS, and more rarely at amniocentesis, AC), are so clearly suggestive of this process (Cassidy et al., 1992; Purvis-Smith et al., 1992).

The mechanism of UPD by "gamete complementation" also implies two cytologenetic errors: a nullisomy and a disomy. These meiotic errors, one occurring in each sex, coincidentally "correct" each other at fertilization without leaving behind a cytologic clue! Gamete complementation was the original argument to build the case for UPD in the 1980 paper (Engel, 1980). The premises were that nullisomic gametes should be abundant, at least for chromosomes 15, 16, 21, 22, and the X, since their disomic counterparts, as judged from so many aborted autosomal trisomies (and X monosomies), appeared in themselves to be so numerous (Kurnit et al., 1978; de la Chapelle et al., 1973; Stephens, 1989; Kajii et al., 1973; Hassold et al., 1978). Consequently, gametes with two or with none of the same chromosome were bound to occasionally meet in the same zygote at fertilization, on the condition that both sexes nurtured a similar type and rate of gametic aneuploidy, which is of course not true.

As will be discussed later, other cytologic causes for UPD have been described, such as the rescue of a monosomy, which may occur through mitotic nondisjunctional duplication of a singly inherited chromosome (Spence et al., 1988); somatic recombination (i.e., somatic crossing-over, the symmetrical trading of a paternal and maternal homologous chromatid segment) may also be a source of segregants producing cell lines with segmental UPD, as seen in some syndromes (e.g., Beckwith-Wiedemann) and tumors (Henry et al., 1991; Junien, 1992).

ISODISOMY AND HETERODISOMY*

The criteria for inclusion of identical, purely homozygous, isodisomic material in a uniparental chromosomal pair or segment are met when the chromosomal segments are copies of each other. This is best illustrated when monosomy gives rise to a

*See also Figure 1a, Chapter 3.

diploid cell line through nondisjunction of the two chromatids of the solitary chromosome, or in the case of acrocentrics, through misdivision of the centromere (Freeman et al., 1993; Fridman et al., 1998) creating an isochromosome with two identical long-arm copies. Isodisomy can also be achieved through nondisjunction at meiosis 2, whenever the chromatids have remained achiasmatic, i.e., not subject to the normal process of crossing-over at Me1. In heterodisomy, two homologous but nonidentical members of a parental pair could result from Me1 nonsegregation. In such cases, the centromeric areas of these homologues are always spared by crossing-over and thus retain their heterozygosity. The rest of these chromosomes could have regions of heterozygosity or homozygosity, depending on the number and location of cross-overs.

Given crossing-over at Me1, the nondisjoined pair usually contains both isodisomic and heterodisomic segments, but what surely persists in such pairs is centromeric and juxta-centromeric heterozygosity (i.e., heterodisomy), the hallmark of an Me1 nondisjunction. By contrast, these centromeric segments of homologous chromosomes after a Me2 (meiosis 2) nondisjunction are isodisomic and, in turn, constitute the hallmark of a Me2 nondisjunction.

So far, more than 20 different disorders have, one or more times, resulted from isodisomy of a uniparental pair carrying a recessive mutant allele (Engel, 1993, 1996 (see also Table 1, Chapter 4). These include, among others, cystic fibrosis (Spence et al., 1988; Voss et al., 1989), beta thalassemia (Beldjord et al., 1992), and the cartilage-hair syndrome (Sulisalo et al., 1994).

IMPRINTING AND UPD IN MAMMALS

Another consequence of UPD, not initially considered in humans (Engel, 1980), was the interference with imprinting mechanisms. UPD, by switching the parental source of one chromosome, may lead to imbalance for certain alleles whose expression in an offspring normally depends on the sex of the parent-of-origin. By 1980, when the UPD concept emerged, research in mice had already indicated the existence of parent-of-origin-dependent gene expression, underscoring the importance of genomic imprinting. As reviewed by Sapienza (1989), the word imprinting was initially applied by Crouse to describe the selective elimination of paternal chromosomes from somatic and germline nuclei of the fly Sciara (Crouse, 1960). In mammals, the display of "functional imprints" was to include, for instance, selective inactivation of the paternal X chromosome in extra-embryonic membranes, already reported in the mouse in the mid-1970s (Takagi and Sasaki, 1975; Lyon and Rastan, 1984).

About this time, evidence that parent-of-origin differential gene expression may also concern autosomes appeared in work such as Johnson's (1974), who reported that a mouse mutation (the hair-pin tail T^{hp}) was viable in the heterozygote when paternally derived, but lethal if maternally derived. Such a parent-of-origin effect on differential gene expression was to become more and more evident in ingenious mouse complementation studies (Searle and Beechey, 1978, 1985) in which heterozygotes for reciprocal or Robertsonian translocations were mated. In such matings, the fusion of complementary unbalanced gametes produced chromosomally balanced zygotes, fully viable for some types of chromosome rearrangements but failing to appear or to survive for others (Searle and Beechey, 1978). Complementation maps were constructed (Ferguson-Smith et al., 1991), based on phenotypic observations in cases of maternal duplication, "correcting" for a corresponding paternal deficiency, and vice-versa. From a historical standpoint, it was George D. Snell (1946) who, 30 years earlier reported the unexpected absence of complementation product from intercrosses of heterozygotes for a translocation. This observation, involving mouse chromosome 7, provided the first example of an imprinting effect in mammals, which was later documented with greater sophistication for that same chromosome (Ferguson-Smith et al., 1991). This kind of work and other multidisciplinary investigations into imprinting have been diligently pursued in the last 20 years (McGrath and Solter, 1984; Surani et al., 1984, 1990; Cattanach and Kirk, 1985; Takagi and Sasaki, 1975; Reik et al., 1987; Solter, 1988; Cattanach and Beechey, 1990; Sapienza, 1990; Barton et al., 1991; Surani, 1991; Cattanach et al., 1992; Cattanach and Jones, 1994) (see Chapters 2 and 10).

Thus by 1979, as the concept of UPD was being mulled over (Engel, 1980), the scenario was already being written in "mouse language": the equivalent of the high spontaneous rate of human gamete aneuploidy, liable to complementations, was being artificially produced in rodents by mating animals with known translocations, whose meiotic segregation generated numerous unbalanced, complementary duplication-deficiency gametes. Intercrosses of the animals were producing offspring complemented through disomy of uniparental segments, the latter identified by homozygosity (isodisomy!) for recessive genes chosen to mark and track the parental source of the genes that remained diploid. Therefore, the understanding of the mouse imprinting was in 1979 much ahead of the human hypotheses. As briefly surveyed above (see also Chapter 10), complementation in mice, resulting in "orthoploid," i.e., balanced offspring with isodisomy, had existed "in the flesh" for some years, and the suggestion of parent-of-origin effects on gene expression was becoming accepted in mice; this idea was still in limbo in the human field, and certainly not predicted in the 1980 paper (Engel, 1980).

During the 1980s the mouse work on genomic imprinting and its bearing on normal development made great strides. Of note among investigations were microtransplantation experiments showing that eggs with genomes made of two pronuclei issued from a same gender animal (i.e., androgenetic or gynogenetic) failed to support complete embryogenesis, a result at variance with control eggs made diploid with one pronucleus transplanted from each sex (Surani et al., 1984; McGrath and Solter, 1984; Mann and Lovell-Badge, 1984). In gynogenetic products, however, the embryonic development appeared more normal than did that of the extra-embryonic membranes and trophoblast; in androgenetic conceptuses studied, the opposite was true (Surani et al., 1984; McGrath and Solter, 1984; Solter, 1988). These observations are indeed reminiscent of the human situation in which endogenous androgenotes result in hydatiform moles, whereas gynogenotes cause embryonic teratomas in women (Kajii et al., 1973; Crouse, 1960). The basic experiments using reconstructed mouse eggs pointed to the very fact that the maternal and paternal legacies to the embryonic genome were not equivalent, at least not in all developmental aspects, and that both parental genomes were required for full prenatal development (McGrath and Solter, 1984; Surani et al., 1984). Obviously, parental genomic contributions could not be, at least entirely, experimentally replaced by one sex for the other, in order to achieve normalcy: Some paternally or maternally derived homologous genes simply did not work the same way during the course of differentiation.

What then could alter the function of such inherently symmetrical material, solely on the basis of its gametal origin, male or female? What could impress pairs of alleles, at certain loci, to perform so distinctly from each other, once present in a conceptus? How many such special loci might undergo sex-of-origin functional modifications?

In this respect, attention was being attracted to superimposed allele-specific methylation as a major mechanism of parental imprinting (Cattanach and Beechey, 1990). Transgenes are foreign genes that have been introduced into zygotes or early embryos to become a permanent part of the genome (Reik et al., 1987; Solter, 1988). Those paternally inherited are (with exceptions) markedly undermethylated in all conceptuses, whereas the same transgenes are highly methylated when inherited maternally (Reik et al., 1987). Depending on the parental origin, methylation of a transgene can be reversed when passed from one generation to another. Whereas the above could be an overall pattern of reaction to foreign or misplaced DNA, it appears to closely mimic the pattern observed for endogenous genes naturally subject to imprinting (Solter, 1988).

In contrast to the experimental mouse studies, it took nearly 10 more years to discover the first cases of autosomal UPD in man: maternal isodisomy for chromosome 7 in a patient with cystic fibrosis (Spence et al., 1988), followed by maternal disomy 15 in a subject with Prader-Willi syndrome (PWS), attributed to a disruption of genomic imprinting (Nicholls et al., 1989). For the human sex chromosomes, XY heterodisomy was described shortly afterward in a boy with hemophilia inherited from his father! (Vidaud et al., 1989).

UPD AND IMPRINTING IN HUMANS

In humans, as in other mammals, such modulation of gene activity at specific loci, as a function of the gender of the transmitting parent, occurs also through reversible epigenetic modifications of the nuclear DNA, both in the germline and somatic cells. In the soma, for some cells and tissues at least, epigenetic changes such as specific methylation at some sites occur (Dittrich et al., 1993; Glenn et al., 1993, 1996), which secondarily modify gene expression (Driscoll et al., 1992). In line with what is now recognized and defined as genomic imprinting, the secondary epigenetic molecular inscriptions added to the basic DNA might be designated "epiglyphes." Thus, by definition, genomic imprinting is the differential modification and expression of alleles of a gene, according to the sex of the parent-of-origin. Obviously, the uniparental origin of a chromosomal pair would prevent whatever differential modifications of expression are normally assigned to a locus or loci, since one parent would not contribute a chromosome to that pair.

Thus, the hallmark of UPD for some human chromosomes has been the interference with the process of genomic imprinting. In humans so far, five chromosome pairs (6, 7, 11, 14, 15) (Donlon et al., 1986; de la Chapelle et al., 1973; Buiting et al., 1993; Dittrich et al., 1993) are known to carry imprinted loci, judging from complementation failures leading to abnormal phenotypes; for two more chromosomes (2 and 16), imprinted genes are suspected (see Chapter 4 Figure 1, and Chapter 5). These chromosomes contain a minimum of nine domains of paternal or maternal expression. The most representative human pathologies associated with these imprinted domains are the Prader-Willi syndrome (PWS) due to maternal disomy 15 (Nicholls et al., 1989), the Angelman syndrome (AS) with paternal disomy 15 (Malcolm et al., 1991; Nicholls et al., 1992), and paternal disomy of chromosome 11p in the Beckwith-Wiedemann syndrome (BWS) (Henry et al., 1991). More recent observations (see Chapter 5) have shown that maternal disomy 7 may cause growth retardation, at times associated with the Russell-Silver syndrome (Kotzot et al., 1995); that paternal UPD6 can be responsible for transient neonatal diabetes mellitus (James et al., 1995; Temple et al., 1996); and that maternal UPD14 may result in multiple anomalies, including abnormal growth and premature sexual development (Antonarakis et al., 1993; Healey et al., 1994), whereas paternal UPD14 is associated with a dwarfing condition (Cotter et al., 1997; Wang et al., 1991). So far, imprinted loci or domains seem to exist for three maternally transmitted chromosomes, i.e., 7, 14, and 15 (Donlon et al., 1986; Christian et al., 1999; Mutirangura et al., 1993) and for four paternally inherited chromosomes, i.e., 6, 14, 15, and 11. Most UPD11 occur through somatic recombination of the chromosome pair (Henry et al., 1991).

Other UPDs under scrutiny for potential genomic imprinting effects include maternal UPD2 (Harrison et al., 1995; Hansen et al., 1997) and maternal UPD16 (Kalousek et al., 1993; Vaughan et al., 1994).

The molecular mechanisms or "epiglyphes" conferring genomic imprinting to loci or domains have been explored in some depth for the AS/PWS critical region in 15q11-q13 (see Chapters 6 and 7), and the 11p13-p15.5 areas of chromosome 11 (Chapter 8). The putative role of imprinting control elements and of the so-called imprinting centers (IC) that may control the imprinting mechanisms are under investigation (Mann and Bartolomei, 1999; Jiang et al., 1998; Nicholls et al., 1998).

Thus far, effects associated with UPD have included disorders of genomic imprinting, chromosomal imbalances, recessive disorders through isodisomy, phenocopies of some chromosomal deletions, and a possible contribution to the stepwise development of the oncogenic process (Engel, 1993, 1996).

In the clinical field, UPD needs to be considered in the context of fetal diagnosis and within the framework of genetic counseling (Engel, 1995), particularly when mosaic aneuploidy has been documented or chromosome rearrangements such as translocations or inversions observed. The same applies to etiologic considerations in the development of certain neoplasias (Cavenee 1991; Rousseau-Merck et al., 1999).

In some ways, the story of UPD is a new and exciting chapter in the emerging field of nontraditional inheritance. This book is an attempt at putting these and related problems into perspective, and is intended for clinicians, genetic counsellors, and laboratory scientists dealing with molecular DNA testing in human disease.

REFERENCES

Antonarakis, S. E. Parental origin of the extra chromosome in trisomy 21 as indicated by analysis of DNA polymorphisms. Down Syndrome Collaborative Group. *N Engl J Med* 324:872–876, 1991.

Antonarakis, S. E., Blouin, J. L., Maher, J., Avramopoulos, D., Thomas, G. and Talbot, C.C.J. Maternal uniparental disomy for human chromosome 14, due to loss of a chromosome 14 from somatic cells with t(13;14) trisomy 14. *Am J Hum Genet* 52:1145–1152, 1993.

Barton, S. C., Ferguson-Smith, A. C., Fundele, R. and Surani, M. A. Influence of paternally imprinted genes on development. *Development* 113:679–687, 1991.

Beldjord, C., Henry, I., Bennani, C., Vanhaeke, D. and Labie, D. Uniparental disomy: a novel mechanism for thalassemia major [letter]. *Blood* 80:287–289, 1992.

Buiting, K., Dittrich, B., Gross, S., et al. Molecular definition of the Prader-Willi syndrome chromosome region and orientation of the SNRPN gene. *Hum Mol Genet* 2:1991–1994, 1993.

Cassidy, S. B., Lai, L. W., Erickson, R. P., et al. Trisomy 15 with loss of the paternal 15 as a cause of Prader-Willi syndrome due to maternal disomy. *Am J Hum Genet* 51:701–708, 1992.

Cattanach, B. M., Barr, J.A., Evans, E. P., et al. A candidate mouse model for Prader-Willi syndrome which shows an absence of Snrpn expression. *Nat Genet* 2:270–274, 1992.

Cattanach, B. M. and Beechey, C. V. Autosomal and X-chromosome imprinting. *Dev Suppl* 63–72, 1990.

Cattanach, B. M. and Jones, J. Genetic imprinting in the mouse: implications for gene regulation. *J Inherit Metab Dis* 17:403–420, 1994.

Cattanach, B. M. and Kirk, M. Differential activity of maternally and paternally derived chromosome regions in mice. *Nature* 315:496–498, 1985.

Cavenee, W. K. Recessive mutations in cancer predisposition and progression. *Basic Life Sci* 57:171–180, 1991.

Christian, S. L., Rich, B. H., Loebl, C., et al. Significance of genetic testing for paternal uniparental disomy of chromosome 6 in neonatal diabetes mellitus [see comments]. *J Pediatr* 134:42–46, 1999.

Cotter, P. D., Kaffe, S., McCurdy, L. D., Jhaveri, M., Willner, J.P. and Hirschhorn, K. Paternal uniparental disomy for chromosome 14: a case report and review. *Am J Med Genet* 70:74–79, 1997.

Crouse, H. V. The controlling element in sex chromosome behaviour in Sciara. *Genetics* 45:1429–1443, 1960.

de la Chapelle, A., Schroder, J. and Kokkonen, J. Cytogenetics of recurrent abortion or unsuccessful pregnancy. *Int J Fertil* 18:215–219, 1973.

Dittrich, B., Buiting, K., Gross, S. and Horsthemke, B. Characterization of a methylation imprint in the Prader-Willi syndrome chromosome region. *Hum Mol Genet* 2:1995–1999, 1993.

Donlon, T. A., Lalande, M., Wyman, A., Bruns, G. and Latt, S. A. Isolation of molecular probes associated with the chromosome 15 instability in the Prader-Willi syndrome [published erratum appears in Proc Natl Acad Sci U S A 1986 Sep;83(18):6964]. *Proc Natl Acad Sci U S A* 83:4408–4412, 1986.

Driscoll, D. J., Waters, M. F., Williams, C. A., et al. A DNA methylation imprint, determined by the sex of the parent, distinguishes the Angelman and Prader-Willi syndromes. *Genomics* 13:917–924, 1992.

Engel, E. A new genetic concept: uniparental disomy and its potential effect, isodisomy. *Am J Med Genet* 6:137–143, 1980.

Engel, E. Uniparental disomy revisited: the first twelve years. *Am J Med Genet* 46:670–674, 1993.

Engel, E. [Uniparental disomy: a review of causes and clinical sequelae]. *Ann Genet* 38:113–136, 1995.

Engel, E. Uniparental disomy and genome imprinting: an overview. *Acta Genet Med Gemellol (Roma)* 45:19–39, 1996.

Ferguson-Smith, A. C., Cattanach, B. M., Barton, S. C., Beechey, C. V. and Surani, M. A. Embryological and molecular investigations of parental imprinting on mouse chromosome 7 [see comments]. *Nature* 351:667–670, 1991.

Freeman, S. B., May, K. M., Pettay, D., Fernhoff, P. M. and Hassold, T. J. Paternal uniparental disomy in a child with a balanced 15;15 translocation and Angelman syndrome. *Am. J. Med. Genet* 45:625–630, 1993.

Fridman, C., Varela, M. C., Nicholls, R. D. and Koiffmann, C. P. Unusual clinical features in an Angelman syndrome patient with unparental disomy due to a translocation 15q15q. *Clin Genet* 54:303–308, 1998.

Glenn, C. C., Nicholls, R. D., Robinson, W. P., et al. Modification of 15q11-q13 DNA methylation imprints in unique Angelman and Prader-Willi patients. *Hum Mol Genet* 2:1377–1382, 1993.

Glenn, C. C., Saitoh, S., Jong, M. T., et al. Gene structure, DNA methylation, and imprinted expression of the human SNRPN gene. *Am J Hum Genet* 58:335–346, 1996.

Hansen, W. F., Bernard, L. E., Langlois, S., et al. Maternal uniparental disomy of chromosome 2 and confined placental mosaicism for trisomy 2 in a fetus with intrauterine growth restriction, hypospadias, and oligohydramnios. *Prenat Diagn* 17:443–450, 1997.

Harrison, K., Eisenger, K., Anyane-Yeboa, K. and Brown, S. Maternal uniparental disomy of chromosome 2 in a baby with trisomy 2 mosaicism in amniotic fluid culture. *Am J Med Genet* 58:147–151, 1995.

Hassold, T. J. A cytogenetic study of repeated spontaneous abortions. *Am J Hum Genet* 32:723–730, 1980.

Hassold, T. J. and Jacobs, P. A. Trisomy in man. *Annu Rev Genet* 18:69–97, 1984.

Hassold, T. J., Matsuyama, A., Newlands, I. M., et al. A cytogenetic study of spontaneous abortions in Hawaii. *Ann Hum Genet* 41:443–454, 1978.

Healey, S., Powell, F., Battersby, M., Chenevix-Trench, G. and McGill, J. Distinct phenotype in maternal uniparental disomy of chromosome 14. *Am J Med Genet* 51:147–149, 1994.

Henry, I., Bonaiti-Pellie, C., Chehensse, V., et al. Uniparental paternal disomy in a genetic cancer-predisposing syndrome [see comments]. *Nature* 351:665–667, 1991.

Jacobs, P. A. and Hassold, T. J. The origin of numerical chromosome abnormalities. *Adv Genet* 33:101–133, 1995.

James, R. S., Temple, I. K., Dennis, N. R. and Crolla, J. A. A search for uniparental disomy in carriers of supernumerary marker chromosomes. *Eur J Hum Genet* 3:21–26, 1995.

Jiang, Y., Tsai, T. F., Bressler, J. and Beaudet, A. L. Imprinting in Angelman and Prader-Willi syndromes. *Curr Opin Genet Dev* 8:334–342, 1998.

Johnson, D. R. Hairpin-tail: a case of post-reductional gene action in the mouse egg. *Genetics* 76:795–805, 1974.

Junien, C. Beckwith-Wiedemann syndrome, tumourigenesis and imprinting [published erratum appears in Curr Opin Genet Dev 1992 Aug;2(4):651]. *Curr Opin Genet Dev* 2:431–438, 1992.

Kajii, T., Oama, K., Niikawa, N., Ferrier, A. and Avirachan, S. Banding analysis of abnormal karyotypes in spontaneous abortion. *Am J Hum Genet* 25:539–547, 1973.

Kalousek, D. K., Langlois, S., Barrett, I., et al. Uniparental disomy for chromosome 16 in humans. *Am J Hum Genet* 52:8–16, 1993.

Knoll, J. H., Nicholls, R. D., Magenis, R. E., et al. Angelman syndrome: three molecular classes identified with chromosome 15q11q13-specific DNA markers. *Am J Hum Genet* 47:149–155, 1990.

Kotzot, D., Schmitt, S., Bernasconi, F., et al. Uniparental disomy 7 in Silver-Russell syndrome and primordial growth retardation. *Hum Mol Genet* 4:583–587, 1995.

Kurnit, D. M., Karp, L. and Hoehn, H. Cytogenetics of recurrent abortions [letter]. *Fertil Steril* 30:366–367, 1978.

Lyon, M. F. and Rastan, S. Parental source of chromosome imprinting and its relevance for X chromosome inactivation. *Differentiation* 26:63–67, 1984.

Malcolm, S., Clayton-Smith, J., Nichols, M., et al. Uniparental paternal disomy in Angelman's syndrome. *Lancet* 337:694–697, 1991.

Mann, J. R. and Lovell-Badge, R. H. Inviability of parthenogenones is determined by pronuclei, not egg cytoplasm. *Nature* 310:66–67, 1984.

Mann, M. R. and Bartolomei, M. S. Towards a molecular understanding of Prader-Willi and Angelman syndromes. *Hum Mol Genet* 8:1867–1873, 1999.

McGrath, J. and Solter, D. Completion of mouse embryogenesis requires both the maternal and paternal genomes. *Cell* 37:179–183, 1984.

Mutirangura, A., Jayakumar, A., Sutcliffe, J. S., et al. A complete YAC contig of the Prader-Willi/Angelman chromosome region (15q11-q13) and refined localization of the SNRPN gene. *Genomics* 18:546–552, 1993.

Nicholls, R. D., Knoll, J. H., Butler, M. G., Karam, S. and Lalande, M. Genetic imprinting suggested by maternal heterodisomy in nondeletion Prader-Willi syndrome. *Nature* 342:281–285, 1989.

Nicholls, R. D., Pai, G. S., Gottlieb, W. and Cantu, E. S. Paternal uniparental disomy of chromosome 15 in a child with Angelman syndrome. *Ann Neurol* 32:512–518, 1992.

Nicholls, R. D., Saitoh, S. and Horsthemke, B. Imprinting in Prader-Willi and Angelman syndromes. *Trends Genet* 14:194–200, 1998.

Purvis-Smith, S. G., Saville, T., Manass, S., et al. Uniparental disomy 15 resulting from "correction" of an initial trisomy 15 [letter]. *Am J Hum Genet* 50:1348–1350, 1992.

Reik, W., Collick, A., Norris, M. L., Barton, S. C. and Surani, M. A. Genomic imprinting determines methylation of parental alleles in transgenic mice. *Nature* 328:248–251, 1987.

Rousseau-Merck, M. F., Versteege, I., Legrand, I., et al. hSNF5/INI1 inactivation is mainly associated with homozygous deletions and mitotic recombinations in rhabdoid tumors. *Cancer Res* 59:352–3156, 1999.

Sapienza, C. Genome imprinting and dominance modification. *Ann N Y Acad Sci* 564:24–38, 1989.

Sapienza, C. Parental imprinting of genes. *Sci Am* 263:52–60, 1990.

Searle, A. G. and Beechey, C. V. Complementation studies with mouse translocations. *Cytogenet Cell Genet* 20:282–303, 1978.

Searle, A. G. and Beechey, C. V. Non-complementation phenomena and their bearing on nondisjunctional effects. In: *Aneuploidy: etiology and mechanisms*, edited by Dellarco, V., Voytek, P. and Hollander, P. 1985,

Snell, G. D. An analysis of translocations in the mouse. *Genetics* 31:157–180, 1946.

Solter, D. Differential imprinting and expression of maternal and paternal genomes. *Annu Rev Genet* 22:127–146, 1988.

Spence, J. E., Perciaccante, R. G., Greig, G. M., et al. Uniparental disomy as a mechanism for human genetic disease. *Am J Hum Genet* 42:217–226, 1988.

Stephens, J. D. Karyotypes of 1142 couples with recurrent abortion [letter; comment]. *Obstet Gynecol* 73:144–145, 1989.

Sulisalo, T., Francomano, C. A., Sistonen, P., et al. High-resolution genetic mapping of the cartilage-hair hypoplasia (CHH) gene in Amish and Finnish families. *Genomics* 20:347–353, 1994.

Surani, M. A. Genomic imprinting: developmental significance and molecular mechanism. *Curr Opin Genet Dev* 1:241–246, 1991.

Surani, M. A., Barton, S. C. and Norris, M. L. Development of reconstituted mouse eggs suggests imprinting of the genome during gametogenesis. *Nature* 308:548–550, 1984.

Surani, M. A., Kothary, R., Allen, N. D., et al. Genome imprinting and development in the mouse. *Dev Suppl* 89–98, 1990.

Takagi, N. and Sasaki, M. Preferential inactivation of the paternally derived X chromosome in the extraembryonic membranes of the mouse. *Nature* 256:640–642, 1975.

Temple, I. K., Gardner, R. J., Robinson, D. O., et al. Further evidence for an imprinted gene for neonatal diabetes localised to chromosome 6q22-q23. *Hum Mol Genet* 5:1117–1121, 1996.

Vaughan, J., Ali, Z., Bower, S., Bennett, P., Chard, T. and Moore, G. Human maternal uniparental disomy for chromosome 16 and fetal development. *Prenat Diagn* 14:751–756, 1994.

Vidaud, D., Vidaud, M., Plassa, F., Gazengel, C., Noel, B. and Goossens, M. Father to son transmission of hemophilia A due to uniparental disomy. *Am J Hum Genet* 59:A2261989.(Abstract)

Voss, R., Ben-Simon, E., Avital, A., et al. Isodisomy of chromosome 7 in a patient with cystic fibrosis: could uniparental disomy be common in humans? *Am J Hum Genet* 45:373–380, 1989.

Wang, J. C., Passage, M. B., Yen, P. H., Shapiro, L. J. and Mohandas, T. K. Uniparental heterodisomy for chromosome 14 in a phenotypically abnormal familial balanced 13/14 Robertsonian translocation carrier. *Am J Hum Genet* 48:1069–1074, 1991.

Chapter 2

Detection of Uniparental Disomy and Imprinting by DNA Analysis

In this chapter, we briefly discuss the different laboratory methods used for the detection of uniparental disomy (UPD) and imprinting (or parent-of-origin allelic expression). Only the common and widely used methods are discussed; for an exhaustive and detailed analysis, the reader is referred to the numerous articles in the relevant literature.

DETECTION OF UNIPARENTAL DISOMY IN DNA

The presence of two copies of a locus originating from one parent only and the absence of contribution from the other can be best documented by the use of DNA polymorphisms. For this analysis, DNA from both parents and the proband is needed in order to determine the parental origin of each chromosome. The necessary DNA polymorphisms, which could serve as molecular markers to distinguish the alleles of a locus in a given chromosome, are heritable and abundant variations of the DNA sequence. There exist several types of DNA polymorphisms.

(i) *Single nucleotide polymorphisms (SNP).* An average 1 in 600–1300 nucleotides varies between two randomly chosen chromosomes (Kan and Dozy, 1978; Antonarakis et al., 1982; Chakravarti et al., 1984; Nickerson et al., 1998; Halushka et al., 1999; Cargill et al., 1999; Venter et al., 2001; Sachidanandam et al., 2001). Millions of SNPs exist in the human genome and could be detected by several methods (Dean, 1995). The sequence of the human genome revealed more than a million of potential SNPs. (Venter et al., 2001; Sachidanandam et al., 2001). A frequently updated database of SNPs has been established at <http://www.ncbi.nlm.nih.gov/SNP/>. These are usually biallelic polymorphisms, i.e., there exist two different alleles in the population. Their usefulness depends on the allele frequency, or heterozygote frequency (Botstein et al., 1980).

(ii) *Variable number of short sequence repeats (SSR) or microsatellites.* The repeat unit is usually two to five nucleotides. The most abundant of these are the dinucleotide repeat polymorphisms, but those due to tri-tetra- or penta-nucleotide repeats are much easier for allele scoring (Litt and Luty, 1989; Weber and May, 1989; Tautz, 1989). The repeat unit could also be a mononucleotide run (Economou et al., 1990). The main advantage of the SSR polymorphisms is that there usually exist more than two alleles per locus and therefore the informativeness of SSRs is higher than that of SNPs. There is usually one polymorphic SSR for every 10–50 Kb of human genomic DNA. A public database for SSRs is available at <http://gdbwww.gdb.org/>. The use of these polymorphic markers was greatly facilitated after the discovery of the polymerase chain reaction (Saiki et al., 1985).

(iia) *Variable number of longer repeats* (formerly called VNTR for variable number of tandem repeats) in which the repeat unit is usually 20–60 nucleotides (Wyman and White, 1980; Jeffreys et al., 1985; Nakamura et al., 1987). These are also highly polymorphic with many alleles at a given locus, but not as common as the SSRs and not easily detectable by PCR.

(iii) *Presence or absence of retrotransposons* such as Alu and LINE sequences (long interspersed repeat element) or processed pseudogenes (Schuler et al., 1983; Anagnou et al., 1984; Woods-Samuels et al., 1989). These polymorphisms also comprise diallelic systems, (i.e., there are only two alleles per locus) and are easy to detect, but much rarer than the previous categories.

The basis for the detection of UPD using DNA polymorphisms is shown schematically in the example of Figure 1. In the pedigree of this figure, the DNA of the father (individual 1) is heterozygous for two different polymorphic alleles A and B at locus X. Mother's DNA (individual 2) contains two yet different alleles C and D of locus X (this is obviously an SSR polymorphic locus with many alleles). The second male offspring of these parents (individual 4) inherited alleles A and B from the father; no maternally inherited alleles at this locus are present. The diagnosis of paternal uniparental disomy (UPD) is obvious (assuming that the probability of *de novo* mutation of one of the maternal alleles to A or B is negligible and there is no laboratory error). In this example, we observe heterodisomy for locus X since both paternal alleles are present in the DNA of individual 4. In the case of the third child (individual 5), there is homozygosity of allele B of locus X and we therefore observe isodisomy for locus X. Because of the meiotic recombination, it is not unusual to detect isodisomy for one portion of the chromosome and heterodisomy for another. It is also possible to detect UPD for only one segment of the chromosome and normal biparental inheritance for the remainder. In summary, the use of DNA markers allows the determination of the inheritance of specific parental alleles and the detection of the absence of parental alleles in the offspring.

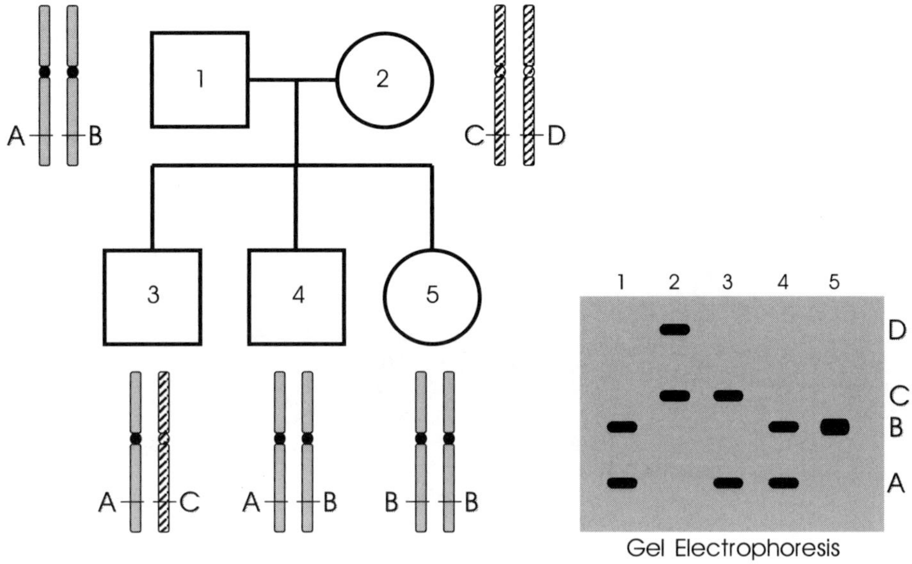

Figure 1 *Schematic representation of the use of DNA polymorphisms in determining the inheritance of parental alleles. A, B, C, and D depict the four different alleles of locus X on the long arm of the chromosomal pair shown next to the symbol of each individual. An image of the polymorphic alleles after gel electrophoresis and autoradiography is shown on the right panel. For the interpretation of the results of the analysis, see the text.*

The detection of the polymorphic alleles of a given locus can be performed using numerous laboratory techniques (Dean, 1995). Almost all use the polymerase chain reaction amplification (Saiki et al., 1985), and subsequently the detection is performed by restriction endonuclease analysis (Kan and Dozy, 1978), allele-specific oligonucleotide hybridization (Conner et al., 1983; Chee et al., 1996), allele-specific amplification (Newton et al., 1989; Wu et al., 1989), single-stranded conformation analysis (Orita et al., 1989), enzymatic or chemical cleavage of nucleotide mismatches (Cotton et al., 1988; Youil et al., 1995), denaturing gradient gel electrophoresis (Myers et al., 1985), denaturing high-pressure liquid chromatography (Underhill et al., 1997), sequence analysis (Sanger et al., 1977); all for SNPs. For the SSRs, the method used is PCR and acrylamide gel electrophoresis for fragment size discrimination that could be done either manually or using the automated nucleotide sequencers.

DETECTION OF DELETIONS BY DNA POLYMORPHISMS ANALYSIS

The detection of a *de novo* (micro)deletion in DNA and its parental origin could be carried out by the use of informative DNA polymorphisms in the DNAs of the proband and its parents, as shown schematically in Figure 2. In the case of a

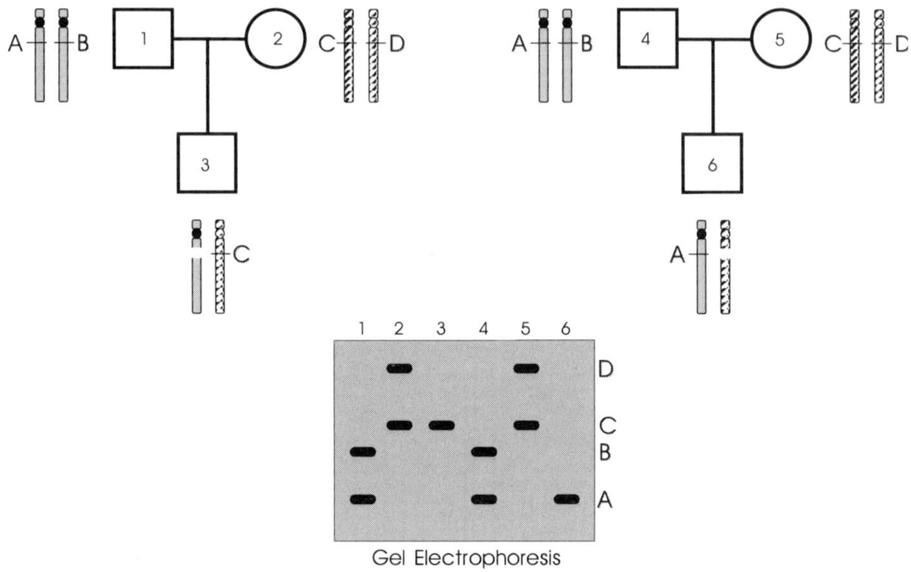

Figure 2 *Schematic representation of the detection of DNA deletions using polymorphisms. In the family on the left, the child inherited a paternal chromosome with a* de novo *deletion. The polymorphism used maps within the deletion and there is therefore no paternal allele seen in the child's DNA. This situation is common in PWS. In contrast, in the family on the right, the child inherited a maternal chromosome with a* de novo *deletion, and consequently, there is no maternal allele in the DNA of the child.*

paternally derived deletion, there is no detectable paternal allele transmitted to the proband; similarly, in the case of a maternally derived deletion, no maternal allele is seen in the DNA of the proband. The limitation of this method is that the polymorphisms used need to be informative in a given family, i.e., there are allelic differences in the parents that permit the detection of the absence of parental contribution. In addition, in the case of parentally derived deletion, nonpaternity needs to be excluded with the use of numerous other polymorphic loci in various chromosomes. Finally, this analysis requires the availability of DNA samples from both parents of the proband, which is not always feasible.

DETECTION OF PARENT-OF-ORIGIN ALLELE-SPECIFIC EXPRESSION

The phenomenon of imprinting can be defined as the specific expression of only one parental allele, but not both, in a specific tissue. For example, expression of gene X is found in blood only from the paternal allele, although a maternal allele is also present in the DNA of the individual.

The polymorphic variability in the coding region and 5′UTR and 3′UTR (untranslated region) of human genes provides ample opportunities for the detection of allele-specific expression. It has been recently shown that there are abundant SNPs in the RNA products of human genes (Cargill et al., 1999; Halushka et al., 1999). The basic principle of the detection of the allele-specific expression is to mark the transcript from each allele with a different polymorphism so that the parent-of-origin specific expression could be determined (Rainier et al., 1993). In the example of Figure 3, the paternal gene is marked with an SNP that could be digested with the restriction enzyme EcoRI, whereas the maternal gene contains the SNP allele that is noncleavable by this enzyme. A reverse transcriptase-PCR (RT-PCR) fragment that

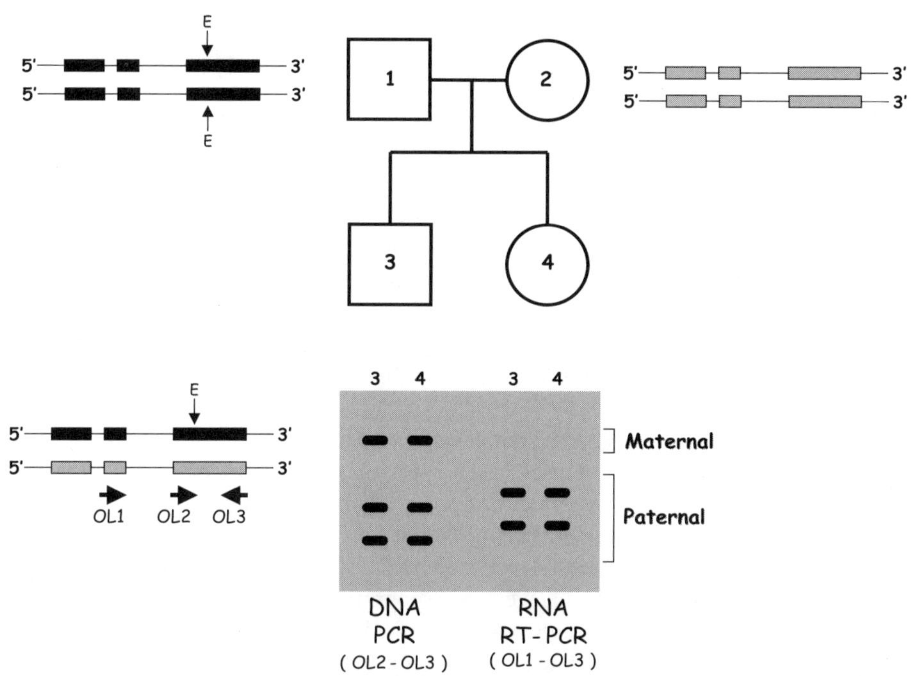

Figure 3 *Schematic representation of the detection of differential gene expression related to the parental origin of the gene. In this example, the paternally derived allele of the gene contains a single nucleotide polymorphism (SNP) that creates a recognition site for the restriction enzyme EcoRI, whereas the maternally derived allele lacks this recognition site. Reverse transcription PCR using oligonucleotide primers OL1 and OL3 results in amplification only from RNA, while PCR using oligonucleotide primers OL2 and OL3 is used to amplify fragments from genomic DNA. Digestion of the RT-PCR products with EcoRI detects transcripts from only the paternal allele since all these amplification products contain the EcoRI restriction site. Analysis of the genomic DNA shows that both the paternal and maternal alleles are present. In this example, the maternal allele is silent, and only the paternal allele is expressed. In this kind of analysis, the investigator should always eliminate the possibility of a deleterious mutation in the nonexpressed allele.*

contains the EcoRI SNP could then be used to discriminate the expression of the two alleles. The oligonucleotide primers for the RT-PCR are placed in different exons, so that the specific amplification originates only from the RNA template and not the genomic DNA. In the example of Figure 3, RT-PCR amplification of only fragments containing the EcoRI site (cleavable) establishes expression of the paternal allele only, whereas RT-PCR amplification of the fragments that do not contain the EcoRI site (noncleavable) fragments documents the expression of the maternal allele only. RT-PCR amplification from both alleles is the common situation of biallelic expression. As with the previous example of detection of uniparental disomy, the existence of the normal variability in the genome provides the tools for the detection of differential allelic expression.

DETECTION OF DIFFERENTIAL METHYLATION BY RESTRICTION ANALYSIS

The parent-of-origin differential allele expression is often associated with differential methylation of cytosine in CpG dinucleotides. Overmethylation has been associated with inactive genes and undermethylation with active, expressed genes (Cedar, 1988). It has been therefore useful to use the detection of differential methylation as a tool for diagnosis of differential allelic expression. The methylation detection is routinely used in the diagnosis of Prader-Willi and Angelman syndromes (see Chapters 6 and 7) (Glenn et al., 1993). In the normal situation, the methylation status of both parental loci on the chromosomal region 15q11-q13 is observed. In Prader-Willi syndrome, only the maternal locus with its specific methylation status is observed; in contrast, in Angelman syndrome only the paternal locus with its specific methylation status is seen. This is shown schematically in Figure 4, in which the differential methylation status is assessed by the use of the methylation-sensitive restriction enzyme HpaII that cleaves its recognition DNA sequence 5′CCGG3′ when it contains unmethylated cytosine, while it does not cleave when its recognition sequence contains a methylated C. In the example of Figure 4, the discrimination between the methylated (maternal) and the unmethylated (paternal) allele is performed after digestion of the DNA with EcoRI and HpaII, Southern blotting of the resulting fragments, and hybridization with a specific probe as shown. The HpaII digestion is only successful in the unmethylated allele, resulting in a small fragment; in contrast, the failure of HpaII to cleave the methylated site results in the large fragment in the Southern blot analysis. In the Prader-Willi syndrome, only the methylated, “larger” maternal fragment is observed; conversely in the Angelman syndrome, only the unmethylated “smaller,” paternal fragment seen by the probe is observed. This study of selected methylated sites within the 15q11-q13 region is a useful diagnostic procedure that is now routinely used in molecular diagnostic laboratories throughout the world (ASHG/ACMG, 1996).

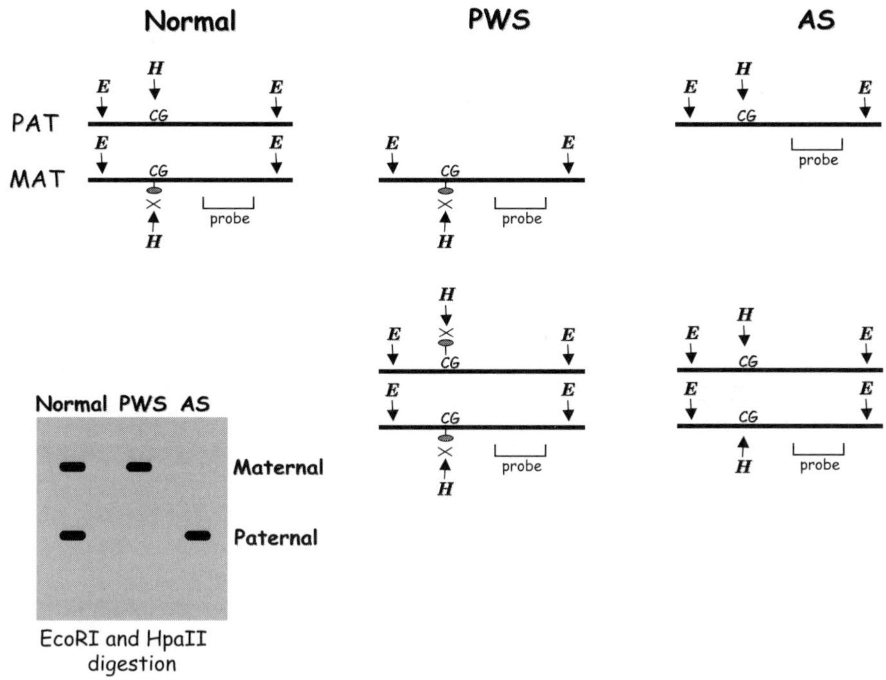

Figure 4 *Schematic representation of the detection of methylation differences using restriction endonuclease analysis. The paternally derived allele is non methylated at the CpG dinucleotide shown between two EcoRI (E) restriction sites and therefore this site is cleaved by HpaII (H) endonuclease. In contrast, the maternally derived allele is* methylated *at this CpG site (shown by the gray circle) and therefore not cleaved by HpaII. The probe for the Southern blot analysis is designated with a bracket. The methylated (maternal) allele is detected as a large band after autoradiography, while the unmethylated (paternal) allele results in a smaller EcoRI-HpaII fragment. In the PWS, the methylation pattern is only maternal due to either a deletion of the paternal allele (top panel) or the abnormal, maternal-type methylation status of the paternal allele. In contrast, in some cases of the AS, the methylation pattern is only paternal due to either a deletion of the maternal allele (top panel) or the abnormal, paternal-type methylation status of the maternal allele.*

PCR DETECTION OF METHYLATION DIFFERENCES

The differences of allele-specific methylation can also be detected by PCR analysis without Southern blotting, although this method is not widely used in routine diagnostic evaluations. The basis of the technique is that DNA treated with sodium bisulfite converts cytosine to uracil except when cytosine is methylated. After this chemical DNA modification, oligonucleotide primers specific for the methylated and unmethylated versions of the CpG clusters are utilized for PCR amplification of specific DNA fragments originating only from the methylated or unmethylated modified templates (Herman et al., 1996; Kubota et al., 1997). The method is shown schematically in Figure 5.

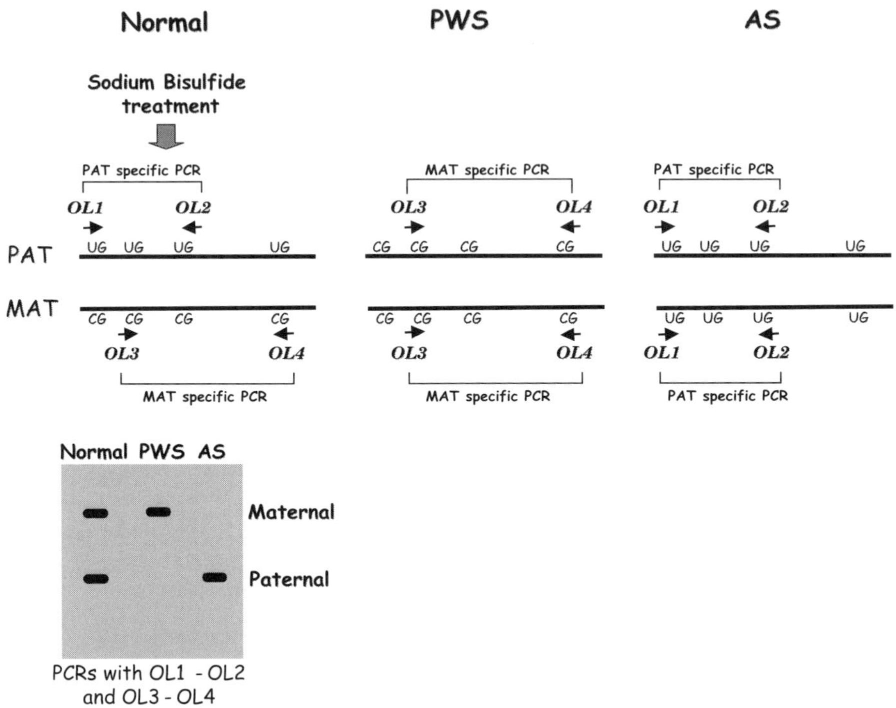

Figure 5 *Schematic representation of the methylation-specific PCR. DNA treated with sodium bisulfite that converts cytosine to uracil except when cytosine is methylated. Oligonucleotide primers OL1 and OL2 were designed to complement "methylated," i.e., unmodified, cytosines and therefore their PCR product is only maternal-specific since these oligonucleotides do not bind the modified DNA. In contrast, oligonucleotides OL3 and OL4 were designed to complement "unmethylated," i.e., modified, cytosine to uracil residues. Their PCR product is only paternal-specific since these oligonucleotide primers do not bind the unmodified DNA. The paternal or maternal allele-specific amplification is recognized by the specific size of the PCR amplification product. In PWS patients, a maternal methylation PCR pattern is only observed, either because there is a deletion of the paternally derived allele or an abnormal, maternal-type methylation status of the paternal allele. In AS patients, a paternal methylation PCR pattern is only observed, either because there is a deletion of the maternally derived allele or the abnormal, paternal-type methylation status of the maternal allele.*

DETECTION OF (MICRO)DELETIONS BY FISH

The technique of *in situ* hybridization is based on the principle that, given the appropriate temperature and salt conditions, single-stranded DNA reanneals with its complementary sequence into double-stranded DNA. A single-stranded labeled probe (usually the DNA insert of a cosmid) is allowed to hybridize with the DNA of metaphase chromosomes fixed on a glass slide (Gerhard et al., 1981; Garson et al., 1987). The probe DNA is labeled with nonisotopic compounds that could then be detected using fluorescence or enzymatic linked reactions. Biotin and digoxigenin

Erratum

Genomic Imprinting and Uniparental Disomy in Medicine
by Eric Engel and Stylianos E. Antonarakis

ISBN 0-471-35126-1

On page 21, figure 6 was incorrectly printed. The correct figure is as follows:

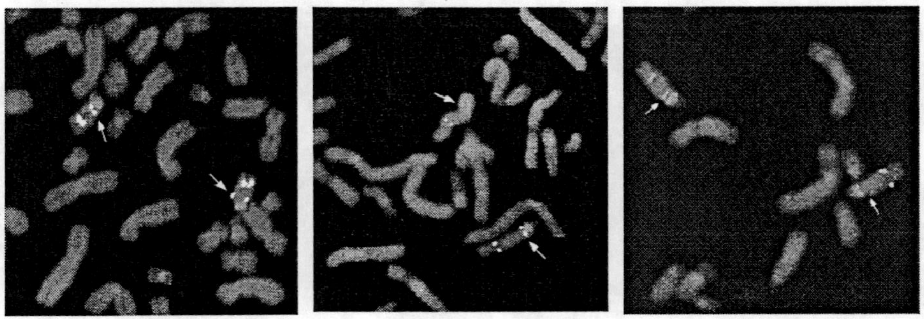

(1) (2) (3)

Figure 6 *FISH (fluorescence* in situ *hybridization) using a P1 probe D15S113 that maps in the 15q11-q13 region. Panel 1 shows a signal in both chromosomes 15 in a normal individual; panels 2 and 3 show microdeletions 15q in one chromosome 15 from patients with PWS and AS, respectively. (This figure was kindly provided by M. A. Morris)*

Figure 6 *FISH (fluorescence* in situ *hybridization) using a cosmid probe XXXX that maps in the 15q11-q13 region. Panel 1 shows a signal in both chromosomes 15 in a normal individual; panels 2 and 3 show microdeletions 15q in one chromosome 15 from patients with PWS and AS, respectively. (This figure was kindly provided by M. A. Morris)*

are two widely used labeling agents and their detection, after binding of the probes to their specific (complementary) regions of the chromosomes, is usually performed with fluorochromes such as FITC (fluorescein isothiocyanate). FISH methods using cosmid probes in commonly deleted chromosomal regions could detect microdeletions not visible with conventional karyotypic analysis. In addition, parental samples or informative DNA polymorphisms are not needed for detecting a deletion in a proband. FISH is therefore a popular method for detecting microdeletion syndromes, including PWS, AS, Di George, Miller-Dieker, and velo-cardiac-facial syndromes (Gopal et al., 1995; Delach et al., 1994; Kuwano et al., 1991, 1992). An example of FISH is shown in Figure 6.

DETECTION OF ALLELE-SPECIFIC EXPRESSION BY FISH (FLUORESCENCE *IN SITU* HYBRIDIZATION)

Several laboratories are now developing methods to visualize the allele-specific RNA transcript by fluorescence *in situ* hybridization in single cells and tissues (Lizardi et al., 1998). These methods are not further discussed here because they are not used widely and have not yet contributed significantly to the elucidation of the biological consequences or improved the diagnostic evaluations of uniparental disomy and imprinting.

REFERENCES

Anagnou, N. P., O'Brien, S. J., Shimada, T., Nash, W. G., Chen, M. J. and Nienhuis, A. W. Chromosomal organization of the human dihydrofolate reductase genes: dispersion, selective amplification, and a novel form of polymorphism. *Proc Natl Acad Sci U S A* 81:5170–5174, 1984.

Antonarakis, S. E., Boehm, C. D., Giardina, P. J. and Kazazian, H. H. J. Nonrandom association of polymorphic restriction sites in the beta-globin gene cluster. *Proc Natl Acad Sci U S A* 79:137–141, 1982.

ASHG/ACMG. Diagnostic testing for Prader-Willi and Angleman syndromes: Report of the ASHG/ACMG Test and Technology Transfer Committee. *Am J Hum Genet* 58:1085–1088, 1996.

Botstein, D., White, R. L., Skolnick, M. and Davis, R. W. Construction of a genetic linkage map in man using restriction fragment length polymorphisms. *Am J Hum Genet* 32:314–331, 1980.

Cargill, M., Altshuler, D., Ireland, J., et al. Characterization of single-nucleotide polymorphisms in coding regions of human genes. *Nat Genet* 22:231–238, 1999.

Cedar, H. DNA methylation and gene activity. *Cell* 53:3–4, 1988.

Chakravarti, A., Buetow, K. H., Antonarakis, S. E., Waber, P. G., Boehm, C. D. and Kazazian, H. H. Nonuniform recombination within the human beta-globin gene cluster. *Am J Hum Genet* 36:1239–1258, 1984.

Chee, M., Yang, R., Hubbell, E., et al. Accessing genetic information with high-density DNA arrays. *Science* 274:610–614, 1996.

Conner, B. J., Reyes, A. A., Morin, C., Itakura, K., Teplitz, R. L. and Wallace, R. B. Detection of sickle cell beta S-globin allele by hybridization with synthetic oligonucleotides. *Proc Natl Acad Sci U S A* 80:278–282, 1983.

Cotton, R. G., Rodrigues, N. R. and Campbell, R. D. Reactivity of cytosine and thymine in single-base-pair mismatches with hydroxylamine and osmium tetroxide and its application to the study of mutations. *Proc Natl Acad Sci U S A* 85:4397–4401, 1988.

Dean, M. Resolving DNA mutations [news]. *Nat Genet* 9:103–104, 1995.

Delach, J. A., Rosengren, S. S., Kaplan, L., Greenstein, R. M., Cassidy, S. B. and Benn, P. A. Comparison of high resolution chromosome banding and fluorescence in situ hybridization (FISH) for the laboratory evaluation of Prader-Willi syndrome and Angelman syndrome [see comments]. *Am J Med Genet* 52:85–91, 1994.

Economou, E. P., Bergen, A. W., Warren, A. C. and Antonarakis, S. E. The polydeoxyadenylate tract of Alu repetitive elements is polymorphic in the human genome. *Proc Natl Acad Sci U S A* 87:2951–2954, 1990.

Garson, J. A., van den Berghe, J. A. and Kemshead, J. T. Novel non-isotopic in situ hybridization technique detects small (1 Kb) unique sequences in routinely G-banded human chromosomes: fine mapping of N-myc and beta-NGF genes. *Nucleic Acids Res* 15:4761–4770, 1987.

Gerhard, D. S., Kawasaki, E. S., Bancroft, F. C. and Szabo, P. Localization of a unique gene by direct hybridization in situ. *Proc Natl Acad Sci U S A* 78:3755–3759, 1981.

Glenn, C. C., Nicholls, R. D., Robinson, W. P., et al. Modification of 15q11-q13 DNA methylation imprints in unique Angelman and Prader-Willi patients. *Hum Mol Genet* 2:1377–1382, 1993.

Gopal, V. V., Roop, H. and Carpenter, N. J. Diagnosis of microdeletion syndromes: high-resolution chromosome analysis versus fluorescence in situ hybridization. *Am J Med Sci* 309:208–212, 1995.

Halushka, M. K., Fan, J. B., Bentley, K., et al. Patterns of single-nucleotide polymorphisms in candidate genes for blood-pressure homeostasis. *Nat Genet* 22:239–247, 1999.

Herman, J. G., Graff, J. R., Myohanen, S., Nelkin, B. D. and Baylin, S. B. Methylation-specific PCR: a novel PCR assay for methylation status of CpG islands. *Proc Natl Acad Sci U S A* 93:9821–9826, 1996.

Jeffreys, A. J., Wilson, V. and Thein, S. L. Hypervariable 'minisatellite' regions in human DNA. *Nature* 314:67–73, 1985.

Kan, Y. W. and Dozy, A. M. Antenatal diagnosis of sickle-cell anaemia by DNA analysis of amniotic-fluid cells. *Lancet* 2:910–912, 1978.

Kan, Y. W. and Dozy, A. M. Polymorphism of DNA sequence adjacent to human beta-globin structural gene: relationship to sickle mutation. *Proc Natl Acad Sci U S A* 75:5631–5635, 1978.

Kubota, T., Das, S., Christian, S. L., Baylin, S. B., Herman, J. G. and Ledbetter, D. H. Methylation-specific PCR simplifies imprinting analysis [letter]. *Nat Genet* 16:16–17, 1997.

Kuwano, A., Ledbetter, S. A., Dobyns, W. B., Emanuel, B. S. and Ledbetter, D. H. Detection of deletions and cryptic translocations in Miller-Dieker syndrome by in situ hybridization. *Am J Hum Genet* 49:707–714, 1991.

Kuwano, A., Mutirangura, A., Dittrich, B., et al. Molecular dissection of the Prader-Willi/Angelman syndrome region (15q11-13) by YAC cloning and FISH analysis [published erratum appears in Hum Mol Genet 1992 Dec;1(9):784]. *Hum Mol Genet* 1:417–425, 1992.

Litt, M. and Luty, J. A. A hypervariable microsatellite revealed by in vitro amplification of a dinucleotide repeat within the cardiac muscle actin gene. *Am J Hum Genet* 44:397–401, 1989.

Lizardi, P. M., Huang, X., Zhu, Z., Bray-Ward, P., Thomas, D. C. and Ward, D. C. Mutation detection and single-molecule counting using isothermal. *Nat Genet* 19:225–232, 1998.

Myers, R. M., Lumelsky, N., Lerman, L. S. and Maniatis, T. Detection of single base substitutions in total genomic DNA. *Nature* 313:495–498, 1985.

Nakamura, Y., Leppert, M., O'Connell, P., et al. Variable number of tandem repeat (VNTR) markers for human gene mapping. *Science* 235:1616–1622, 1987.

Newton, C. R., Graham, A., Heptinstall, L. E., et al. Analysis of any point mutation in DNA. The amplification refractory mutation system (ARMS). *Nucleic Acids Res* 17:2503–2516, 1989.

Nickerson, D. A., Taylor, S. L., Weiss, K. M., et al. DNA sequence diversity in a 9.7-kb region of the human lipoprotein lipase gene [see comments]. *Nat Genet* 19:233–240, 1998.

Orita, M., Iwahana, H., Kanazawa, H., Hayashi, K. and Sekiya, T. Detection of polymorphisms of human DNA by gel electrophoresis as single-strand conformation polymorphisms. *Proc Natl Acad Sci U S A* 86:2766–2770, 1989.

Rainier, S., Johnson, L. A., Dobry, C. J., Ping, A. J., Grundy, P. E. and Feinberg, A. P. Relaxation of imprinted genes in human cancer. *Nature* 362:747–749, 1993.

Sachidanandam, R., Weissman, D., Schmidt, S. C., et al. A map of human genome sequence variation containing 1.42 million single nucleotide polymorphisms. *Nature* 2001; 409:928–933.

Saiki, R. K., Scharf, S., Faloona, F., et al. Enzymatic amplification of beta-globin genomic sequences and restriction site analysis for diagnosis of sickle cell anemia. *Science* 230:1350–1354, 1985.

Sanger, F., Nicklen, S. and Coulson, A. R. DNA sequencing with chain-terminating inhibitors. *Proc Natl Acad Sci U S A* 74:5463–5467, 1977.

Schuler, L.A., Weber, J.L. and Gorski, J. Polymorphism near the rat prolactin gene caused by insertion of an Alu-like element. *Nature* 305:159–160, 1983.

Tautz, D. Hypervariability of simple sequences as a general source for polymorphic DNA markers. *Nucleic Acids Res* 17:6463–6471, 1989.

Underhill, P. A., Jin, L., Lin, A. A., et al. Detection of numerous Y chromosome biallelic polymorphisms by denaturing high-performance liquid chromatography [letter] [see comments]. *Genome Res* 7:996–1005, 1997.

Venter, J. C., Adams, M. D., Myers, E. W., et al. The sequence of the human genome. *Science* 2001; 291:1304–1351.

Weber, J. L. and May, P. E. Abundant class of human DNA polymorphisms which can be typed using the polymerase chain reaction. *Am J Hum Genet* 44:388–396, 1989.

Woods-Samuels, P., Wong, C., Mathias, S. L., Scott, A. F., Kazazian, H. H. J. and Antonarakis, S. E. Characterization of a nondeleterious L1 insertion in an intron of the human factor VIII gene and further evidence of open reading frames in functional L1 elements. *Genomics* 4:290–296, 1989.

Wu, D. Y., Ugozzoli, L., Pal, B. K. and Wallace, R. B. Allele-specific enzymatic amplification of beta-globin genomic DNA for diagnosis of sickle cell anemia. *Proc Natl Acad Sci U S A* 86:2757–2760, 1989.

Wyman, A. R. and White, R. A highly polymorphic locus in human DNA. *Proc Natl Acad Sci U S A* 77:6754–6758, 1980.

Youil, R., Kemper, B. W. and Cotton, R. G. Screening for mutations by enzyme mismatch cleavage with T4 endonuclease VII. *Proc Natl Acad Sci U S A* 92:87–91, 1995.

Chapter 3

Mechanisms Generating Uniparental Disomy and Genomic Imprinting Disorders

In this chapter, we attempt to list the potential mechanisms that lead to uniparental disomy (UPD).

1. TRISOMY RESCUE

Trisomies are frequent occurrences in human zygotes (Hassold and Jacobs, 1984). In all recognized pregnancies, there is a 4% incidence of trisomies (the most frequent of which is trisomy 16). A considerable number of the trisomy pregnancies result in spontaneous abortions in the first trimester. As much as 26% of spontaneous abortions show trisomies for almost all chromosomes (Hassold and Jacobs, 1984). There are numerous indications that UPD could result from the loss of one chromosome of an initial trisomy; this mechanism is termed trisomy rescue. Figure 1 shows the three possible outcomes of the single chromosomal loss in trisomy rescue. In two of the three outcomes, the chromosomal loss will result in the restoration of biparental disomy, because one of the chromosomes involved in the initial nondisjunction is lost. In one of the three outcomes, however, the chromosomal loss involves the chromosome derived from the normal parental meiosis and there will be uniparental disomy. Because all chromosomes are involved in trisomies (most of which are not viable to term), there exist, as expected, UPDs for the majority of chromosomal pairs. In addition, since advanced maternal age remains the only well-documented risk for maternal meiotic nondisjunction (Nicolaidis and Petersen, 1998), there also exists a correlation between maternal age and UPDs due to trisomy rescue (Robinson et al., 1998). The UPD could be isodisomy or heterodisomy, depending on the mechanism of nondisjunction and the occurrence of meiotic crossing-over (Figure 1*a*).

Prenatal diagnosis by CVS or amniocentesis is best suited for detecting reduction of a trisomic to a disomic line, in cases of documented mosaicism for trisomy/euploid cell lines.

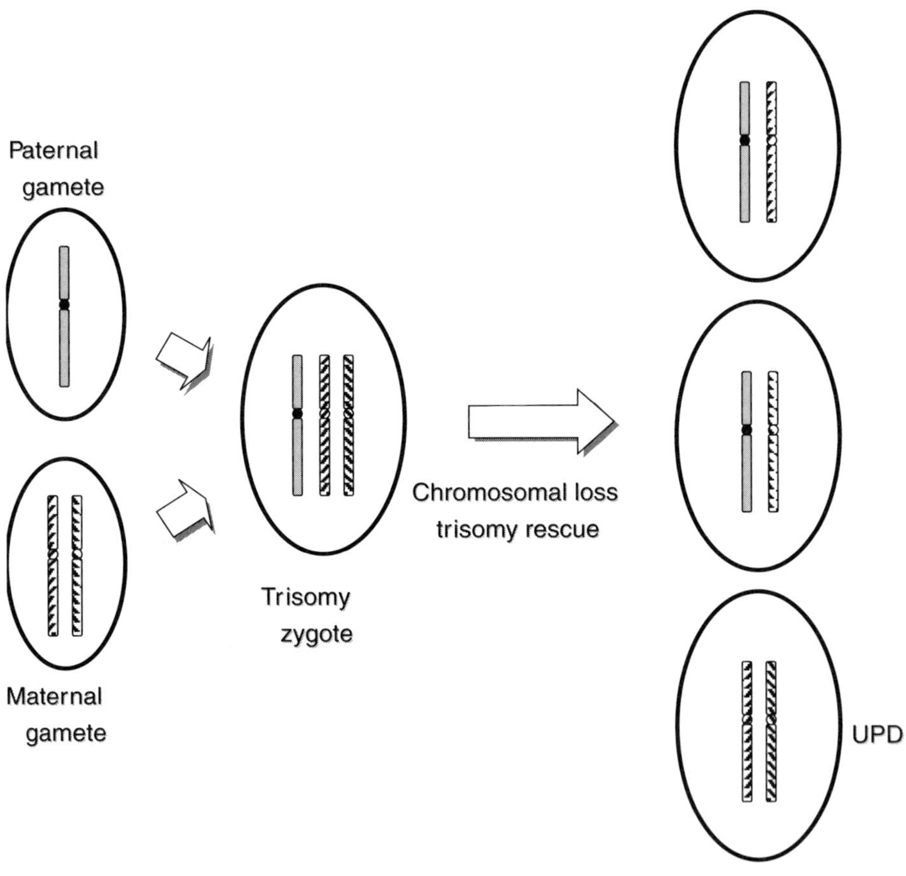

Figure 1 *Schematic representation of the trisomy rescue and the resulting outcomes including the 1/3 possibility of UPD.*

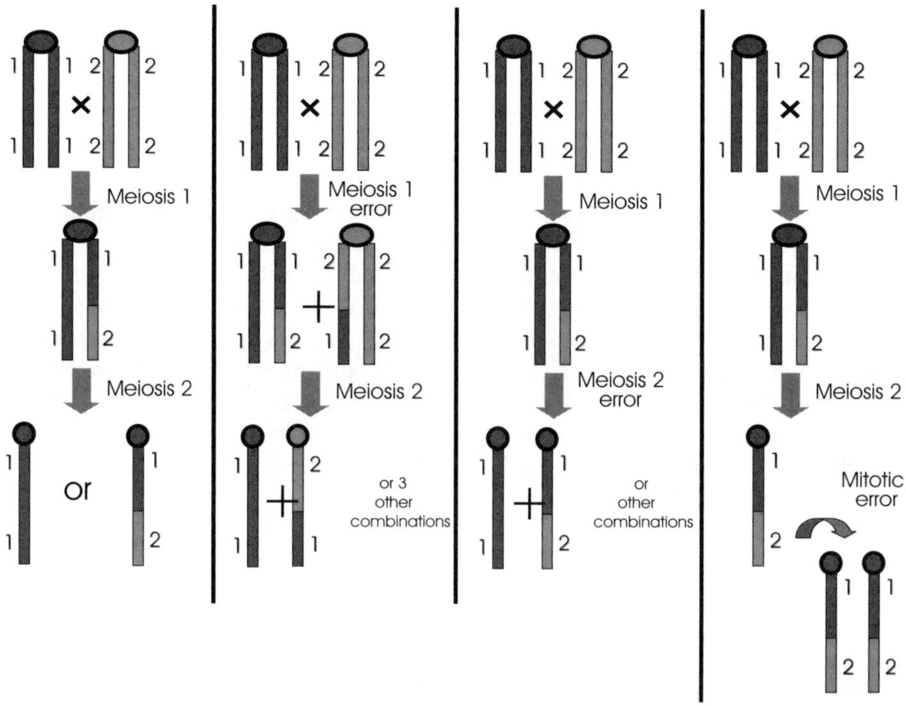

Figure 1a *Schematic representation of the normal meiosis, and the errors in meiosis 1 and 2 and mitosis leading to aneuploidy. The regions of homozygosity and heterozygosity are shown by different shades. In meiosis 1 errors, the centromeric areas of the missegregated chromosomes remain heterodisomic (heterozygous for polymorphic markers). In meiosis 2 errors, these areas are isodisomic (homozygous for polymorphic markers). The situation changes distal to recombination or cross-overs. In mitotic nondisjunction, the entire length of chromosomes is isodisomic. Numbers refer to polymorphic alleles of specific loci along the DNA.*

Mechanisms of Chromosomal Loss

Given a trisomy in a zygote, three major mechanisms may result in a chromosomal loss, namely, nondisjunction, anaphase lag during chromosomal migration on the spindle, or eventually, the breakage or destruction of one chromosome.

The stage at which, cells of an early conception may undergo the secondary loss of a member of the trisomy, and the mechanism causing such a loss, can occur at the first mitotic cleavage of the zygote, or at subsequent cell divisions.

(1) At the First Cleavage of the Zygote This should lead to two blastomeres of distinct chromosomal make-up initiating a mosaic pattern with either two (euploid) or three chromosomes (trisomy) for the specific pair.

(a) *Nondisjunction.* There is an abnormal situation in which two sister chromatids of the trisomic zygote migrate jointly to only one of the daughter cells, while four others separate and migrate to the other daughter cell. Thus, of two resulting blastomeres, one will contain four, the other two homologous chromosomes. Tetrasomy is mostly lethal, except for the sex chromosomes and, from then on, the congruent diploidy would be restored. As such, embryogenesis would proceed from the disomic blastomere. If so, the trisomic derivation of such a case of UPD will entirely escape cytogenetic or molecular detection.

(b) *Anaphase lag.* There are instances of anaphase lag in which the migration of one of the chromatids toward one pole during anaphase is delayed. At completion of cell division, the delayed chromatid is usually found in a micronucleus resulting in a mosaic situation of a trisomic and a newly formed disomic cell. The survival of the trisomy or disomy cells depends on the chromosomes involved and other factors related to imprinting and isodisomy. Certainly, the mosaicism is tolerated for trisomies of chromosomes such as 13, 16, 18, and 21. For certain chromosomes, however, mosaicism with UPD may not be a viable outcome; for example, UPD 18 is perhaps lethal since no such case has yet been described.

(c) *Intrinsic accidents* that damage and destroy one chromosome, or result in smaller structures with or without a centromere. Often unstable, such chromosomal remnants usually vanish from a proportion of cells, thus being only retrieved as part of a mosaicism.

(2) After the First Cleavage If a chromosomal loss occurs at the second or subsequent divisions of a zygote, at least one cell line would retain the original pattern of trisomy, depending on its potential viability. The mechanisms of chromosomal loss are the same as above, i.e., nondisjunction, anaphase lag or chromosomal wreckage. These cases result in persistent mosaicism (if we assume appropriate fitness of the aneuploid or the uniparentally disomic cell lines), exhibiting a mixture of cell populations with trisomy and disomy (UPD or not) for a particular chromosome.

The proof of loss of a supernumerary chromosome is seen only in cases where the original trisomy line could be recognized. Trisomy "rescue" has been best illustrated in cases with initial trisomies 16 or 15 (Bennett et al., 1992; Kalousek et al., 1993; Dworniczak et al., 1992; Cassidy et al., 1992; Purvis-Smith et al., 1992; Markovic et al., 1996; Milunsky et al., 1996), but also in trisomies of other chromosomes (Antonarakis et al., 1993).

The mosaicism for trisomy/disomy cell lines could also be due to the alternative mechanism in which the initial 2n biparental zygote or cell line might give rise to a trisomic clone by mitotic duplication of one of its chromosomes. In this case, there is no possibility for UPD unless there are further rearrangements. What is the proportion of mosaics (trisomy/disomy) originating from chromosomal loss following initial trisomy, versus mitotic duplication following initial disomy? In 32 mosaic cases examined by polymorphic markers (Robinson et al., 1995), it was observed that: (i) of 14 mosaics of the 45,X/46,XX type, chromosome loss following a normal, disomic fertilization product occurred in 12 cases; (ii) disomy/trisomy mosaicism for chromosome 13, 18, 21, and X also frequently originated with the somatic loss of one chromosome after a trisomic fertilization, (iii) in contrast, four of five trisomy 8 cases were consistent with the somatic gain of chromosome 8 following an initially normal zygote.

As has been further learned from experience in chorionic tissue studies, the meiotic origin of the trisomic line is very likely for certain chromosomes, whereas a mitotic conversion to trisomy is more commonly involved for other chromosomes. The somatic origin of the trisomy (duplication of one chromosome following normal disomy) accounts for a majority of the cell lines carrying an extra 2, 7, 8, 10, 12, while many cases of CPM, (confined placental mosaicism) involving trisomy 9, 16, and 22, are meiotically determined (Robinson et al., 1997). It therefore follows that there is more risk for these latter chromosomes to be involved in UPD. At the phenotypic level intrauterine growth retardation (IUGR), IU death and other abnormalities are almost exclusively seen in cases with a meiotically determined trisomy. In addition, in cultured chorion villus sampling (CVS), more trisomic cells of meiotic origin as compared to those of somatic origin have been observed. In somatic trisomies, the frequencies of the trisomic cell lines are significantly lower in cultured CVS stroma and in term trophoblast, while they are not significantly different in term chorionic stroma and chorionic plate.

The cytogenetic studies of the placenta, ideally, should include both the highly mitotic cytotrophoblast cells karyotypically characterized from direct or very short-term cultures, and the extraembryonic mesenchymatous villus (extraembryonic mesoderm EEM), where the chromosomal make-up is revealed by mitoses derived from long-term cultures. The EEM tissue is made up of cells derived from the inner cell mass, i.e., from the embryonic progenitor cells. Paradoxically, the latter "mass" at the 64 cell blastocyst stage is only comprised of three or four cells. Given these ontogenic differences and the potentially different developmental stages at which secondary chromosome mutations may occur, the recording of aneuploid results may conveniently fall into three types: (i) abnormal cells in direct, i.e., cytotrophoblastic, preparations only; (ii) Abnormal cells in extra-embryonic mesoderm (EEM) cultures only; (iii) abnormal cells in both lineages (Wolstenholme, 1996; Kalousek et al., 1992; Simoni and Sirchia, 1994).

The relevant studies of Wolstenholme (1996) resulted in the following conclusions:

(1) Placental trisomies for chromosomes 2, 3, 7, 8, 9 are likely to be due to mitotic nondisjunction; (2) mitotic errors producing CPM for trisomies 7 and 9 contribute to either type I or II anomalous distributions; (3) trisomy 3 of mitotic origin is mostly found as type I; (4) trisomies 2 and 8 of the same origin colonize preferentially the EEM lineage (type II) and are better diagnosed from long-term cultures; (5) meiotic errors are more common for trisomy 2 (20%), trisomy 9 (1 in 3), trisomy 16 (9 in 10) while trisomy 22 results from meiotic errors in two of three occurrences; (6) type III CPM with 100% involvement of cytotrophoblast cells is mostly meiotic in origin.

2. MONOSOMY RESCUE

As schematically illustrated in Figure 2, there is a duplication of a chromosome to "correct" a zygote monosomic for this particular chromosome. In this case, such a correction occurs to make up for a loss affecting the other parent's chromosome. The monosomy could have resulted from a meiotic error and the subsequent fertilization of a nullisomic germ cell by a normal haploid gamete (Figure 2). The mechanism of the duplication of the monosomic chromosome could be either due to mitotic nondisjunction or endo-reduplication of the solitary chromosome. Mosaicism is most unlikely since cells with autosomal monosomies are usually not viable. In monosomy rescue, there is uniparental isodisomy for the entire chromosome.

Acrocentric chromosomes display an additional mode of monosomy rescue. A duplication can result from misdivision, i.e., transverse splitting of the centromere, giving rise to an isochromosome that effectively preserves, in a single chromosome, the content of the long arms of a chromatid pair; here again, the duplicated structure is entirely isodisomic.

Monosomy "correction" is best illustrated by cases of paternal UPD15, often with i[15q] compensating for a lack of maternal chromosome 15 in a zygote (Freeman et al., 1993; Fridman et al., 1998; Tonk et al., 1996; Ramsden et al., 1996; Robinson et al., 1996) or by cases of maternal UPD7, making up for paternal absence of this chromosome (Voss et al., 1989; Spence et al., 1988). Correction of partial monosomy has been hypothesized in cases of monosomy 21 (Petersen et al., 1992).

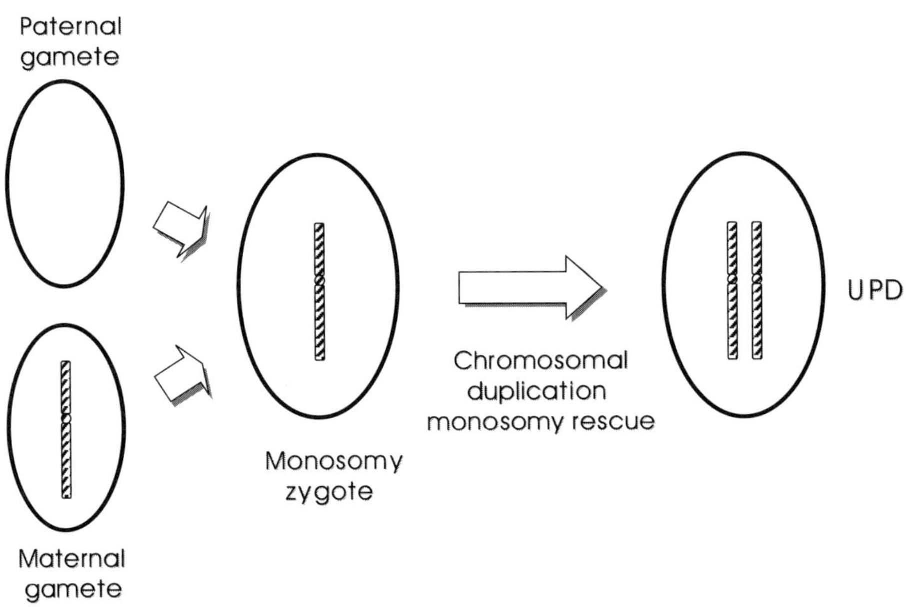

Figure 2 Schematic representation of the monosomy rescue and the resulting outcome of UPD.

3. GAMETE COMPLEMENTATION

Fusion of chromosomally complementary gametes (one with duplication of a chromosome and the other with a deletion) could also be a mechanism for UPD in humans. This has been used experimentally in mice to discover the genomic regions that are likely to contain parentally imprinted genes (Ferguson-Smith et al., 1991; Cattanach and Kirk, 1985).

In humans, the closest approximation to the experimental animal situation for considering complementation as the likely cause of a UPD would be to identify parents, each with a same translocation or both with involvement of the same chromosome in a translocation. This has been observed twice to our knowledge both in cases resulting in paternal UPD14 (Wang et al., 1991; Cotter et al., 1997) (these cases are discussed in Chapter 4 on pages 93 to 96 as UPD14 cases 1 and 4, respectively). In the first case, the mother of the UPD14 proband had a t(1;14) and the father t(13;14). In the second case, the mother had t(14;21) and the father t(13;14). Aside from finding translocations in both parents, complementary gametes, resulting from unidentifiable nondisjunction of intact chromosomes (as shown in Figure 3), cannot be suspected, so that a total bias of ascertainment against the recognition of this mechanism exists. This stands in contrast to trisomies converting to disomies, which are detected in a proportion of cases. Thus, the exact contribution of gamete complementation to human UPD cannot be appreciated.

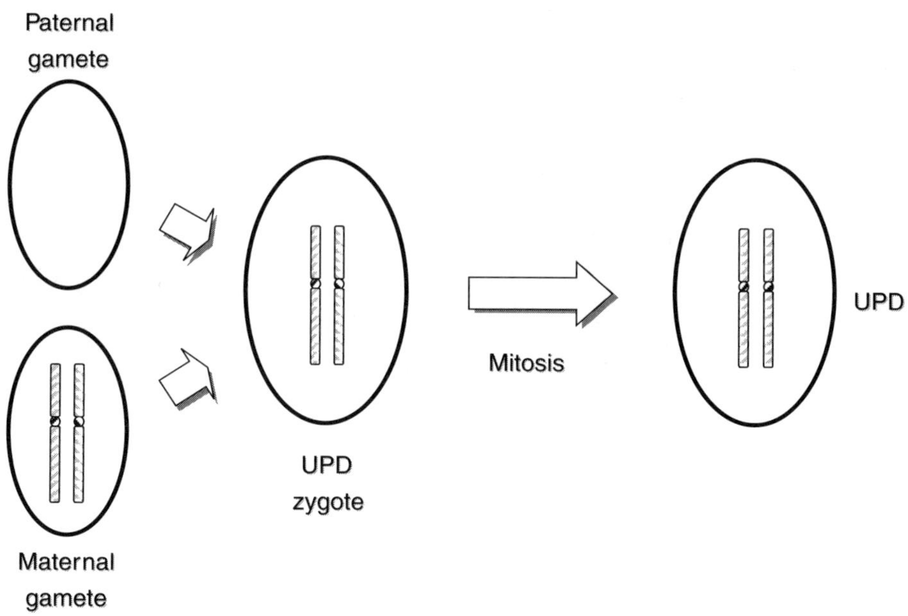

Figure 3 *Schematic representation of one possibility of gamete complementation and the resulting outcome of UPD.*

4. CHROMOSOME TRANSLOCATIONS AND UPD

Chromosomal translocations, particularly of the acrocentric chromosomes, have been found in numerous cases of UPD and they are involved in its etiology. The importance of translocations has been recognized since the early 1980s for chromosome 15 and PWS (Mattei et al., 1983, 1984). By 1989, when the role of genomic imprinting emerged in this condition, 14 instances of Robertsonian translocations had been compiled in PWS patients (Butler, 1990); 12 were t(15;15), one t(13;15), and one t(14;15). All the above appeared balanced, i.e., without 15q11-q13 deletions. Additional translocation 15 cases have been subsequently identified (Engel, 1995).

(A) UPD in the Various Types of Centric fusion of Acrocentric Chromosomes

(1) Familial Heterologous Robertsonian Translocations The mechanism of UPD after trisomy rescue in a heterologous familial Robertsonian translocation of acrocentric chromosomes is shown in Figure 4. Cases of maternal or paternal t(13;14) and, more rarely, a maternal t(13;15) or t(14;15) have been described to cause paternal or maternal UPD14 or maternal UPD15 in the offspring. In such cases, adjacent segregations of a trivalent should have induced nullisomic or disomic

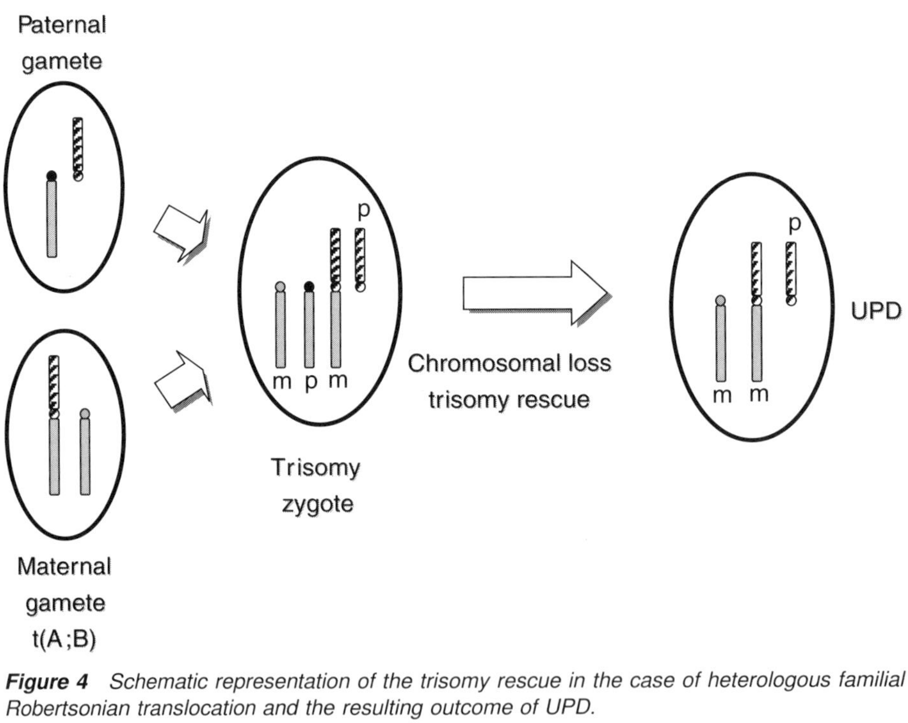

Figure 4 *Schematic representation of the trisomy rescue in the case of heterologous familial Robertsonian translocation and the resulting outcome of UPD.*

gametes to occur, resulting after fertilization in monosomic or trisomic embryos. Subsequently, a secondary event occurred to correct the aneuploidy, either by loss (in trisomy) or gain (in monosomy) of a member in the involved pair.

(2) Familial Homologous Robertsonian Translocation The mechanism of UPD after trisomy rescue in a homologous familial Robertsonian translocation of an acrocentric chromosome is shown in Figure 5. When the translocation chromosome is transmitted to a balanced offspring, it is safe to conclude that such an event has generated UPD due to the loss of the homologous chromosome originating from the other parent (trisomy rescue). This is one of the very rare instances where UPD can be ascertained without a need for molecular studies to prove it (Palmer et al. 1980; Kirkels et al. 1980). Indeed, genitors passing down these translocations can only form disomic or nullisomic gametes, resulting in trisomic or monosomic zygotes, after fertilization. It is also possible that UPD could be generated by the monosomy rescue mechanism described earlier. As a hallmark, also, such families could be ascertained through the occurrence of a long string of spontaneous abortions or births of infants with abnormalities.

(3) Heterologous* de novo *Centric Fusions There are some nonfamilial cases with either UPD14 or UPD15 and *de novo* t(13;14) or t(14;15) respectively. In one illustrative *de novo* t(13;14) case with maternal UPD14, a trisomy 14 lineage was

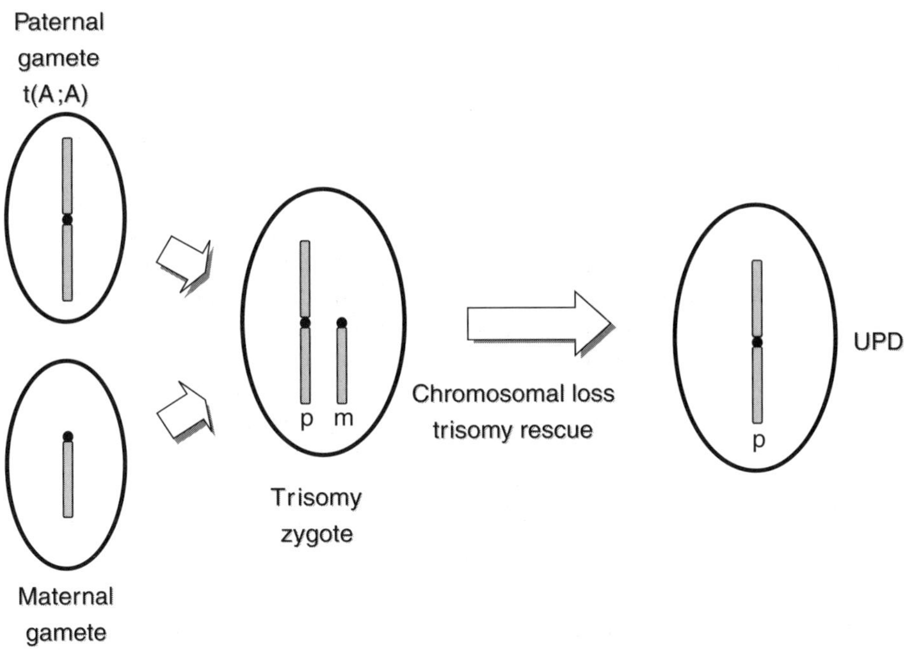

Figure 5 *Schematic representation of the trisomy rescue in the case of homologous familial Robertsonian translocation and the resulting outcome of UPD.*

uncovered in blood (5% of cells) both by molecular analysis and karyotype. Studies with polymorphic markers indicated crossing-over between the free and attached maternal chromosome 14. It was concluded that a 13;14 translocation had occurred during meiosis, leading to a zygote with 2 maternal 14, one free, one fused to a 13; the paternal contributions for 13 and 14 were normal. This presumed trisomy translocation 14 conceptus subsequently underwent the early loss of a parental 14, thus resulting in maternal UPD14 for most cells. (Antonarakis et al., 1993).

In a series involving *de novo* t(13;14) and t(14;21), the translocations apparently originated through common breakpoint regions, during maternal meiosis. These breakpoints were almost constantly specific to the short-arm regions, suggesting that exchanges at these areas were favored by the recombination of homologous DNA sequences common to the proximal short arm of chromosomes 13, 14, 21. Of 16 *de novo* nonhomologous Robertsonian translocations, 14 were derived from the maternal chromosome, a ratio significantly different from that expected for random occurrence. Furthermore, the 10 t(13;14) had consistent breakpoints, compatible with recombination within specific short-arm regions of chromosomes 13 and 14; a similar mechanism seemed to apply to the few cases of t(14;21) also studied (Page et al., 1996). The *de novo* centric fusions thus display some very remarkable characteristics: (i) In the great majority of cases, they are of maternal origin; (ii) they appear to be formed at meiosis, (iii) the prevalent ones, rob(13q14q) and (14q21q), have consistent short-arm breakpoints, suggesting recombination at hot spots between homologous DNA sequences in proximal short arms of chromosomes 13, 14, 21.

(4) Homologous* do novo *Centric Fusions These cases represent the most numerous group, diagnosed with maternal and paternal UPD14, UPD15, and UPD21, and maternalUPD 22 for a total of 19 cases (see Table 1). In these instances, isodisomy is the rule and the fusion products represent the union of two isochromatids through transverse splitting across the centromere or across the short-arm material. A trisomy rescue via the loss of the homologous chromosome from the other parent would result in UPD. This mechanism is schematically shown in Figure 6. Alternatively, the failure of chromatid separation of an achiasmatic translocated chromosome in meiosis 2 could also produce similar isodisomy.

The eight cases studied by Robinson et al. (1994) involved all acrocentric chromosomes but for number 13. In four of the eight, UPD was noted, and homozygosity throughout the involved chromosome supported the mechanism of postmeiotic origin, implying misdivision and resulting in isochromosome formation (probably compensating for an initial monosomy). In the four other cases, there was biparental origin of the fused homologues.

Table 1 shows some examples of cases of Robertsonian translocations of acrocentric chromosomes associated with UPD.

(B) Reciprocal Translocations (Balanced Interchanges) and Imprinting Disorders

The contribution of balanced translocations that are associated with imprinting disorders (and not necessarily with UPD) is discussed here. In the medical literature,

TABLE 1 Translocations by Centric Fusions, Inherited or *de novo*, Heterologous or Homologous, Observed in UPD Cases

Types	CHR	Origins	UPD	References
Heterologous (inherited)	13;14	Maternal	14	(Temple et al., 1991; Barton et al., 1996)
	13;14	Paternal	14	(Cotter et al., 1997; Wang et al., 1991)
	13;15	Maternal	15	(Nicholls et al., 1989; Smith et al., 1993)
	14;15	Maternal	15	(Trent et al., 1991)
Homologous (inherited)	13;13	Maternal	13	(Stallard et al., 1995; Slater et al., 1994)
	13;13	Paternal	13	(Slater et al., 1995)
	22;22	Maternal	22	(Palmer et al., 1980; Kirkels et al., 1980)
Heterologous (*de novo*)	13;14	Maternal	14	(Coviello et al., 1996; Linck et al., 1996; Antonarakis et al., 1993; Healey et al., 1994)
	14;15	Maternal	15	(Toth-Fejel et al., 1995)
Homologous (*de novo*)	14;14	Maternal	14	(Robinson et al., 1994; Pentao et al., 1992; Tomkins et al., 1996)
	14;14	Paternal	14	(Walter et al., 1996; Papenhausen et al., 1995)
	15;15	Maternal	15	(Bettio et al., 1995; Freund et al., 2000; Hamabe et al., 1991; Robinson et al., 1994; Saitoh et al., 1994; Narahara et al., 1992)
	15;15	Paternal	15	(Tomkins et al., 1996; Fridman et al., 1998; Freeman et al., 1993; Ramsden et al., 1996; Tonk et al., 1996)
	21;21	Maternal	21	(Creau-Goldberg et al., 1987)
	21;21	Paternal	21	(Blouin et al., 1993)

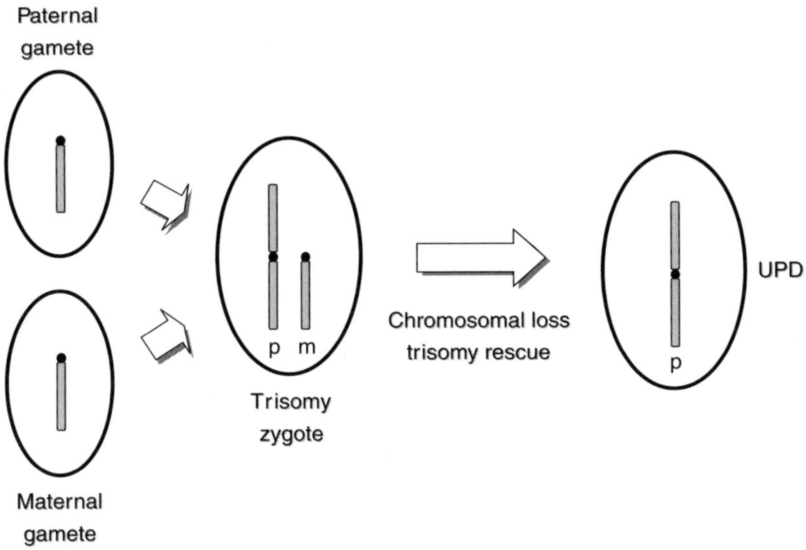

Figure 6 *Schematic representation of the trisomy rescue in the case of homologous* de novo Robertsonian translocation and the resulting outcome of UPD.

between 1978 and 1986, six cases of balanced translocations had been collected in patients with the PWS (Butler, 1990). Unlike the centric fusions, translocations of this type rarely lead to UPD. They are more often noted for causing deletions (or occasionally duplications) as a result of uneven crossing-over or for other unknown reasons. The most notorious have been responsible for, or associated with AS, PWS, and BWS. Some cases published in the 1990s associated with UPD, deletions, or more cryptic lesions, causing abnormal phenotypes due to imprinted genomic regions, are now listed in Table 2 and briefly discussed. Although the examples of translocations discussed below are not many, each one serves to illustrate a particular facet of its theoretical or clinical interest.

Case 1

When unbalanced segregation of the t(15;22) of case 1 included the der(22) with deletion 15 for a segment pter-q13 [without the complementary der(15)], the resulting imbalance, when originating from the father caused PWS in two children, while it produced AS when transmitted from his balanced daughter to her child. This pedigree provided compelling proof of the imprinting process as a function of the parental sex of origin (Hulten et al., 1991).

Case 2

The balanced t(6;15) led to different meiotic accidents, i.e., unequal crossing-over and 15q11-q13 deletion, accounting for the PWS offspring of one brother, and nondisjunction and UPD15, causing AS in the child born to the other brother! (Smeets et al., 1992).

Cases 3–5

The balanced translocations involved 11p15.4 or 11p15.5 as one breakpoint in women (Tommerup et al., 1993; Weksberg et al., 1993). When transmitted apparently unchanged, these translocations caused BWS, presumably by relaxation of the normal imprint of a maternal locus at 11p15.5, causing biparental expression.

Cases 6–7

These examples are of importance (Horsthemke et al., 1996) in prenatal diagnosis. In both situations, amniocentesis, performed because of the paternal translocations involving chromosome 15, revealed a normal 46,XY karyotype and absence of the translocation, which in both cases had involved the PWS critical region of 15q. When it appeared, after birth, that both children had the PWS, FISH and microsatellite analyses with markers mapping at the PWS/AS region showed lack of alleles for which the fathers were heterozygous, thus demonstrating a typical deletion. This suggested that, at paternal meiosis, unequal crossing-over between the derivative chromosomes 15 and their normal counterparts had occurred (Horsthemke et al., 1996). On the practical side, these cases

TABLE 2 Reciprocal Translocations and Genomic Imprinting

Case	Type	Reference	Disease	Carriers
1. Familial Balanced Autosomal Interchanges				
1	46,XY, t(15;22) (q13;q11)	(Hulten et al., 1991)	PWS, AS	Father and daughter
2	46,XY, t(6;15) (p25.3;q11.1)	(Smeets et al., 1992)	PWS, AS	Two brothers
3	46,XY, t(9;11) (p11.2;p15.5)	(Tommerup et al., 1993)	PWS	Father and daughter
4	46,XX, t(11;22) (p15.4;q11.2)	(Weksberg et al., 1993)	PWS	Mother
5	46,XX, t(11;16) (p15.5;q12)	(Weksberg et al., 1993)	PWS	Mother
6	46,XY, t(15; 18) (q12;q12)	(Horsthemke et al., 1996)	PWS	Father
7	46,XY, t(8;15) (q11.2;q11.2. q13)	(Horsthemke et al., 1996)	PWS	Father
8	46,XX, t(14;15) (q11.2;q11.2)	(Burke et al., 1996)	AS	Mother, sister
9	46,XX, t(13;15)(q12.3;q12-13)	(Tepperberg et al., 1993)	AS	Mother and half sister and brother
10	46,XX, t(1;14) (q32;q32)	(Wang et al., 1991)	UPD14	Mother
2. de novo *Balanced Autosomal Interchanges*				
11	46,XY, t(2;15) (q37.2;q11.2)	(Conroy et al., 1997)	PWS	*De novo*
12	46,XY t(15;19) (q12;q13.4)	(Sun et al., 1996)	PWS	
3. Familial or de novo*, Y; Autosome Interchanges, Balanced or not*				
13	46XY, t(Y;15) (q12; p11)	(Saitoh et al., 1997)	PWS	Father
14	45XY, t(Y;15) (q12; q11.2)	(Qumsiyeh et al., 1992)	PWS	
15	46,X ,t(Y;15) (p11.3; q11.3)	(Eliez et al., 1997)	PWS	

clearly indicated that prenatal diagnosis in such cases should include FISH and microsatellite analysis, to rule out deletion or UPD.

Cases 8–9

They feature short-arm and juxtacentromeric long-arm interchanges between acrocentrics 13 and 15 or 14 and 15 (Burke et al., 1996; Tepperberg et al., 1993) but represent also cases of cryptic translocations, which could only be demonstrated by FISH and microsatellite analyses. In case 8 (Burke et al., 1996), the AS proband had inherited an unbalanced form of the translocation, having received the abnormal 15 whose pericentromeric portion had come from the top of 14; there was therefore monosomy from 15pter to SNRPN. The methylation pattern was solely paternal in type at SNRPN, PW71, and ZNF127 loci. The mother and an unaffected sister, both with the balanced translocation, had a normal methylation test at these loci. This case suggested that the AS locus lies proximal to D15S10. This was later corroborated by the fact that the UBE3A gene is centromeric and adjacent to the 15 breakpoint of this cryptic rearrangement of the AS critical region (Matsuura et al., 1997; Kishino et al., 1997; Burke et al., 1996).

Case 10

This illustrates the extraordinary coincidence, in an unrelated couple, of translocations involving chromosome 14 in both partners; a centric fusion t(13;14) in the man and an interchange t(1;14) in the spouse (Wang et al., 1991). For the mechanism see case 1 of paternal UPD14 (Chapter 4).

Cases 11–15

All involve translocations of chromosome Y in a breakage–reunion partnership with proximal 15q, as a cause of deletions in PWS (Sun et al., 1996; Saitoh et al., 1997; Conroy et al., 1997; Vickers et al., 1994; Qumsiyeh et al., 1992; Eliez et al., 1997). Interestingly, in the case of Eliez et al. (1997), the translocation involved 15p, and the 15q deletion causing the PWS syndrome was unexpectedly *de novo*; it is unknown if the translocation predisposed to a 15q deletion.

5. MITOTIC HOMOLOGOUS INTERCHANGES (*de novo* SOMATIC RECOMBINATION)

They result from symmetrical homologous (but not sister) chromatid exchanges occurring in mitosis for causes as of yet unidentified. Examples of this mechanism were first demonstrated in Bloom's syndrome (German, 1993) and

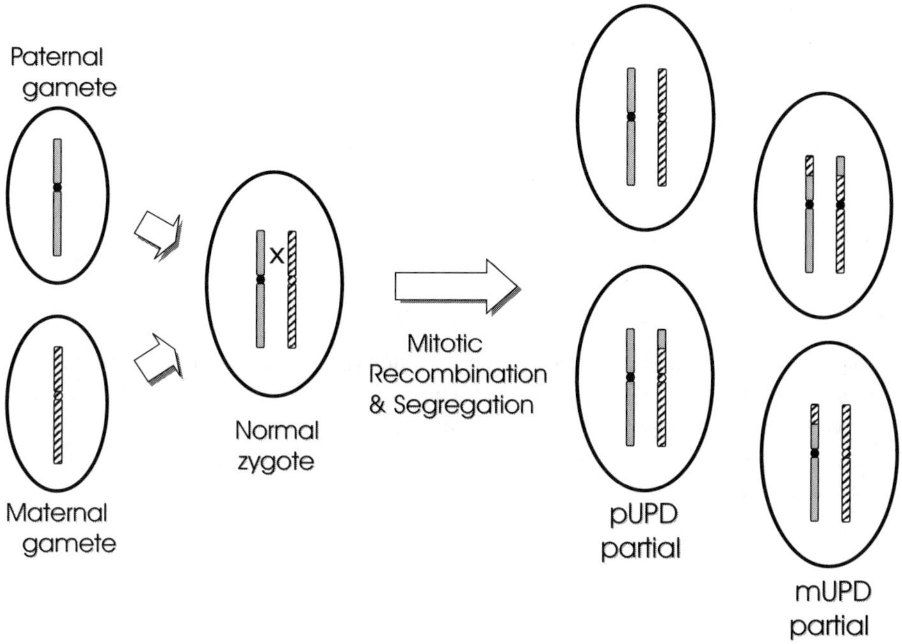

Figure 7 *Schematic representation of mitotic recombination and the resulting outcomes including a partial UPD for maternal or paternal alleles.*

Retinoblastoma (Cavenee, 1991). In the case of such a mitotic recombinational event, just as happens at normal meiosis, segments are symmetrically traded between one maternal and one paternal chromatid. Thus, the pair involved now includes a paternal segment on a maternal chromatid and vice-versa; these segments could be terminal (Henry et al., 1991) or interstitial (Stephens et al., 1996) and result in partial UPDs.

The end result of such individual cellular events will depend on the mitotic segregation in daughter cells of the four strands of both chromosomes involved. There are two modified and two intact, one of each for each chromosome (Figure 7). The theoretical results of the segregation are as follows: 50% biparental inheritance (25% identical to the parental chromosomes, 25% with reciprocal exchanges), 25% partial paternal UPD (isodisomy), and 25% partial maternal UPD (also isodisomy). Importantly, such events will introduce mosaic cell lineages, causing tissue confined or more widespread mosaicism, depending on the stage of occurrence. It is also of note that mitotic recombination leads to loss of heterozygosity in certain cell lineages. This is precisely what has now been documented in the BWS and a number of other pathologic conditions particularly with predisposition to tumors (Henry et al., 1991; Cavenee, 1991; Stephens et al., 1996).

6. GENOMIC IMPRINTING SYNDROMES AND PERI- OR PARACENTRIC INVERSIONS IN IMPRINTED DOMAINS

It is assumed that crossing-over in an inversion loop is the cause of interstitial deletions or rearrangements. It is therefore not surprising that, in the case of inversions of chromosome 15, segregating in the same family, more than one relative could be found affected with either AS or PWS, depending on the sex of the parent, whose inversion predisposed to a secondary deletion.

(A) Pericentric Inversions of Chromosome 15

A 15 inv(p11q13) was transmitted from a father to both his sons and a daughter without immediate ill-effects; however, when the same inversion was transmitted from the daughter to her son, it resulted in a deletion of chromosome 15, presumably the cause of AS in this boy (Webb et al., 1992). Analysis of polymorphisms revealed a *de novo* deletion in the proband, encompassing loci D15S9 to D15S13, while locus D15S24 maintained heterozygosity. As the boy's mother was heterozygous for several of these loci, it was concluded that the deletion in 15q12 had arisen *de novo*, most probably as a result of incorrect pairing at maternal meiosis; this maternally derived deletion was thus the explanation of the AS in the proband.

In another family, two cases of partial *de novo* 15q12 deletion occurred, derived from a chromosome 15 carrying a pericentric inversion (Chan et al., 1993). The cousins had AS and PWS, respectively, as the secondary deletion had arisen in the inverted maternal chromosome in the former and the paternal one in the latter! These cases also confirmed the sex-dependent imprinting of the genomic domains involved.

(B) Paracentric Inversions

These inversions could similarly lead to a secondary rearrangement with deletion in the inverted interstitial segment. Such a case was documented in two first cousins affected by AS and PWS, respectively, with the karyotype 46,XY, del(15)(pter-q11.2:q13-qter). These deletions extended from 15p34 to 15q to include locus CMW1 distally. This deletion was maternal in origin in the AS patient and paternal in the PWS. Methylation analysis of the DN34 region corroborated these results; the pattern was solely paternal in type in the AS and only maternal in the PWS patient (Clayton-Smith et al., 1993).

It is therefore safe, whenever an inversion in the chromosome 15 (or other chromosomes) is noted, pericentric or paracentric, familial or *de novo*, to use molecular markers and FISH analysis to rule out a deletion (and/or duplication) in the same region. Recurrence of cases within a same family confirms the relative frequency of such events, which might be more common in paracentric than pericentric inversions. These situations have been dealt with at some length because they not only cause interchromosomal rearrangements but also favor nondisjunction and possible UPD by interfering with synapsis at meiosis.

7. UPD IMPRINTING DISORDERS WITH SMALL MARKER CHROMOSOMES

At times, trisomic lines with small marker chromosomes, often in mosaic patterns, and occasionally small markers within pseudodipoid lineages can be found in somatic or extraembryonic tissues. In some of these cases, there is a cell population with UPD for the chromosome from which the marker was derived. Attention to this situation was attracted by several cases of invdup(15) and tiny rings. Thus, some cases of PWS or AS have been associated with invdup(15) (Webb, 1994; Robinson et al., 1993). Examples of isodisomy in cases with chromosomal rings and other minute chromosomes have also been identified, such as paternal UPD6 in cell lines with a small r(6) maternal (James et al., 1995) or UPD21 in cases of r(21) and a 21q-marker (Petersen et al., 1992). In the latter cases, the cells with abnormal chromosomes were replaced over time with those with a full complement of chromosomes.

(A) Invdup(15) or Idic(15)

They are among the most commonly observed marker chromosomes and their characteristics have been discussed by Webb (1994). These isodicentric members are derived from proximal 15q and have two centromeres separated by a variable amount of chromatin. When documentation is available, most are derived from a maternal chromosome. They are subcategorized into three classes:

Type 1. Small invdup, with breakpoints in 15q11 often appearing metacentric, and smaller than chromosomes 21–22.

Type 2. Medium-sized invdup, with breakpoints in 15q12, featuring a larger duplication of proximal 15q, and eventually the shape of a submetacentric. One of the two distinct centromeres is inactive.

Type 3. Large invdup with breakpoints in 15q13 or more telomeric, with a shape mimicking an acrocentric chromosome with satellites at both ends and a duplication likely to encompass the whole of the PWS/AS critical region.

As a rule, the larger the duplication, the more frequent an altered phenotype, often with mental retardation and seizure activity, is documented.

In PWS and AS, the invdup(15) has been of type 1 in all eight cases reviewed (Webb, 1994). The origin was documented in five cases: The marker chromosome was paternal and there was maternal UPD15 (Spinner et al., 1995; Christian et al., 1998; Robinson et al., 1993).

(B) Other Marker Chromosomes Found in UPD

The association of other marker chromosomes and UPD is less clear. The concomitant finding of a small r(6) maternal and a paternal UPD6 was observed in a study (James et al., 1995) to determine the association of UPD and supernumerary marker chromosomes (SMCs). The UPD6 was the only one observed in this study of 39 cases with SMCs, 17 originating from chromosome 15 and the others from a number of other autosomes. There are two cases with PWS and concomitant sex chromosomal anomalies: one with a minute marker derived from chromosome X (Bettio et al., 1997) and the other with XXY (Clayton-Smith et al., 1992). It seems that this is a coincidence without any etiologic significance.

8. UPD SUSPECTED BY CHROMOSOMAL POLYMORPHISMS OR HETEROMORPHISMS

Chromosomal polymorphic variations include: (i) larger than usual heterochromation blocks such as 1qh+, 9qh+, 16qh+; (ii) small pericentric inversions of the same areas, particularly for chromosome 9; (iii) short-arm, stalk and satellite configurations for some acrocentrics, including the presence of large or duplicated satellite knobs. Chromosomal heteromorphisms could include rearrangements such as inversions or occasional fragile sites. When both parental karyotypes are available, detailed analysis of the inherited heteromorphisms in the offspring may uncover some cases of UPD. This will occur, for example, when a heteromorphism of one parental chromosome is observed in two chromosomes of the offspring, or when there is no inheritance of a heteromorphism from a homozygous parent. Some suggestive heteromorphisms for chromosomes 3, 4, 9 and the acrocentrics have been observed in cases of suspected (Betz et al., 1974; Carpenter et al., 1982) or proven UPD (Willatt et al., 1992), respectively.

REFERENCES

Antonarakis, S. E., Blouin, J. L., Maher, J., Avramopoulos, D., Thomas, G. and Talbot, C.C.J. Maternal uniparental disomy for human chromosome 14, due to loss of a chromosome 14 from somatic cells with t(13;14) trisomy 14. *Am J Hum Genet* 52:1145–1152, 1993.

Barton, D. E., McQuaid, S., Stallings, R., Griffin, E. and Geraghty, M. Further evidence for an emerging maternal uniparental disomy chromosome 14 syndrome: Analysis of a phenotypically abnormal de novo Robertsonian translocation t(13;14) carrier. *Am J Hum Genet* 59:A125,1996.(Abstract)

Bennett, P., Vaughan, J., Henderson, D., Loughna, S. and Moore, G. Association between confined placental trisomy, fetal uniparental disomy, and early intrauterine growth retardation [letter]. *Lancet* 340:1284–1285, 1992.

Bettio, D., Rizzi, N., Giardino, D., et al. FISH analysis in Prader-Willi and Angelman syndrome patients. *Am J Med Genet* 56:224–228, 1995.

Bettio, D., Rizzi, N., Giardino, D., et al. FISH characterization of small supernumerary marker chromosomes in two Prader-Willi patients. *Am J Med Genet* 68:99–104, 1997.

Betz, A., Turleau, C. and de Grouchy, J. Heterozygosity and homozygosity for a pericentric inversion of human chromosone 3. *Ann Genet* 17:79–80, 1974.

Blouin, J. L., Avramopoulos, D., Pangalos, C. and Antonarakis, S. E. Normal phenotype with paternal uniparental isodisomy for chromosome 21. *Am J Hum Genet* 53:1074–1078, 1993.

Burke, L. W., Wiley, J. E., Glenn, C. C., et al. Familial cryptic translocation resulting in Angelman syndrome:implications for imprinting or location of the Angelman gene? *Am J Hum Genet* 58:777–784, 1996.

Butler, M. G. Prader-Willi syndrome: current understanding of cause and diagnosis. *Am J Med Genet* 35:319–332, 1990.

Carpenter, N. J., Say, B. and Barber, N. D. A homozygote for pericentric inversion of chromosome 4. *J Med Genet* 19:469–471, 1982.

Cassidy, S. B., Lai, L. W., Erickson, R. P., et al. Trisomy 15 with loss of the paternal 15 as a cause of Prader-Willi syndrome due to maternal disomy. *Am J Hum Genet* 51:701–708, 1992.

Cattanach, B. M. and Kirk, M. Differential activity of maternally and paternally derived chromosome regions in mice. *Nature* 315:496–498, 1985.

Cavenee, W. K. Recessive mutations in cancer predisposition and progression. *Basic Life Sci* 57:171–180, 1991.

Chan, C. T., Clayton-Smith, J., Cheng, X. J., et al. Molecular mechanisms in Angelman syndrome: a survey of 93 patients. *J Med Genet* 30:895–902, 1993.

Christian, S. L., Mills, P., Das, S. and Ledbetter, D. High risk of uniparental disomy 15 associated with amniotic fluid containing de novo small supernumerary marker 15 chromosomes. *Am J Hum Genet* 63:A11–A111998. (Abstract)

Clayton-Smith, J., Driscoll, D. J., Waters, M. F., et al. Difference in methylation patterns within the D15S9 region of chromosome 15q11-13 in first cousins with Angelman syndrome and Prader- Willi syndrome. *Am J Med Genet* 47:683–686, 1993.

Clayton-Smith, J., Webb, T., Robb, S. A., et al. Further evidence for dominant inheritance at the chromosome 15q11-13 locus in familial Angelman syndrome. *Am J Med Genet* 44:256–260, 1992.

Conroy, J. M., Grebe, T. A., Becker, L. A., et al. Balanced translocation 46,XY,t(2;15)(q37.2;q11.2) associated with atypical Prader-Willi syndrome. *Am J Hum Genet* 61:388–394, 1997.

Cotter, P. D., Kaffe, S., McCurdy, L. D., Jhaveri, M., Willner, J. P. and Hirschhorn, K. Paternal uniparental disomy for chromosome 14: a case report and review. *Am J Med Genet* 70:74–79, 1997.

Coviello, D. A., Panucci, E., Mantero, M. M., et al. Maternal uniparental disomy for chromosome 14. *Acta Genet Med Gemellol (Roma)* 45:169–172, 1996.

Creau-Goldberg, N., Gegonne, A., Delabar, J., et al. Maternal origin of a de novo balanced t(21q21q) identified by ets-2 polymorphism. *Hum Genet* 76:396–398, 1987.

Dworniczak, B., Koppers, B., Kurlemann, G., Holzgreve, W., Horst, J. and Miny, P. Uniparental disomy with normal phenotype [letter]. *Lancet* 340:12851992.

Eliez, S., Morris, M. A., Dahoun-Hadorn, S., et al. Familial translocation t(Y;15)(q12;p11) and de novo deletion of the Prader-Willi syndrome (PWS) critical region on 15q11-q13. *Am J Med Genet* 70:222–228, 1997.

Engel, E. [Uniparental disomy: a review of causes and clinical sequelae]. *Ann Genet* 38:113–136, 1995.

Ferguson-Smith, A. C., Cattanach, B. M., Barton, S. C., Beechey, C. V. and Surani, M. A. Embryological and molecular investigations of parental imprinting on mouse chromosome 7 [see comments]. *Nature* 351:667–670, 1991.

Freeman, S. B., May, K. M., Pettay, D., Fernhoff, P. M. and Hassold, T. J. Paternal uniparental disomy in a child with a balanced 15;15 translocation and Angelman syndrome. *Am J Med Genet* 45:625–630, 1993.

Freund, M., Dahan, K., Liebaers, I., Hubinont, C. and Verellen-Dumoulin, C. Prenatal diagnosis of the Prader-Willi syndrome through a 15;15 isochromosome and maternal isodisomy. *Eur J Hum Genet* 8:50–50, 2000. (Abstract)

Fridman, C., Varela, M. C., Nicholls, R. D. and Koiffmann, C. P. Unusual clinical features in an Angelman syndrome patient with uniparental disomy due to a translocation 15q15q. *Clin Genet* 54:303–308, 1998.

German, J. Bloom syndrome: a mendelian prototype of somatic mutational disease. *Medicine (Baltimore)* 72:393–406, 1993.

Hamabe, J., Fukushima, Y., Harada, N., et al. Molecular study of the Prader-Willi syndrome: deletion, RFLP, and phenotype analyses of 50 patients. *Am J Med Genet* 41:54–63, 1991.

Hassold, T. J. and Jacobs, P. A. Trisomy in man. *Annu Rev Genet* 18:69–97, 1984.

Healey, S., Powell, F., Battersby, M., Chenevix-Trench, G. and McGill, J. Distinct phenotype in maternal uniparental disomy of chromosome 14. *Am J Med Genet* 51:147–149, 1994.

Henry, I., Bonaiti-Pellie, C., Chehensse, V., et al. Uniparental paternal disomy in a genetic cancer-predisposing syndrome [see comments]. *Nature* 351:665–667, 1991.

Horsthemke, B., Maat-Kievit, A., Sleegers, E., et al. Familial translocations involving 15q11-q13 can give rise to interstitial deletions causing Prader-Willi or Angelman syndrome. *J Med Genet* 33:848–851, 1996.

Hulten, M., Armstrong, S., Challinor, P., et al. Genomic imprinting in an Angelman and Prader-Willi translocation family [letter] [see comments]. *Lancet* 338:638–639, 1991.

James, R. S., Temple, I. K., Dennis, N. R. and Crolla, J. A. A search for uniparental disomy in carriers of supernumerary marker chromosomes. *Eur J Hum Genet* 3:21–26, 1995.

Kalousek, D. K., Barrett, I. J. and Gartner, A. B. Spontaneous abortion and confined chromosomal mosaicism. *Hum Genet* 88:642–646, 1992.

Kalousek, D. K., Langlois, S., Barrett, I., et al. Uniparental disomy for chromosome 16 in humans. *Am J Hum Genet* 52:8–16, 1993.

Kirkels, V. G., Hustinx, T. W. and Scheres, J. M. Habitual abortion and translocation (22q;22q): unexpected transmission from a mother to her phenotypically normal daughter. *Clin Genet* 18:456–461, 1980.

Kishino, T., Lalande, M. and Wagstaff, J. UBE3A/E6-AP mutations cause Angelman syndrome [published erratum appears in Nat Genet 1997 Apr;15(4):411]. *Nat Genet* 15:70–73, 1997.

Linck, L., McMilin, K., Popovich, B. and Magenis, R. E. Maternal uniparental disomy for chromosome 14. *Am J Hum Genet* 59:A124, 1996. (Abstract)

Markovic, V. D., Chitayat, D. A., Ritchie, S. M., Chodakowski, B. A. and Hutton, E. M. Trisomy 15 mosaic derived from trisomic conceptus: report of a case and a review. *Am J Med Genet* 61:363–370, 1996.

Matsuura, T., Sutcliffe, J. S., Fang, P., et al. De novo truncating mutations in E6-AP ubiquitin-protein ligase gene (UBE3A) in Angelman syndrome. *Nat Genet* 15:74–77, 1997.

Mattei, J. F., Mattei, M. G. and Giraud, F. Prader-Willi syndrome and chromosome 15. A clinical discussion of 20 cases. *Hum Genet* 64:356–362, 1983.

Mattei, M. G., Souiah, N. and Mattei, J. F. Chromosome 15 anomalies and the Prader-Willi syndrome: cytogenetic analysis. *Hum Genet* 66:313–334, 1984.

Milunsky, J. M., Wyandt, H. E., Huang, X. L., Kang, X. Z., Elias, E. R. and Milunsky, A. Trisomy 15 mosaicism and uniparental disomy (UPD) in a liveborn infant. *Am J Med Genet* 61:269–273, 1996.

Narahara, K., Hiramoto, K., Murakami, M., et al. Unique karyotypes in two patients with Prader-Willi syndrome. *Am J Med Genet* 42:671–677, 1992.

Nicholls, R. D., Knoll, J. H., Butler, M. G., Karam, S. and Lalande, M. Genetic imprinting suggested by maternal heterodisomy in nondeletion Prader-Willi syndrome. *Nature* 342:281–285, 1989.

Nicolaidis, P. and Petersen, M. B. Origin and mechanisms of non-disjunction in human autosomal trisomies. *Hum Reprod* 13:313–319, 1998.

Page, S. L., Shin, J. C., Han, J. Y., Choo, K. H. and Shaffer, L. G. Breakpoint diversity illustrates distinct mechanisms for Robertsonian translocation formation. *Hum Mol Genet* 5:1279–1288, 1996.

Palmer, C. G., Schwartz, S. and Hodes, M. E. Transmission of a balanced homologous t(22q;22q) translocation from mother to normal daughter. *Clin Genet* 17:418–422, 1980.

Papenhausen, P. R., Mueller, O. T., Johnson, V. P., Sutcliffe, M., Diamond, T. M. and Kousseff, B. G. Uniparental isodisomy of chromosome 14 in two cases: an abnormal child and a normal adult [see comments]. *Am J Med Genet* 59:271–275, 1995.

Pentao, L., Lewis, R. A., Ledbetter, D. H., Patel, P. I. and Lupski, J. R. Maternal uniparental isodisomy of chromosome 14: association with autosomal recessive rod monochromacy. *Am J Hum Genet* 50:690–699, 1992.

Petersen, M. B., Bartsch, O., Adelsberger, P. A., Mikkelsen, M., Schwinger, E. and Antonarakis, S. E. Uniparental isodisomy due to duplication of chromosome 21 occurring in somatic cells monosomic for chromosome 21. *Genomics* 13:269–274, 1992.

Purvis-Smith, S. G., Saville, T., Manass, S., et al. Uniparental disomy 15 resulting from "correction" of an initial trisomy 15 [letter]. *Am J Hum Genet* 50:1348–1350, 1992.

Qumsiyeh, M. B., Dalton, J. D., Gordon, P. L., Wilroy, R. S. and Tharapel, A. T. Deletion of chromosome 15pter->q11.2 due to t(Y;15) in a boy with Prader-Willi syndrome. *Am J Med Genet* 42:109–111, 1992.

Ramsden, S., Gaunt, L., Seres-Santamaria, A. and Clayton-Smith, J. A case of Angelman syndrome arising as a result of a de novo Robertsonian translocation. *Acta Genet Med Gemellol (Roma)* 45:255–261, 1996.

Robinson, W. P., Barrett, I. J., Bernard, L., et al. Meiotic origin of trisomy in confined placental mosaicism is correlated with presence of fetal uniparental disomy, high levels of trisomy in trophoblast, and increased risk of fetal intrauterine growth restriction. *Am J Hum Genet* 60:917–927, 1997.

Robinson, W. P., Bernasconi, F., Basaran, S., et al. A somatic origin of homologous Robertsonian translocations and isochromosomes. *Am J Hum Genet* 54:290–302, 1994.

Robinson, W. P., Binkert, F., Bernasconi, F., Lorda-Sanchez, I., Werder, E. A. and Schinzel, A.A. Molecular studies of chromosomal mosaicism: relative frequency of chromosome gain or loss and possible role of cell selection. *Am J Hum Genet* 56:444–451, 1995.

Robinson, W. P., Kuchinka, B. D., Bernasconi, F., et al. Maternal meiosis I non-disjunction of chromosome 15: dependence of the maternal age effect on level of recombination. *Hum Mol Genet* 7:1011–1019, 1998.

Robinson, W. P., Langlois, S., Schuffenhauer, S., et al. Cytogenetic and age-dependent risk factors associated with uniparental disomy 15. *Prenat Diagn* 16:837–844, 1996.

Robinson, W. P., Wagstaff, J., Bernasconi, F., et al. Uniparental disomy explains the occurrence of the Angelman or Prader-Willi syndrome in patients with an additional small inv dup(15) chromosome. *J Med Genet* 30:756–760, 1993.

Saitoh, S., Buiting, K., Cassidy, S. B., et al. Clinical spectrum and molecular diagnosis of Angelman and Prader-Willi syndrome patients with an imprinting mutation. *Am J Med Genet* 68:195–206, 1997.

Saitoh, S., Mutirangura, A., Kuwano, A., Ledbetter, D. H. and Niikawa, N. Isochromosome 15q of maternal origin in two Prader-Willi syndrome patients previously diagnosed erroneously as cytogenetic deletions. *Am J Med Genet* 50:64–67, 1994.

Simoni, G. and Sirchia, S. M. Confined placental mosaicism. *Prenat Diagn* 14:1185–1189, 1994.

Slater, H., Shaw, J. H., Bankier, A., Forrest, S. M. and Dawson, G. UPD 13: no indication of maternal or paternal imprinting of genes on chromosome 13 [letter; comment]. *J Med Genet* 32:4931995.

Slater, H., Shaw, J. H., Dawson, G., Bankier, A. and Forrest, S. M. Maternal uniparental disomy of chromosome 13 in a phenotypically normal child [see comments]. *J Med Genet* 31:644–646, 1994.

Smeets, D. F., Hamel, B. C., Nelen, M. R., et al. Prader-Willi syndrome and Angelman syndrome in cousins from a family with a translocation between chromosomes 6 and 15 [see comments]. *N Engl J Med* 326:807–811, 1992.

Smith, A., Robson, L., Neumann, A., et al. Fluorescence in-situ hybridisation and molecular studies used in the characterisation of a Robertsonian translocation (13q15q) in Prader-Willi syndrome. *Clin Genet* 43:5–8, 1993.

Spence, J. E., Perciaccante, R. G., Greig, G. M., et al. Uniparental disomy as a mechanism for human genetic disease. *Am J Hum Genet* 42:217–226, 1988.

Spinner, N. B., Zackai, E., Cheng, S. D. and Knoll, J. H. Supernumerary invdup(15) in a patient with Angelman syndrome and a deletion of 15q11-q13. *Am J Med Genet* 57:61–65, 1995.

Stallard, R., Krueger, S., James, R. S. and Schwartz, S. Uniparental isodisomy 13 in a normal female due to transmission of a maternal t(13q13q). *Am J Med Genet* 57:14–18, 1995.

Stephens, K., Weaver, D. D., Leppig, K., Side, L. E., Shannon, K. M. and Maruyama, K. Somatic confined uniparental disomy of the NF1 gene region in myeloid leukemic cells of an NF1 patient mimics a loss of heterozygosity due to deletion. *Am J Hum Genet* 59:A51996. (Abstract)

Sun, Y., Nicholls, R. D., Butler, M. G., Saitoh, S., Hainline, B. E. and Palmer, C. G. Breakage in the SNRPN locus in a balanced 46,XY,t(15;19) Prader-Willi syndrome patient. *Hum Mol Genet* 5:517–524, 1996.

Temple, I. K., Cockwell, A., Hassold, T., Pettay, D. and Jacobs, P. Maternal uniparental disomy for chromosome 14. *J Med Genet* 28:511–514, 1991.

Tepperberg, J. H., Tennison M. B., Kaiser-Rogers K., Albright S. G., Aylsworth A. S. and Rao K. W. An inherited cryptic translocation between chromosomes 13 and 15, detected by FISH, in a child with Angelman syndrome. *Am J Hum Genet* 53:609–609, 1993.(Abstract)

Tomkins, D. J., Roux, A. F., Waye, J., Freeman, V. C., Cox, D. W. and Whelan, D. T. Maternal uniparental isodisomy of human chromosome 14 associated with a paternal t(13q14q) and precocious puberty. *Eur J Hum Genet* 4:153–159, 1996.

Tommerup, N., Brandt, C. A., Pedersen, S., Bolund, L. and Kamper, J. Sex dependent transmission of Beckwith-Wiedemann syndrome associated with a reciprocal translocation t(9;11)(p11.2;p15.5). *J Med Genet* 30:958–961, 1993.

Tonk, V., Schultz, R. A., Christian, S. L., Kubota, T., Ledbetter, D. H. and Wilson, G. N. Robertsonian (15q;15q) translocation in a child with Angelman syndrome: evidence of uniparental disomy. *Am J Med Genet* 66:426–428, 1996.

Toth-Fejel, S., Magenis, R. E., Leff, S., et al. Prenatal diagnosis of chromosome 15 abnormalities in the Prader-Willi/Angelman syndrome region by traditional and molecular cytogenetics. *Am J Med Genet* 55:444–452, 1995.

Trent, R. J., Volpato, F., Smith, A., et al. Molecular and cytogenetic studies of the Prader-Willi syndrome. *J Med Genet* 28:649–654, 1991.

Vickers, S., Dahlitz, M., Hardy, C., Kilpatrick, M. and Webb, T. A male with a de novo translocation involving loss of 15q11q13 material and Prader-Willi syndrome. *J Med Genet* 31:478–481, 1994.

Voss, R., Ben-Simon, E., Avital, A., et al. Isodisomy of chromosome 7 in a patient with cystic fibrosis: could uniparental disomy be common in humans? *Am J Hum Genet* 45:373–380, 1989.

Walter, C. A., Shaffer, L. G., Kaye, C. I., et al. Short-limb dwarfism and hypertrophic cardiomyopathy in a patient with paternal isodisomy 14: 45,XY,idic(14)(p11). *Am J Med Genet* 65:259–265, 1996.

Wang, J. C., Passage, M. B., Yen, P. H., Shapiro, L. J. and Mohandas, T. K. Uniparental heterodisomy for chromosome 14 in a phenotypically abnormal familial balanced 13/14 Robertsonian translocation carrier. *Am J Hum Genet* 48:1069–1074, 1991.

Webb, T. Invdup(15) supernumerary marker chromosomes. *J Med Genet* 31:585–594, 1994.

Webb, T., Clayton-Smith, J., Cheng, X. J., et al. Angelman syndrome with a chromosomal inversion 15 inv(p11q13) accompanied by a deletion in 15q11q13. *J Med Genet* 29:921–924, 1992.

Weksberg, R., Teshima, I., Williams, B. R., et al. Molecular characterization of cytogenetic alterations associated with the Beckwith-Wiedemann syndrome (BWS) phenotype refines the localization and suggests the gene for BWS is imprinted. *Hum Mol Genet* 2:549–556, 1993.

Willatt, L. R., Davison, B. C., Goudie, D., et al. A male with trisomy 9 mosaicism and maternal uniparental disomy for chromosome 9 in the euploid cell line. *J Med Genet* 29:742–744, 1992.

Wolstenholme, J. Confined placental mosaicism for trisomies 2, 3, 7, 8, 9, 16, and 22: their incidence, likely origins, and mechanisms for cell lineage compartmentalization. *Prenat Diagn* 16:511–524, 1996.

Chapter 4

Uniparental Disomy for Individual Human Chromosomes: Review of Cases

There are 47 possible uniparental chromosomal pairs, each of which could, in theory, substitute for a biparental pair, resulting from a Me1 or Me2 error or a mitotic duplication. These 47 pairs could be formed by the doubling of either a maternal or paternal chromosome, thus offering the theoretical possibility of 46 chromosomal doublets to which the potential of a monoparental XY should be added (substitute for the usual XY biparental pair). In this chapter, the vast majority of published cases of UPD are presented and discussed. Figure 1 summarizes the chromosomes for which UPD has been reported as of December 2000. Table 1 lists the autosomal recessive phenotypes due to homozygosity of mutant alleles because of UPD.

CHROMOSOME 1 UPD

Maternal UPD1

Case 1

A male patient with junctional epidermolysis bullosa of the Herlitz type (H-JEB; OMIM 226700) was described who was homozygous for a nonsense mutation, Q243X, in the LAMB3 (beta3 subunit of laminin 5) gene on chromosome 1 and

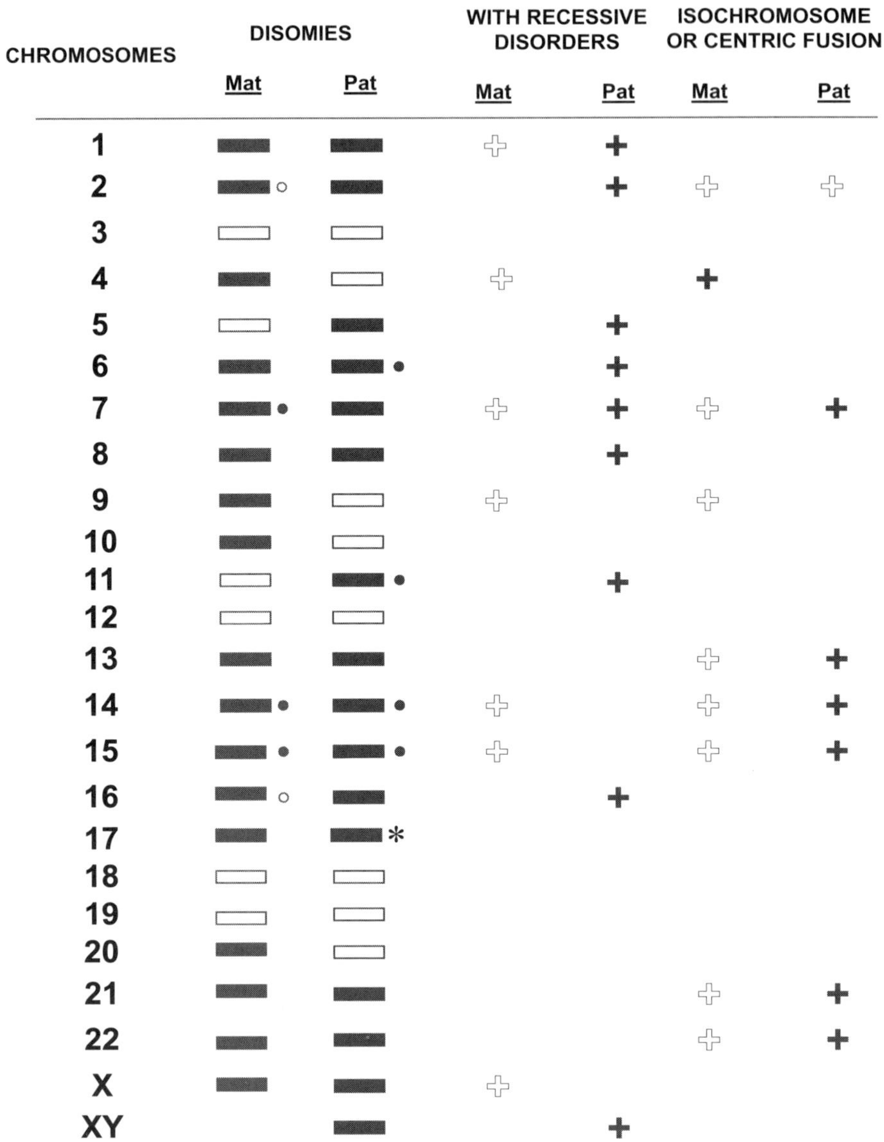

Figure 1 Overview of reported cases of UPD. Gray indicates cases of paternal and maternal origins in the appropriate columns. In the column "disomies," filled rectangles show chromosomes for which UPD has been reported at least once, empty rectangles represent those members for which no case of UPD has yet been described. The circles beside the rectangles indicate a confirmed (filled circles) or potential (open circles) association with genomic imprinting. Under the column "with recessive disorders," a + sign indicates that at least one patient has been described with a recessive condition caused by UPD for this chromosome. In the column "with isochromosome or centric fusion," the + signs indicate that such chromosomal rearrangements have been observed with UPD. (*Paternal UPD17 was described only once, and only involving an interstitial q segment of chromosome 17.)

TABLE 1 Autosomal Resessive Diseases due to Uniparental Disomy

Disorder	OMIM	Gene	UPD	References
Junctional epidermolysis bullosa; Herlitz type	226700	LAMB3	Mat UPD1	(Pulkkinen et al., 1997)
Junctional epidermolysis bullosa; Herlitz type	226700	LAMC2	Pat UPD1	(Takizawa et al., 2000)
Chediak-Higashi syndrome	214500	LYST	Mat UPD1	(Dufourcq-Lagelouse et al., 1999)
Pycnodysostosis	265800 601105	CTSK	Pat UPD1	(Gelb et al., 1998)
Congenital pain insensitivity with anhidrosis (CIPA)	256800	TRKA	Pat UPD1	(Miura et al., 2000)
5alpha-reductase 2 deficiency	264600	SRD5A2	Pat UPD2	(Chavez et al., 2000)
Abetalipoproteinemia	157147 200100	MTP	Mat UPD4	(Yang et al., 1999)
Spinal muscular atrophy	253300	SMA	Pat UPD5	(Brzustowicz et al., 1994)
Complement C4 deficiency	120810	C4	Pat UPD6	(Welch et al., 1990)
Methylmalonic acidemia	251000	MUT	Pat UPD6	(Abramowicz et al., 1994)
21-hydroxylase deficiency	201910	CYP21	Pat UPD6	(Lopez-Gutierrez et al., 1998)
Cystic fibrosis	602421 219700	CFTR	Mat UPD7	(Spence et al., 1988) (Voss et al., 1989)
Osteogenesis imperfecta	120160	COL1A2	Mat UPD7	(Spotila et al., 1992)
Congenital chloride diarrhea	214700 126650	DRA	Pat UPD7	(Hoglund et al., 1994)
Cystic fibrosis and immotile cilia syndrome	602421 219700 242650	CFTR	Pat UPD7	(Pan et al., 1998)
Lipoprotein lipase deficiency	238600	LPL	Pat UPD8	(Benlian et al., 1996)
Cartilage-hair hypoplasia	250250 157660	RMRP	Mat UPD9	(Sulisalo et al., 1997) (Ridanpaa et al., 2000)
Leigh syndrome	256000 185620	SURF1	Mat UPD9	(Tiranti et al., 1999)
Beta thalassemia	141900	HBB	Pat UPD11	(Beldjord et al., 1992)
Rod monochromacy	603096		Mat UPD14	(Pentao et al., 1992)
Bloom syndrome	10900	BLM	Mat UPD15	(Woodage et al., 1994)
Alpha thalassemia	141800	HBA	Pat UPD16	(Ngo et al., 1993)

who had normal karyotype 46,XY. This is a lethal autosomal recessive disorder characterized by blister formation at the level of the lamina lucida within the cutaneous basement-membrane zone. The proband was a newborn male with unrelated healthy parents who were of Scottish-Irish (mother) and German-Irish (father) descent and had no family history of blistering skin diseases. At birth, he was noted to have abnormal nails, and, within a few days of birth, he developed blisters first in the mouth and then covering the entire body. The proband died at the age of 2 months from complications of the disease. The mother was found to be a carrier of the Q243X mutation, whereas the father had two normal LAMB3 alleles. Nonpaternity was excluded by use of 11 microsatellite markers from six different chromosomes. The use of informative microsatellite markers spanning the entire chromosome 1 revealed that the patient had both maternal uniparental isodisomy of a 35-cM region on 1q containing the maternal LAMB3 mutation and maternal uniparental heterodisomy of other regions of chromosome 1 (Pulkkinen et al., 1997; Takizawa et al., 1998). The authors concluded that the H-JEB phenotype in this patient was due to a reduction to homozygosity of the 1q region containing the maternal LAMB3 mutation (Table 1). The patient was normally developed at term and did not show overt dysmorphisms or malformations.

Case 2

During a genome screening of 77 families having at least two children affected with type 1 (insulin-dependent) diabetes, in order to identify by linkage analysis genes predisposing to this disorder, a British family was noted to have a child with maternal uniparental disomy for loci on chromosome 1 (Field et al., 1998). Nonpaternity was excluded. The pattern of the molecular analysis was consistent with maternal uniparental heterodisomy (probably arising from nondisjunction during meiosis I), with an embedded region of homozygosity (secondary isodisomy) on the short arm created by a double exchange event. The isodisomic region within the double-exchange included markers D1S159, D1S410, D1S1665, D1S550, D1S1728, D1S551, D1S1159, which have all been cytogenetically localized between 1p21 and 1p32. The mother and both of the two children in this family have type 1 diabetes, and all three individuals have HLA genotypes associated with a high risk of developing diabetes. The authors assumed that the presence of chromosome 1 UPD in one of the diabetic children is unrelated to her IDDM. Apart from her diabetes, she had no other unusual conditions, and there was no evidence of dysmorphic features at birth. She had a full-term birthweight of 2930 gm with no indication of intrauterine growth retardation. Subsequently (the patient was 23 years old at the time of publication), she showed no signs of mental or developmental retardation or precocious puberty. This case of maternal chromosome 1 UPD in a developmentally normal adult provides valuable additional evidence that there are probably no imprinted genes on chromosome 1 with major phenotypic effects.

Case 3

Chediak-Higashi syndrome (CHS) is a rare autosomal recessive disorder (incidence around 1 in 10^6 births), characterized by complex immunologic defects, reduced pigmentation, and presence of giant granules in many different cell types. It most likely results from defective organellar trafficking or protein sorting. The causative gene (LYST) is homologous to the beige locus in the mouse. A unique patient with CHS was reported, who was homozygous for one nucleotide deletion, leading to frameshift and premature stop codon in the LYST gene on chromosome 1, and who had a normal 46,XY karyotype. The mother was found to be a carrier of the mutation, whereas the father had two normal LYST alleles. Nonpaternity was excluded by the analysis of microsatellite markers from different chromosomes. The results of 13 informative microsatellite markers spanning the entire chromosome 1 revealed that the proband had a maternal UPD1 (isodisomy) encompassing the LYST mutant allele (Dufourcq-Lagelouse et al., 1999). The authors concluded that the proband's clinical presentation confirms the absence of imprinted genes on chromosome 1 (Dufourcq-Lagelouse et al., 1999).

Paternal UPD1

Case 1

Molecular analysis of a patient affected by the autosomal recessive skeletal dysplasia, pycnodysostosis (MIM 265800), revealed homozygosity for a missense mutation A277V in the cathepsin K gene. Since the A277V mutation was carried by the patient's father but not by his mother, who had two normal cathepsin K alleles, paternal uniparental disomy of chromosome 1, to which cathepsin K is mapped, was suspected (Gelb et al., 1998). Karyotyping of the patient and of both parents was normal, and high-resolution cytogenetic analyses of chromosome 1 revealed no abnormalities. Evaluation of polymorphic DNA markers spanning chromosome 1 demonstrated that the patient had inherited two paternal chromosome 1 homologues, whereas alleles for markers from other chromosomes were inherited in a Mendelian fashion. The patient was homoallelic for informative markers mapping near the chromosome 1 centromere, but he was heteroallelic for markers near both telomeres, establishing that the paternal uniparental disomy with partial isodisomy was probably caused by a meiosis 2 nondisjunction event. Phenotypically, the patient had normal height and weight at birth, had normal psychomotor development at age 7 years, and had only the usual features of pycnodysostosis. The phenotype of this patient was due to the homozygosity for a recessive allele, and therefore this case provides conclusive evidence that paternally derived genes on human chromosome 1 are not imprinted.

Case 2

A 43-year-old woman was reported who was referred for evaluation because of minor facial anomalies, myopathy, sterility, short stature, hearing loss, downward slant of palpebral fissures, bilateral ptosis, severe micro/retrognathia, high arched palate, and scoliosis (Chen et al., 1999). Cytogenetic analyses showed the presence of one i(1p) and one i(1q) without normal chromosome 1 homologues. Fluorescence *in situ* hybridization analysis showed hybridization to only two chromosomes, consistent with the G-banded interpretation of i(1p) and i(1q). Molecular investigations using markers for chromosome 1 showed inheritance of only one set of paternal alleles and absence of any maternal alleles in the patient. The authors concluded that the adverse phenotype of the patient may be due to one or more recessive mutations, genomic imprinting, or a combination of both. However, it is also possible that the isochromosomes 1p and 1q contain undetectable chromosomal gains or losses.

Case 3

A male patient with CIPA (congenital insensitivity to pain with anhidrosis) (MIM 256800) was described who developed normally at term and did not show overt dysmorphisms or malformations. The human TRKA gene on chromosome 1q21-q22 encodes a receptor tyrosine kinase for nerve growth factor and is responsible for this phenotype. This case was shown to be due to a homozygous mutation (deletion of C at nt 1726 in exon 14 that caused a frameshift and premature termination codon) in the TRKA gene. The patient had only the usual features of CIPA. Haplotype analysis of the TRKA locus and genotypes for different polymorphisms on the entire chromosome 1 revealed that the chromosome pair was exclusively derived from his father (Miura et al., 2000). Nonmaternity was excluded by analyses of autosomes other than chromosome 1. Thus, there was a complete paternal isodisomy for chromosome 1 causing reduction to homozygosity of the TRKA gene mutation, leading to CIPA.

Case 4

A patient of Japanese origin with Herlitz junctional epidermolysis bullosa (MIM 226700), who died at the age of 8 months from complications of the disease, was reported. The mutation analysis revealed that the proband was homozygous for a nonsense mutation C553X in the LAMC2 gene encoding the gamma2 chain of laminin 5. The father was a heterozygous carrier of this mutation, whereas the mother had two normal alleles of this gene. The patient showed homozygosity for 15 known intragenic polymorphisms in the LAMC2 gene. Furthermore, genotype analysis, performed from the parents and the proband, using 16 microsatellite markers spanning the entire chromosome 1, revealed that the patient was homozygous for all markers tested, and that these alleles originated from the father. Among the 16 markers, eight were fully informative for the absence of the maternal chromosome 1 in the proband, suggesting that the patient had complete paternal isodisomy of this chromosome. Thus, the Herlitz junctional epidermolysis bullosa phenotype in this patient was caused by homozygous LAMC2 mutation that was of paternal origin and probably resulted from nondisjunction and uniparental disomy involving monosomy rescue (Takizawa et al., 2000).

CHROMOSOME 2 UPD

Paternal UPD2

To our knowledge, no paternal UPD2 has been described yet (data up to December 2000). The following case is, however, compatible with paternal UPD2.

Case 1

DNA analyses of the steroid 5alpha-reductase 2 gene (SRD5A2) were performed in two unrelated subjects bearing the enzyme deficiency and revealed differences in the mode of transmission for the disease. In both families, the fathers were carriers for an E197D mutation, whereas the mothers were carriers for a P212R mutation. Patient 1 was identified as compound heterozygote because he had both alterations (E197D/P212R). On the contrary, patient 2 was found to be homozygous, but only for the paternal mutation. Because this finding could not be explained on the basis of nonpaternity or a chromosomal abnormality, the presence of uniparental disomy was suggested (Chavez et al., 2000). The reduction to homozygosity for the E197D mutation, as confirmed by restriction analysis, supported this view. The results of this study give evidence of the first case of 5alpha-reductase deficiency resulting from paternal UPD2, since the SRD5A2 gene map on 2p23.

Maternal UPD2

Case 1

Case 1 (Figure 5 in Chapter 5) was that of a male whose UPD2 was suspected from the result of an amniocentesis, 46,XY/ 47,XY,+2, carried out at 17 weeks for abnormal serum HCG levels, in a 31-year-old woman (Harrison et al., 1995). Intrauterine growth retardation (IUGR) and oligohydramnios were present and a cesarean section was performed at 36 weeks. The baby, although vigorous, was of small weight (10th centile) and head circumference (10–25th centile). There were no malformations or major dysmorphic features. Cytogenetic examination showed diploid cord blood cells (100/100) and diploid buccal epithelial cells by interphase FISH (50/50), but trisomy 2 was observed in 10–30% of chorionic villi and amnion cells. Over a 14-month period, he developed hyaline membrane disease complicated by bronchopulmonary dyplasia and a need for oxygen therapy until 11 months. He was also treated for primary hypothyroidism in the neonatal period. At 31 months, he was alert, speaking, interactive but his height and weight remained in the 5%.

Analysis of polymorphic markers showed segmental maternal iso- and -heterodisomy of chromosome 2, in a pattern suggestive of Me1 maternal non-disjunction. Trisomy 2 was not observed in blood lymphocytes or buccal mucous membrane cells, but no other tissues were available.

In summary, the child had normal development, aside from lasting growth failure, treated hypothyroidism, bronchopulmonary problems and maternal UPD2, with the detectable trisomy 2 cell line being found only in extra-embryonic tissues. The authors noted that human chromosome 2 has homologous segments on mouse chromosomes 1, 2, 6, 8, 11, 12, 17, the latter showing imprinted areas on 1, 2, 6, 11. They also indicated that human chromosome 2 carries the surfactant associated protein 3 and thyroid peroxydase loci. The phenotype was discussed as a function of genomic imprinting, isodisomy and undetected somatic trisomy mosaicism, and placental dysfunction caused by trisomy 2 mosaicism (Harrison et al., 1995).

Case 2

This case of UPD2 was initially suspected following a placental biopsy done at 23 weeks' gestation, for severe IUGR and oligohydramnios in a 23-year-old woman (Hansen et al., 1997). Although chorionic villi showed trisomy 2/euploid mosaicism, the karyotype was diploid on amniotic fluid. The fetus was delivered at 36 weeks and died during the neonate period. At autopsy, severe IUGR, hypospadias, and undescended testes were noted, along with effects secondary to oligohydramnios including pulmonary hypoplasia, nasal deviation, ankle deformities, and joint contractures. While trisomy 2 mosaicism was confirmed in placental biopsies, four different fetal tissues failed to show it. UPD2 was diagnosed using microsatellite markers, with two maternal alleles only in the fetus and three alleles, including a paternal contribution, in villi. In the fetus, centromere markers disclosed heterodisomy, indicating that the uniparental fetal pair had originated from a conceptus trisomic because of Me1 maternal chromosome 2 nondisjunction. The authors suggested that either the maternal UPD2 or the placental malfunction secondary to confined placental mosaicism (CPM) for chromosome 2 was responsible for the fetal phenotype.

Case 3

This case was identified after the detection of two isochromosomes for 2p and 2q, in a child with growth retardation and perineoscrotal hypospadias (Shaffer et al., 1997).

The proband was the 785 gm product of a 41-week gestation complicated by pre-eclampsia and marked oligohydramnios. The newborn developed severe bronchopulmonary dysplasia. At 8 years of age, height was 116 cm (−2.5 DS), with a normal growth velocity and normal bone age. He presented pectus carinatum and bilateral external earpits but normal intellectual development was reported. Maternal isodisomy was demonstrated using eight highly polymorphic chromosome 2 markers, consistent with the homozygosity of each chromosomal arm. The authors emphasized that this patient shares the features of IUGR, oligohydramnios, short stature, hypospadias, and bronchopulmonary hypoplasia with the other cases of maternal UPD2, and they postulated that the mat UPD2 phenotype is more likely to be caused by a chromosome 2 maternal imprinting effect than by homozygosity for a recessive allele or fetal damage due to confined placental mosaicism (Shaffer et al., 1997).

Case 4

This case was the result of the 4th pregnancy of a 43-year-old mother and 50-year-old father (Webb et al., 1996). On CVS, at 11 weeks, both direct and cultured specimens showed uniform trisomy 2. At 14 weeks, the fetal size and amniotic fluid volume were normal; AF culture revealed a mosaic euploid/trisomy 2 female karyotype. Oligohydramnios was observed at 20 weeks; fetal growth declined as of the late second trimester and IUGR was present at 29 weeks preceded by anhydramnios. Fetal blood sampling around 39 weeks of gestation showed a normal karyotype, but stagnant fetal growth and central circulatory problem led to the C-section delivery of a 765-gm female. She developed renal problems, and had surgery for patent ductus arteriosus at 1 month, pylorotomy, and hiatus hernia operation at 3 months. From 4–8 months, weight declined from −3 to −5 SD, but no dysmorphic features or other developmental problems were observed. The study of DNA polymorphisms at various loci demonstrated that the trisomy 2 originated from meiotic nondisjunction at oogenesis, with subsequent maternal UPD2, by loss of the paternal chromosome 2. Segments of both heterodisomy and isodisomy were present in the nondisjoined maternal pair; there was isodisomy at the most centromeric of the loci, which suggested a Me2 nondisjunction.

The authors, comparing this case to the others with proven maternal UPD2, noted that they all showed IUGR and oligohydramnios, without consistency of other clinical features. This left as unanswered the possible contributions of placental trisomy, UPD effects related to imprinting, and homozygosity for recessive alleles. They underlined the significant homologies of chromosome 2p and proximal 2q with regions of mouse chromosomes 2, 6, 11, 12, 17 containing imprinted segments.

Case 5

This 36-year-old, normal healthy female was studied because of five pregnancies ending in spontaneous abortion (Bernasconi et al., 1996). Similar to the subject in case 3 (Shaffer et al., 1997), she had isochromosomes of 2pi(2)(p10)] and 2q [i(2)(q10)] substituting for the normal pair of chromosomes 2. On molecular investigation, the isochromosomes showed complete homozygosity for maternal polymorphic alleles. She had been born to a 25-year-old mother, after a normal pregnancy, with a birthweight of 3.0 kg. Postnatal physical and mental development was normal as was puberty. Achievements in school were comparable to those of her siblings. Her weight was 55 kg and her height 155 cm.

Given that the normal proband presented none of the phenotypic signs noted in other cases of maternal UPD2, the authors expressed the view that such anomalies could have been due to undetected trisomy mosaicism, suggesting that there might be no imprinted genes on maternal chromosome 2 (Bernasconi et al., 1996).

Case 6

Maternal UPD2 was discovered during analysis for paternity testing. Serological and molecular analysis of a paternity case demonstrated exclusion of paternity of the presumptive father in two markers (ACP and Apo B, localized on chromosome 2p25.2 and 2p23-24, respectively) in a phenotypically normal girl with a normal karyotype. The index of paternity calculated for other serological (seven erythrocyte antigens, six serum protein systems, and seven isozymes, as well as the A- and B-HLA loci) and nine DNA markers gives a very high (virtually certain) degree of paternity for the presumptive father. Maternal uniparental disomy (UPD) for chromosome 2 was suspected. Evaluation of polymorphic DNA markers spanning chromosome 2 of the child, mother, and presumptive father demonstrated that the girl had inherited two maternal chromosome 2 homologues, whereas alleles for markers from other chromosomes were inherited from the father in a Mendelian fashion (Heide et al., 2000). The girl was homoallelic for informative markers mapping to the chromosomal regions 2p23-25, but she was heteroallelic for informative markers on the long arm of chromosome 2, establishing that the maternal UPD with partial isodisomy of the short arm was probably caused by a meiosis 1 nondisjunction event with genetic recombination (chiasmata in the region 2p23-25) during oogenesis.

CHROMOSOME 3 UPD

There is to our knowledge no proven case of UPD3 reported, although several investigators have examined this possibility in many cases of Brachmann-de Lange syndrome (De Marchi et al., 1994; Shaffer et al., 1993).

Possible Maternal UPD3

The only potential case, reported in 1974, is suggested by the duplication in an offspring of the chromosomal heteromorphism of a mother heterozygous for chromosome 3 pericentric inversion, inv(3)(p12;q14), inherited from her own father (Betz et al., 1974). This slightly retarded woman had two nondysmorphic, mentally retarded daughters; one was heterozygous but the other homozygous for this chromosome 3 inversion. The father was unavailable for study. To explain this homozygosity, the authors considered the (consanguineous) mating of two closely related carriers of this inversion that was also present in several other nonretarded, family members. Since no DNA polymorphisms were available at the time, there is no formal demonstration of maternal UPD3 in this case.

CHROMOSOME 4 UPD

Maternal UPD4

Case 1

In one case reported in 1982, either an incestuous mating or a chromosome 4 maternal isodisomy could account for homozygosity for a familial chromosome 4 pericentric inversion [inv(4)(p15.2→q12)], in a female whose mother was a heterozygous carrier (Carpenter et al., 1982). This African-American girl, weighing 1930 gm at birth, showed at 4 years 8 months, height and weight at the third percentile, normal head circumference, craniofacial dysmorphism (prominent forehead, mild dolichocephaly, large earlobes, deep-set eyes, broad nasal bridge, pointed chin), long slender extremities, poor muscle mass, jerky movements, and a flat-footed gait. Gross and fine motor skills were delayed; she had walked at 3 years and her language was that of a 1-year-old. Moderate sensorineural deafness was noted. On both her chromosomes 4, there was an inversion of segment p15.2-q12, inherited from the heterozygous mother. The father was not available for study but consanguinity was denied. The authors considered the homozygosity of recessive alleles or the existence of a submicroscopic deletion at the inversion breakpoint(s) (Carpenter et al., 1982).

Case 2

A phenotypically normal woman, the tallest of five siblings, was studied because of repeated abortions; her karyotype showed both an isochromosome 4p and an isochromosome 4q substituting for the chromosome 4 pair (Lindenbaum et al., 1991). Two biochemical polymorphisms suggested inheritance of only maternal material for chromosome 4. Polymorphic chromosome 4 microsatellite markers indicated that this individual was disomic at all tested loci on chromosome 4, possibly isodisomic. This case suggests that there are no paternally imprinted loci on chromosome 4.

Case 3

A patient was described with abetalipoproteinemia, a rare autosomal recessive disorder, that was homozygous for an intron 9 splice acceptor (-1) G-to-A mutation of the microsomal triglyceride transfer protein (MTP) gene (Yang et al., 1999). This mutation alters the splicing of the mRNA, in an in-frame deletion of sequences encoded by exon 10. Analysis of chromosome 4 short tandem repeat polymorphic markers revealed that the proband had only maternal alleles for chromosome 4q markers spanning a 150-cM (centimorgan) region; i.e., there was segmental maternal isodisomy 4q21-35. Nonpaternity was excluded using polymorphic markers from different chromosomes (paternity probability, 0.999). Maternal isodisomy (maternal UPD 4q) was therefore the basis for homozygosity of the MTP gene mutation in this patient.

Paternal UPD4

To our knowledge, no paternal UPD4 has been described yet (data up to December 2000).

CHROMOSOME 5 UPD

Maternal UPD5

To our knowledge, no maternal UPD5 has been described yet (data up to December 2000)

Paternal UPD5

Case 1

While performing routine prenatal tests for families with a previous child affected with spinal muscular atrophy (SMA), a case of SMA, associated with paternal isodisomy for chromosome 5, was identified (Brzustowicz et al., 1994).

The male proband, seen at 2.5 years of age, was born at 35 weeks' gestation with a weight of 5 lb 13 oz to a 29-year-old primiparous mother and a 36-year-old father. When first seen, the child's height was in the 90th centile and weight and head circumference were at the 75th centile. Neurologic features were consistent with the diagnosis of SMA type III. Lack of biparental inheritance for short-sequence repeat polymorphic markers at three informative loci in the SMA region of chromosome 5 prompted further testing, which confirmed the presence of only paternal alleles at each of 11 additional loci distributed along this chromosome. FISH analysis with probes closely flanking the SMA locus in the critical 5q11.2-q13.3 region ruled out deletion hemizygosity around that locus. It was thus established that both chromosomes 5 were identical, homozygous for their entire length, without the contribution of maternal alleles from 5p15.1-15.3 to 5q33.3-qter. The authors stated that the chromosomal region containing the SMA locus is homologous to mouse chromosome 13, which appears to be free of genomic imprinting. In conclusion, paternal isodisomy for human chromosome 5, as judged from this unique case, did not appear to cause other anomalies aside from homozygosity for a mutant paternal SMA gene (Brzustowicz et al., 1994).

CHROMOSOME 6 UPD

Maternal UPD6

Case 1

HLA analysis of the family of a renal transplant patient from the Netherlands revealed that the only demonstrable HLA antigens shown in the propositus were from the maternal haplotype (HLA-A11,-B46,-CW1,-DR14,-DQ1) (van den Berg-Loonen et al., 1996). No paternal antigens could be demonstrated either by serologic or DNA-typing methods. A paternity investigation was carried out to exclude the possibility of the legal father not being the biological father. The results of this investigation showed a fatherhood probability of 99.995%. A karyotype of the patient showed two normal chromosomes 6 and no other chromosomal abnormalities. Maternal isodisomy was demonstrated from the analysis of polymorphic DNA markers on 6p and 6q. These data were consistent with this patient having uniparental maternal disomy 6. It is likely that maternal UPD6 is not associated with an abnormal phenotype and that there are no maternally imprinted genes on chromosome 6.

Paternal UPD6

Case 1

A 9-year-old girl, with systematic lupus erythematosus (SLE), had complete complement C4 deficiency and showed homozygosity for all HLA markers studied. Complete C4 deficiency (for both C4A and C4B), extremely rare, is almost always associated with the development of SLE and is, in most cases, seen in the offspring of consanguinous unions. This was not so in the proband who was proven to have UPD for paternal chromosome 6 (Welch et al., 1990). She was initially evaluated for malar erythema, vasculitic skin lesions, and photosensitivity. Both parents and a sister were healthy. She expressed only a paternal HLA type and did not possess the glyoxylase-1-allotype of her homozygous mother, thus showing the absence of maternal contribution from HLA-A to glyoxylase. Restriction fragment analysis, including an informative 6q telomeric marker, gave no evidence of maternal DNA on either arm of her chromosome 6 pair. The paternal MHC (major histocompatibility complex) haplotype, singly expressed at all loci examined, showed however a hybridization intensity consistent with two copies of these paternal genes.

The authors emphasized that this case provided unequivocal evidence for the existence of this novel mechanism for the expression of recessive genetic traits, and that homozygosity of HLA markers need not always imply consanguinity (Welch et al., 1990), two tenets that, since this case was initially reported in 1990, have been verified repeatedly over the last 10 years.

Case 2

A female, small for gestational age and born of nonconsanguinous parents, died at 16 days with methylmalonic acidemia and diabetes mellitus. Both HLA class I serotyping and DNA analyses of the HLA-B and HLA-class II genes showed apparent homozygosity and lack of maternal allele inheritance over a 25-cM (centimorgan) length encompassing the HLA loci (Abramowicz et al., 1994). The authors concluded that the methylmalonic acidemia in this patient was due to homozygosity for a mutant recessive MUT allele that was the result of paternal isodisomy over at least 25 cM of chromosome 6.

At autopsy, immunohistochemistry, electron microscopy and *in vitro* hybridization studies with insulin mRNA probes showed the absence of beta-cells, in apparently normal islets of the pancreas.

The authors mentioned that, in transgenic mice models, congenital absence of beta-cells could result from beta-cell targeted ectopic expression of histocompatibility antigens. Such a mechanism, if extrapolated to this human case, would imply conjectural alteration of prenatal HLA gene expression in the developing islets, owing to isodisomy of chromosome 6. As an alternative hypothesis, even though there was an apparent lack of congenital diabetes in the other published case of paternal UPD6, case 1 (Welch et al., 1990), the authors proposed that there might be "... a developmental gene on chromosome 6 implicated in pancreatic beta-cell differentiation." They considered that case 2 could be explained by the loss of such a putative gene, either by a paternal mutation made homozygous by isodisomy or because of paternal imprinting of a locus, active only on the maternal side (Abramowicz et al., 1994).

Case 3

This little girl had transient neonatal diabetes (TNDM) and a small maternally derived chromosome 6 ring in about 3/4 of her cells [46,XX/47,XX, + r(6)] (Temple et al., 1995; James et al., 1995). Term birthweight was 1.8 kg (below 3rd centile). Hyperglycemia and dehydration occurred within 24 hours with a low endogenous insulin level. The patient recovered from diabetes at 4 months and showed normal mental development at 16 months; her subsequent height was at the 10th centile but she had mild dysmorphic features. At 3 years of age, although normoglycemic, she had "evidence of insulin resistence reflected by a raised C-peptide level" (Temple et al., 1995). Children with TNDM often develop an adult type II diabetes later in life. Analysis of DNA polymorphisms revealed that the two intact chromosomes 6 were paternal in origin and isodisomic. All 16 informative loci on chromosome 6, which were heterozygous in the father, had been reduced to homozygosity in the proband. The small ring chromosome 6 that contained material from only 6p11-6p21 was maternally derived. Evidently, this made the child trisomic for the segment of ring chromosome 6p. Because of case 2 (Abramowicz et al., 1994), which had diabetes mellitus and complete absence of pancreatic-islet beta-cells, the authors proposed that an imprinted gene(s) important for cellular development could exist on chromosome 6 and proceeded to test for UPD in two patients with TNDM (Temple et al., 1995).

Case 4

The proband was one of the two above referred patients with TNDM and showed UPD for chromosome 6 (the other patient had a biparental inheritance of chromosome 6). The proband weighed 1.7 kg at term (below 3rd centile), immediately developed hyperglycemia with dehydration, had undetectable insulin levels, but recovered from diabetes by 6 months, and showed normal growth and intelligence at 13 years of age (Temple et al., 1995). The uniparental paternal pair 6 of the child was isodisomic and/or homozygous for 16 loci spanning chromosome 6 from 6p24-p25 to 6q27. It was concluded that paternal UPD6 was revealing an imprinted gene rather than unmasking a rare recessive mutant gene since, by then, all three isodisomies in TNDM cases were paternal, and affected individuals were unrelated (Temple et al., 1995). In addition, imprinting had been shown in mouse chromosomal regions syntenic (homologous) to human 6.

With the addition of subsequent cases, it is likely that paternal UPD6 does not deprive the genome of a maternally active gene essential to "... development and maturation of early fetal and neonatal pancreatic beta- cell function ..." (Temple et al., 1995, 1996). This became apparent when other cases of TNDM were studied in two more families (Temple et al., 1996). In one of these families, PCR analysis of dinucleotide repeats along 6q, detected a duplication of four loci, all of which mapped to 6q22.33-q23.3. FISH analysis showed an insertion of chromosome 6 material into chromosome 2p. This duplication was present in the female TNDM proband, as well as in her normal father and paternal grandmother.

It thus appears that increased dosage of a paternally expressed locus, due in this instance to a duplication inserted in 2p, could result in TNDM, whereas maternal inheritance of the same duplication would not. If this is true, then the unaffected carrier grandmother should have also inherited her q22-23 insertion duplication from her mother. It is of interest to note that an imprinted imbalance, in this gene "overdose" context, has a parallel with the situation resulting from 11p paternal isodisomy or paternal duplication in WBS (Henry et al., 1991).

In another family, one brother had three sons with TNDM, and a second brother had a daughter affected at age 28 with type 2 adult diabetes; she had been an infant of low birthweight with perhaps undiagnosed TNDM. One of her three TNDM first cousins had, in turn, fathered a similarly affected child. In this family, there was no detectable 6q duplication, but the pedigree suggested dominant modification and paternal-only-derived penetrance of a putative "mutated allele/cryptic duplication" that was linked to a locus assigned to 6q23.3 (Temple et al., 1996).

To conclude with more evidence for the paternal expression of a maternally imprinted gene, overexpressed in neonatal diabetes, several cases of 6q duplication were collected from the literature, one of which was found to have confirmed neonatal diabetes. In line with paternally fostered dominance, the example of an unaffected man who had three children with TNDM, each born from a different mother, was quoted (Temple et al., 1996).

Case 5

This was the report of a further patient with TNDM and paternal UPD6 (Whiteford et al., 1997). The term-born infant was growth-retarded, with weight, length, and occipitofrontal circumference below the 3rd centile. Insulin infusion was started on the first day, after hyperglycemia and hyperglucosuria were noted; subcutaneous injections were given after 2 weeks, insulin requirements decreasing by 6 weeks and discontinued at 8 weeks of age. Imaging, by ultrasound and computerized tomography, twice failed to reveal pancreatic tissue, but stool chemotrysin activity (from exocrine glands) and glucagon (from alpha-cells) had been tested as normal and, at 14 weeks, a pancreas of normal size was visualized. At 7 months, weight and length were at the 3rd centile, with head circumference at the 10th centile. DNA analysis using short-sequence repeats mapping to 6p22-23 showed isodisomy for one of the two alleles present in the father and no inheritance from the mother. Dinucleotide markers mapping at 6p11-12, 6q21, 6q23, 6q24 all confirmed paternal isodisomy for chromosome 6. The authors proposed that a prenatal anomaly of pancreatic development in NTDM may be due to paternal UPD6 (Whiteford et al., 1997).

Case 6

UPD6 was fortuitously uncovered in a 9-year-old female with beta-thalassemia major, in the course of HLA typing to identify potential marrow transplantation donors (Bittencourt et al., 1997). Birthweight had been 2800 gm (25th centile) and height 50 cm (75th centile). Typing HLA class I and II loci with specific oligonucleotides revealed that the patient was uniformly homozygous for one paternal haplotype, with no maternal alleles at five fully informative loci. Analysis of other (erythrocyte) markers on six different chromosomes was compatible with biparental origin, as was microsatellite analysis at the chromosome 11 HBB locus. Serum C4 levels were normal in the patient. Five other highly informative microsatellite polymorphisms on chromosome 6 were compatible with paternal UPD6. No TNDM had been observed in the neonatal period, but retrospective consideration on this possibility did not permit one to rule it out either.

In the absence of apparent phenotypic anomaly other than beta-thalassemia major, the authors stated that the existence of an imprinted domain on chromosome 6 was not supported by their case (Bittencourt et al., 1997).

Case 7

A further patient who presented at birth with transient neonatal diabetes mellitus (TNDM) and paternal UPD6 was described (Christian et al., 1999). This child presented with low birthweight, macroglossia, hypertelorism, and club foot in addition to neonatal diabetes (see Figure 1, Chapter 5). Hyperglycemia was transient, and insulin treatment was discontinued at 4 months of age. Analysis with polymorphic DNA markers for chromosome 6 indicated the presence of paternal UPD6. The authors discussed that there were three cases with paternal UPD6 that also included additional anomalies, such as macroglossia. Therefore, the simultaneous finding of NDM and macroglossia should be a strong indicator for genetic testing (Christian et al., 1999).

Case 8

A Mexican-American female with transient neonatal diabetes (TNDM) was diagnosed. She required insulin injections for the first 3 months of life, then remained euglycemic at last evaluation at 7 months. There was a low birthweight, borderline macroglossia, narrow pyloric outlet, and mild tricuspid regurgitation. Cytogenetic studies were normal. DNA analysis was performed on samples from the infant and mother (father was unavailable). There was a single allele at loci D6S314, D6S435, D6S292, D6S311, D6S613, D6S409, and D6S1684; none of these alleles were shared by the mother consistent with paternal UPD6 (Palmer et al., 1998).

Case 9

Samples from a cohort of patients with transient neonatal diabetes mellitus for UPD6 were analysed using polymorphic microsatellite repeats. An additional case of paternal UPD6 associated with classic transient neonatal diabetes mellitus was discovered (Gardner et al., 1998). The authors estimated that UPD6 accounts for approximately one-fifth of cases of transient neonatal diabetes mellitus. By analyzing this case and two other cases with duplications, this group has defined the so-called TNDM critical region (Gardner et al., 1999). In one patient, polymorphic microsatellite analysis identified a duplicated region of chromosome 6, flanked by markers D6S472 and D6S311. Using markers within the region and P1-derived artificial chromosomes (PACs), the authors localized the distal breakpoints of the two duplications. The results of these mapping studies indicated that an imprinted gene for TNDM lies within an 18.72 cR (approximately 5.4 Mb) interval on chromosome 6q24.1-q24.3 between markers D6S1699 and D6S1010 (Gardner et al., 1999).

Case 10

During a molecular characterization study of a group of 47 Mexican families with 21-hydroxylase deficiency, one patient was identified with exclusive paternal inheritance of all eight markers tested on chromosome 6p, despite normal maternal and paternal contributions for eight additional markers on three different chromosomes (Lopez-Gutierrez et al., 1998). No duplication of paternal alleles for markers in the 6q region was found, consistent with a lack of expression of transient neonatal diabetes. The authors concluded that the 21-hydroxylase deficiency in this patient was due to homozygosity for a recessive allele secondary to the paternal UPD6 (Lopez-Gutierrez et al., 1998).

Case 11

A patient was described with neonatal diabetes, macroglossia, and craniofacial abnormalities, with partial paternal uniparental disomy of chromosome 6 involving the distal portion of 6q, from 6q24-qter (Das et al., 2000). The patient died on day 14 of sepsis. This observation demonstrates that mitotic recombination of chromosome 6 can also give rise to uniparental disomy and neonatal diabetes, a situation similar to that observed in Beckwith-Wiedemann syndrome, another imprinted disorder. The somatic recombination event was localized between markers D6S292 and D6S1009, which map to chromosome band 6q23.2 and 6q24.1, respectively. This finding has clinical implications, since somatic mosaicism for uniparental disomy of chromosome 6 should also be considered in patients with transient neonatal diabetes mellitus.

CHROMOSOME 7 UPD

This chromosome along with chromosomes 15 (Nicholls et al., 1989; Malcolm et al., 1991), and 11 (Henry et al., 1991) comprise the majority of chromosomes involved in uniparental disomy. It was also the first chromosome for which UPD was recognized (Spence et al., 1988).

Maternal UPD7

Case 1

This female patient has made medical history by being reported twice. The first time was at seven years of age because of the combination of short stature, cystic fibrosis (CF), and growth hormone deficiency (Hubbard et al., 1980), the second time as the first known human case of UPD (Spence et al., 1988). At 16, she measured 52 in. (130 cm), with slight leg-length discrepancy, and corporal asymmetry suggesting the Russel-Silver syndrome, but without other features. Menses had appeared at 14. She was of normal intelligence. Her mother, 36 years of age at the time of her birth, had died of lymphoma at 39. She had two siblings.

Results of chromosome analysis were normal but segregation analysis of three RFLPs at two loci closely linked to the CF locus on chromosome 7 contradicted Mendelian rules of inheritance by showing that the father was homozygous for alleles distinct from those for which the affected daughter was homozygous. In the absence of weakly hybridizing fragments from paternal alleles for these loci in Southern blots, mosaicism was ruled out, and this was further substantiated by the study of 100 metaphases of both blood and skin cultures. Nonpaternity and *de novo* chromosome 7 deletion were excluded. Dosage techniques indicated diploidy for the MET and J3 11 polymorphic markers in the proband; HLA markers were used to rule out an incestuous relationship in the patient's family. The use of 19 additional RFLPs assigned to chromosome 7 (from 7pter to 7q22) confirmed the absence of paternal alleles; in addition, all polymorphisms examined were homozygous in the patient's DNA. This led to the hypothesis that the patient must have received two copies of maternal alleles from chromosome 7 and none from the father for all investigated loci. As the mother was deceased, alphoid probes were used to trace the origin of the proband's chromosome 7 centromeric pair and RFLPs, detected with alpha-satellite DNA probes, showed homozygosity for one centromere of the maternal grandmother's chromosome 7. Taken together, the data were interpreted to show isodisomy for much of a chromosome 7 inherited only from the mother.

In conclusion,"... the uniparental origin of the centromere, the lack of heterozygosity on chromosome 7 and failure to detect mosaicism led (the authors) to favor monosomic conception for the findings in this case . . .". Obviously then, mitotic nondisjunction or replication of a solitary chromosome 7 could result in disomy 7 and residual monosomy could not be found. As to the phenotype, the cystic fibrosis was explained as a homozygosity for a maternal recessive allele; the short stature was assigned to either embryonic chromosomal mosaicism or a second recessive genetic disorder on chromosome 7, the last hypothesis probably influenced by the growth hormone deficiency stated in the first case report (Hubbard et al., 1980). However, the child had poorly responded to growth hormone therapy. It was later discovered that the patient was homozygous for a nonsense Gly542Ter mutation in the CFTR gene (Beaudet et al., 1991).

This article not only describes the various aspects of the first human case of UPD, but it also provides the framework regarding cytologic mechanisms of this then novel type of nontraditional inheritance, the etiopathology of phenotypic features in UPD, the clues to detecting such cases, and the laboratory techniques necessary to clarify them.

Case 2

A male patient was diagnosed with isodisomy of maternal chromosome 7. His family was one of 55 families examined with DNA markers for the diagnosis of CF. He was born to a 25-year-old mother and 30-year-old father, the second of four children, at the low weight of 1770 gm for a 38-week pregnancy (Voss et al., 1989). He remained small and, at 4 years, his stature was 87 cm, the average height of a 2-year-old child. Cystic fibrosis had already been diagnosed in infancy.

As with case 1, six DNA polymorphisms in the CF region were homozygous in the child, and four of these showed only maternal and no paternal contribution. Five additional polymorphisms, spanning the whole length of chromosome 7, all elicited a homozygous pattern, and these alleles were not present in the father. On chromosome studies, there was no mosaicism but a chromosome 7 maternal centromeric heteromorphism was found to be homozygous in the child. Again, as in case 1, nonpaternity or an irregular chromosome transmission, leading to isodisomy of chromosome 7, was considered. Nonpaternity was excluded by the study of RFLP markers on chromosomes 11 and Y and by several VNTR (variable number of tandem repeats) polymorphisms. As the patient was homozygous for maternal alleles at 11 loci along chromosome 7 and for the centromeric heteromorphism, it was concluded that there were two identical chromosomes 7 inherited from the mother (Voss et al., 1989). This must have occurred with a minimum of two abnormal events such as: (1) mitotic duplication of a maternal chromosome 7, made initially monosomic in a zygote produced by fertilization with a sperm nullisomic for 7, as a result of a paternal meiotic error. (2) nondisjunction of a meiotically nonrecombinant chromosome 7 at Me2 with an isodisomic ovum being fertilized by a nullisomic 7 sperm. (3) As above, but with normal haploid sperm fertilization and subsequent early mitotic loss of paternal 7. (4) Concomitant mitotic loss of a paternal chromosome 7 and compensatory or coincidental nondisjunction of a maternal 7, i.e., double mitotic nondisjunction. Some of these mechanisms could have resulted in mosaicism with trisomy or monosomy 7; such mosaicism was not detected. Of course, as in case 1 above, CF resulted from homozygosity of a mutation for which the mother was a carrier. A most perceptive explanation was set forth for the short stature, attributed to parental imprinting, i.e., to the lack of paternal contribution of alleles of certain gene(s) on chromosome 7.

Case 3

A male patient, with short stature and severe osteoporosis, had UPD, with partial isodisomy for chromosome 7; (Spotila et al., 1992). His mother had postmenopausal osteoporosis associated with a mutation of one allele at the COL1A2 locus. Even though the father did not have a mutation, the proband was found to be homozygous for the Gly661Ser COL1A2 mutation and was shown to have UPD7 (Spotila et al., 1992).

At 30 years, his height was 143.7 cm and weight 36.6 kg. Intelligence and body proportions were normal. Birthweight had been around 2025 gm. At age 1, a "slightly pointed facies was apparent." At 9 years, he was hardly taller than his 4-year-old brother and growth hormone therapy was undertaken for 9 years. Bone age was consistently retarded by 2 years. Fractures occurred at 4 and 14 years. An anteroposterior increase in chest diameter and rounded shoulders with laterally placed scapulae were noted. At 30, bone mineral density was below the 2nd centile, sclerae were slightly blue, hearing was normal, and body segment ratios appropriate.

Molecular analysis showed homozygosity for the Gly661Ser codon of the COL1A2 gene. Again, nonpaternity and deletion monosomy for the homozygous segment were ruled out. A total of 12 polymorphic markers, mapping to chromosome 7, were analyzed. Four were dinucleotide repeat polymorphisms. D7S435 and EGFR loci were maternally derived and homozygous. Of the eight RFLPs, three in the long arm, ELN, COL1A2, and L281, were homozygous and derived from the mother. The remaining were on 7p and the centromeric regions. The proband was homozygous at four loci for maternal specific alleles. However, the proband was heterozygous at the IGBP-1 locus mapping at the 7p13-7p12, making him heterodisomic for a segment bracketed proximally and distally by a maternally specific homozygous area. Obviously, recombination had occurred in the proband's maternally derived chromosomes 7. It is likely that the chromosome 7 pair was derived from an ovum disomic for 7, fertilized either by a nullisomic sperm (complementation), or a haploid one, with subsequent loss of paternal 7. Furthermore, if we assume that the 7p heterodisomy resulted from a double recombination at Me1, the isodisomic state of the remaining chromosome would point to a Me2 maternal nondisjunction event.

The patient suffered from a form of recessive osteogenesis imperfecta (OI) (light blue sclerae, pronounced low mineral bone density, two fractures), presumably due to the homozygosity for a "recessive" COL1A2 allele. The short stature, however, was not of the OI type and was similar to that observed in cases 1 and 2. Imprinting of genes on chromosome 7 was favored as the cause of this growth retardation, with lack of paternal alleles required for a growth regulation. IGBP-1 (insulinlike growth factor-binding protein 1) was quoted as a potential candidate gene for imprinting at 7p13-p12. This region is syntenic with chromosomes 6, 11, and 13 in the mouse, in which 6 and 11 may contain parentally imprinted genes.

Case 4

Oddly enough, this case had UPD with paternal isodisomy for 7p and maternal isodisomy for 7q, as a result of an isochromosome formation of each arm, from different parental sources (Eggerding et al., 1994).

The baby, born of a 39-year-old mother and 37-year-old father, was of normal size for gestational age (weight 3 kg, length 48 cm, OFC 35 cm.) An amniocentesis at 16 weeks, on maternal age indication, had revealed an apparently balanced karyotype, 46XX,−7, −7, +i(7p), + i(7q). There was postnatal growth retardation, weight and length below the 5th centile at 6 months, but head circumference was normal. The occiput was rather prominent, facies somewhat triangular, with narrow palate and there was slight leg-length asymmetry. Development at 1 year of age showed normal social and language skills, and slightly delayed gross motor activity. At $2\frac{1}{2}$ years, head size remained normal. Psychomotor development continued normally but growth remained well below the 5th centile. Parents and two siblings were of average height. Chromosome studies failed to reveal any mosaicism. PCR amplification of microsatellite polymorphisms was used to study 18 loci on chromosome 7. All but three were fully or partially informative and all 18 loci were homozygous. Loci from chromosome 7p showed no maternal allele contribution. By contrast, failure of transmission of paternal alleles was demonstrated for 7q markers. There was no evidence for mosaicism. It was concluded that the proband's i(7p) and i(7q), substituting for normal 7 homologues, were true isochromosomes of paternal and maternal origin, respectively.

One possible explanation of this case was a mitotic accident, i.e., a concomitant and probably interrelated break of chromosome 7 and sister chromatid reunion of, respectively, 7p paternal and 7q maternal, in one zygote. Alternately, this mitotic accident could have occurred later in embryonic development. If we assume that this uniparental disomy was confined to the embryonic disk, a normal diploid placenta might then explain the normal intrauterine growth observed in the case.

At the phenotypic level, the authors emphasized a distinction between the intrauterine and postnatal growth failures. Lack of paternally expressed genes on 7p causing intrauterine growth failure and loss of paternal 7q alleles, as in the case here, responsible for postnatal growth failure.

This is the 5th example of true isochromosome formation of p and q arms observed in cases of isodisomy, the others being for chromosome 1, (case 2) (Chen et al., 1999), chromosome 2 (cases 3 and 5) (Bernasconi et al., 1996; Shaffer et al., 1997), and chromosome 4 (case 2) (Lindenbaum et al., 1991).

Maternal UPD7 Cases in the Russel-Silver Syndrome.

It is worth noting that some of the above cases with maternal UPD7 have some features of the Russel-Silver syndrome (asymmetry, triangular face, pointed chin, clinodactyly) (Spence et al., 1988).

Individuals with RSS show intrauterine and postnatal growth retardation plus some of the following signs: prominent forehead (with late closure of anterior fontanelle), triangular face, downturned corners of the mouth, faciotruncular and limb asymmetry, 5th finger brachymesophalangy and clinodactyly, cutaneous syndactyly of toes 2 and 3, areas of hypo- and hyperpigmentation of the skin, genital hypoplasia in males, decreased subcutaneous tissue, delayed bone age, and excessive sweating. Intelligence is usually within the normal range.

A systematic study of 35 families was undertaken to search for UPD7 among 25 sporadic cases of RSS (with prenatal and postnatal growth retardation) and 10 more with idiopathic primordial (both pre- and postnatal) growth retardation, PGR (Kotzot et al., 1995). Probands ranged from 14 months to $16\frac{3}{4}$ years of age at examination.

In three of the 25 RSS individuals, maternal UPD7 was documented using polymorphic markers. Two had apparently complete isodisomy for several maternally derived markers, and one had heterodisomy at 15 maternal loci. All three children displayed the full spectrum of the RSS. In one out of ten individuals with the diagnosis of PGR, maternal UPD7 (isodisomy) was detected using informative microsatellite markers. On reexamination, this patient had a broad forehead, triangular face, prominent lips and philtrum, short clinodactyly of 5th fingers, excessive sweating of the forehead, mild limb asymmetry, and retarded bone age.

These results prompted the authors (Kotzot et al., 1995) to reach the following conclusions and considerations: (1) The three RSS cases with UPD7 (isodisomy) were the result of postzygotic mitotic nondisjunction of maternal chromosome 7 and loss of paternal 7 that could occur before, during, or after meiosis. (2) Complete maternal heterodisomy 7 in one of the four patients would indicate maternal Me1 nondisjunction, possibly with lack of pairing and meiotic recombination. Maternal age at birth was 39 and paternal age was 35 years. (3) There exists at least one maternally imprinted (i.e., paternally expressed) allele that influences intrauterine and postnatal growth. In case 4 (Eggerding et al., 1994), however, paternal 7p loci were present, which potentially implicates 7q loci in intrauterine and postnatal growth. (4) Approximately 10% of RSS cases might arise from maternal UPD7. (5) Most interestingly, the documentation of maternal UPD7 in RSS led to further hypotheses as to the heterogenous etiology of the syndrome, including: (a) microdeletion of the paternally derived chromosome 7; (b) mutation of a paternally expressed gene in this segment; (c) mutation or microdeletion of a chromosomal region regulating imprinting at other loci on chromosome 7.

After this study, several other groups contributed to the molecular analysis of RSS. In a prospective study of 33 patients with sporadic RSS using microsatellite repeat markers, two patients with maternal UPD7 were identified (Preece et al., 1997). There were no gross clinical differences in probands with UPD7 versus those without. One RSS patient was also found with maternal UPD7 in an initial subset of 14 cases (Shuman et al., 1996). In another study of 20 additional cases with RSS phenotype, one further case with maternal UPD 7 was detected (Penaherrera et al., 1996). In another study from Germany, 37 RSS families were typed with short tandem repeat markers from chromosomes 2, 7, 9, 14, and 16. Maternal UPD7 was detected in three of these patients (Eggermann et al., 1997). The allelic distribution in one of the three families indicated complete isodisomy, whereas allelic patterns in the other two families were consistent with partial and complete heterodisomy, respectively. There was no UPD of chromosomes 2, 9, 14, and 16. In a different study of 28 French RSS nuclear families, maternal UPD7 was found in two cases (Dupond et al., 1998). In one of these cases, there was also a balanced t(7;16)mat. In a study of 18 Italian RSS patients, maternal UPD7 was found in one case (Cogliati et al., 1998).

Taken together, there is a total number of 14 cases of maternal UPD7 out of 185 RSS cases analyzed. The overall frequency of UPD7 in RSS is therefore approximately 7.5%.

Monk et al. (2000) have identified a *de novo* duplication of 7p11.2-p13 in a proband with features characteristic of RSS (Monk et al., 2000). FISH confirmed the presence of a tandem duplication encompassing the genes for growth factor receptor-binding protein 10 (GRB10) and insulinlike growth factor-binding proteins 1 and 3 (IGFBP1 and -3), but not that for epidermal growth factor-receptor (EGFR). Microsatellite markers showed that the duplication was of maternal origin. These findings provide the first evidence that RSS may result from overexpression of a maternally expressed imprinted gene, rather than from absent expression of a paternally expressed gene. GRB10 lies within the duplicated region and is a strong candidate, since it is a known growth suppressor. Furthermore, the mouse homologue (Grb10/Meg1) is reported to be maternally expressed and maps to the imprinted region of proximal mouse chromosome 11 that demonstrates prenatal growth failure when it is maternally disomic. The authors have demonstrated that the GRB10 genomic interval replicates asynchronously in human lymphocytes, suggestive of imprinting. An additional 36 RSS probands were investigated for duplication of GRB10, but none were found. However, it remains possible that GRB10 and/or other genes within 7p11.2-p13 are responsible for some cases of RSS.

The GRB10 gene of chromosome 7 (a homologue of the mouse imprinted gene Grb10) was also shown to be monoallelically expressed in human fetal brain tissues and is transcribed from the maternally derived allele in somatic-cell hybrids. Mutation analysis of GRB10 in 58 unrelated RSS patients identified in two patients, a P95S substitution within the N-terminal domain of the protein. In these two cases, the mutant allele was inherited from the mother (Yoshihashi et al., 2000). The fact that monoallelic GRB10 expression was observed from the maternal allele in this study suggests but does not prove that these maternally transmitted mutant alleles contribute to the RSS phenotype.

Case 5

A case of prenatal and postnatal growth retardation associated with a prenatal diagnosis of mosaicism for trisomy 7 confined to the placenta was reported (Langlois et al., 1995). Genotyping of polymorphic markers on chromosome 7 has established that the zygote originated as a trisomy 7 with two maternal and one paternal chromosome 7 with subsequent loss of the paternal chromosome resulting in a disomic child with maternal UPD7 (heterodisomy). The growth failure observed in this child with heterodisomy 7 strongly supports the hypothesis of imprinted gene(s) on chromosome 7. The authors also studied the effect of confined placental mosaicism with trisomy 7 in 13 cases with biparental inheritance of the two chromosomes 7 in the newborn (Kalousek et al., 1996). It appeared that intrauterine fetal growth was not greatly affected by the presence of a trisomy 7 cell line in the placenta.

Paternal UPD7

Case 1

This 23-year-old Finnish female, born to a 39-year-old mother and 44-year-old father, the last of five children, was delivered at 34 weeks' gestation, with a weight of 1900 gm and length of 45 cm (both appropriate for gestational age). Hydramnios had been noted and the placenta was large, weighing 550 gm (Hoglund et al., 1994). Marked hyperbilirubinemia, a distended abdomen, and watery diarrhea led, at 8 days, to the diagnosis of congenital chloride diarrhea (CLD) for which NaCl and KCl substitution therapy were successful. Development was subsequently normal, with bouts of urinary infections up to age 6 years, at which time a high-frequency sensorineural hearing loss was diagnosed.

Menarche occurred at 12 years; skeletal maturation had been found to be normal at 4, 6, and 9 years. At 22 years of age, height was 159 cm and weight 60 kg, and both were similar to those of three female siblings. Intelligence was normal and she attended a business school. UPD7 was suspected when a linkage study of the extended family using polymorphisms near the CLD locus, known to map to chromosome 7, uncovered the lack of maternal contribution for CLD-linked microsatellite markers. The determination of genotypes of polymorphic loci, spanning chromosome 7, confirmed the absence of a maternal contribution for 10 informative markers. For an additional 10 markers, paternal and maternal alleles could not be distinguished, but the proband was always homozygous (Hoglund et al., 1994). Paternity was confirmed by the assessment of informative loci on the other chromosomes. It was thus concluded that this woman had paternal UPD7 (isodisomy), and that her CLD was caused by reduction to homozygosity of a recessive mutant allele at the CLD locus; an analogous mechanism was also considered for the sensorineural deafness. Although the mutation in the DRA gene responsible for CLD had not been determined in the patient, it is likely that homozygosity for the private Finnish mutation Val317Del is responsible for this phenotype.

The authors favored the hypothesis of a maternal nullisomy 7 compensated by early mitotic duplication of the paternal chromosome 7 in the conceptus. Alternative explanations could also be hypothesized. Mouse regions syntenic to human 7 are multiple (on chromosomes 2, 5, 6, 10–13); four of these display imprinting in the mouse. It is possible that duplication of the paternally derived genes could have contributed to the overweight (~30%) of the placenta observed at birth (Hoglund et al., 1994).

In summary, paternal isodisomy 7 was compatible with normal growth and development of the affected individual, but had brought about homozygosity at the congenital chloride diarrhea locus and also perhaps at another recessive locus involved in an undefined form of sensorineural deafness.

Case 2

This male infant of Hispanic origin had cystic fibrosis (homozygosity for the Phe508del mutation) and complete situs inversus with structurally normal dextrocardia. Bronchial cilia, immotile on light microscopy, were structurally normal by electron microscopy. The mother was normal, and the father was unavailable. Nineteen polymorphic markers spanning chromosome 7 demonstrated paternal UPD7 (isodisomy), whereas maternity was verified by 10 markers on other autosomes. DNA testing showed the patient to be homozygous for the F508del CFTR mutation, with his mother lacking any detectable alteration. Concomitant X-linked heterotaxy was ruled out both by absence of structural heart disease and by absence of Xq26 deletion, as tested by nine DNA sequences in the 1.2 megabase critical region. It was concluded that a locus influencing body plan laterality and ciliary motility resides on chromosome 7 and part of the phenotype in the patient was due to homozygosity for a recessive maternal allele (Pan et al., 1998).

CHROMOSOME 8 UPD

Maternal UPD8

Case 1

A female child with mild dysmorphisms, and motor and mental retardation had a 45,XX,−8,−8,+psudic(8)(p23.3) karyotype in blood lymphocytes, skin fibroblasts, and a lymphoblastoid cell line. DNA analysis showed that the proposita was nullisomic for the 8pter region distal to D8S264, which is less than 1 cM from the 8p telomere. Analysis of DNA polymorphisms of 38 loci spread along the entire chromosome 8 revealed that only maternal alleles were present, distributed in four heterozygous and four homozygous regions (Piantanida et al., 1997). This finding indicated that the rearrangement occurred during maternal meiosis in a chromosome with a minimum of seven cross-overs. The authors concluded that there was maternal UPD8 and nullisomy for the distal

1-cM portion of 8p. The available data support the assumption that no imprinted genes are present on chromosome 8 (Piantanida et al., 1997). Thus, dysmorphisms, and motor and mental retardation of the proposita are likely to be caused by the nullisomy for the region distal to D8S264, a region in which a recessive gene for epilepsy with progressive mental retardation is known to be located.

Case 2

An uncommon case of maternal UPD8 due to pseudodicentric chromosome diagnosed prenatally was described (Turleau et al., 1998). The fetal karyotype was interpreted as 45,XX,−8,−8,+psudic(8;8)(p23.1;p23.3). Only one constriction was present, suggesting inactivation of one of the two centromeres. A YAC probe on 8p23.1 revealed only one FISH signal, suggesting partial monosomy 8. This was confirmed by analysis of DNA polymorphisms. There was also maternal origin of the pseudodicentric chromosome without paternal contribution, 8p23.3-pter nullisomy, 8p23.1-p23.3 monosomy, and UPD8 that switched between iso- and heterodisomy for the remaining portion of chromosome 8. The authors proposed that these data were compatible with an unequal U-type exchange between two sister chromatids of one maternal chromosome associated with two cross-overs between homologous chromosomes. The fetal phenotype could be due to the partial monosomy/nullisomy for chromosomes 8 or to a homozygosity for a recessive allele (Turleau et al., 1998).

Case 3

An individual with apparently nonmosaic complete maternal isodisomy of chromosome 8 was reported. This individual was identified during routine genotyping in a genomewide search for type 2 diabetes susceptibility genes, although he does not have diabetes. He is of normal appearance, stature, and intelligence, but there is an unusual history of early onset ileal carcinoid (Karanjawala et al., 2000). The discovery of other maternal UPD8 cases will be necessary to define whether this condition causes a distinct phenotype.

Paternal UPD8

Case 1

A thorough molecular study of the lipoprotein lipase (LPL) gene in a child with LPL deficiency revealed that "... the homozygous variant in the proband was present in a heterozygous fashion in the father and was not found in the mother" (Benlian et al., 1996).

This nontraditional pattern of inheritance led to suspect paternal UPD8 as the mechanism for LPL deficiency in the child. This female baby was the first of a 24-year-old mother and 33-year-old father, born by C-section for breech presentation at 35 weeks' gestation, measuring 49 cm and weighing 3250 gm. An episode of acute pancreatitis at 1 month of age led to the detection of chylomicronemia and LPL deficiency. An appropriate diet was successful and the child was symptom-free until her current age of $5\frac{1}{2}$ years. Statural and ponderal growth was within normal limits. Intellectual, social, and developmental milestones were normal. Mutation analysis of the LPL gene has shown a frameshift mutation by an insertion at codon Gly35 of the second exon, causing a downstream premature stop codon at residue 62. Furthermore, all nine polymorphisms analyzed at the LPL locus and an additional 15 markers spanning chromosome 8 from 8p23.2−>pter to 8q24 revealed that the child was homozygous for polymorphic alleles originating from the father and absent in the mother. Normal segregation of informative markers was observed from both parents for four other chromosomes, blood karyotypes were normal, and no faint maternal bands for chromosome 8 were detected on DNA blot analyses, making trisomy 8 mosaicism unlikely. It was concluded that paternal UPD8 (isodisomy) was causing reduction to homozygosity of the paternal LPL mutant gene and LPL deficiency in a developmentally normal child. The authors (Benlian et al., 1996) therefore proposed that there are no paternally imprinted genes on chromosome 8. They added that chromosome 8 contains genes syntenic to mouse chromosomes that do not show imprinting either.

CHROMOSOME 9 UPD

Maternal UPD9

Case 1

The first described case of maternal UPD9 also displayed a residual trisomy 9 cell line in blood. A 17-year-old male was referred for evaluation of mental delay and speech abnormality (Willatt et al., 1992). Both height and weight were in the 3rd centile but head circumference was normal (50th centile). Facial asymmetry, short neck, micrognathia and low-set ears, kyphoscoliosis, limited elbow extension, tight hamstrings, and knock-knees were noted. There were 13 ribs and abnormalities of dorsal vertebrae 6 to 9, tibio-femoral subluxation, patellar lateral dislocation, and osteochondritis. Paternal age at birth was 35 and maternal age was 30 years.

On karyotype analysis, 7% of blood cells (100 counted) showed trisomy 9 and two of the three chromosomes 9 had a pericentric inversion (p11q12) of the heterochromatic regions, also present in euploid cells. Another blood study confirmed this finding, while cultures of muscle and skin—obtained at knee surgery—did not display any trisomic cells (100 cells study per sample).

DNA analysis of markers along chromosome 9, from 9p to 9q34 (both dinucleotide repeats and RFLPs), showed reduced intensity or absence of alleles derived from paternal DNA. The heterodisomy of maternally derived alleles, along with the cytogenetic findings, indicated that recombination must have occurred between 9q31 and the centromere at Me1. It was concluded that there was maternal UPD9 in the euploid cell line that had arisen after postzygotic loss of the paternal chromosome 9 from the trisomic cell line.

The authors concluded that the boy's phenotype might have resulted from homozygosity for a recessive maternal allele, an imprinting effect of chromosome 9 gene(s), or the presence of the trisomic cell line in a mosaic state (Willatt et al., 1992).

This case is an example of incomplete trisomy rescue that precludes any firm conclusion about the potential role of imprinting and recessive genes from the maternal UPD9 on the phenotype. It is possible that there might also be cryptic cases where somatic mosaicism has been phased out (dead mosaicism), having presumably impacted the earlier development but becoming undetectable later in life.

Case 2

The pregnancy of a 38-year-old woman, tested for an increased maternal age, was voluntarily interrupted on parental decision after short-term cultures of CVS showed trisomy 9/euploid mosaicism that could not be demonstrated in long-term villus cultures. Abortion products such as cartilage, eye, cord, and skin tissues were studied. Interphase cells were studied by FISH for signals with a cosmid cloned from the pericentromeric region of chromosome 9, and PCR amplified microsatellite repeat markers at four loci from 9p23-22 to 9q32-33. FISH did not elicit trisomic signals above the normal control background (in 20 of 500 cells) in the four fetal tissues. An additional maternal chromosome in extraembryonic tissues, villi, and amnion was found, and both maternal alleles were more intense than the single paternal allele. In the fetal tissues, only the two maternal alleles were detected (paternal alleles were not observed). Heterozygosity of maternal alleles was present at three of the four loci tested, the fourth one being homozygous, indicating that maternal meiotic nondisjunction occurred after recombination took place and that maternal UPD9 resulted from the loss of the paternal chromosome from an initial trisomy (Wilkinson et al., 1996).

The authors had thus shown that an extensive search did not substantiate trisomic cells in the four fetal tissues (all with UPD 9), thus suggesting that the trisomy 9 had segregated only in the extraembryonic tissues, amnion, and trophoblast (confined placental mosaicism; CPM) (Wilkinson et al., 1996).

Cases 3 and 4

These two cases are presented together, as they were documented from a cohort of 54 uniplex families with cartilage-hair hypoplasia (CHH) (Sulisalo et al., 1997), an autosomal recessive form of chondrodysplasia, unusually frequent in Finland, where the locus was mapped to 9p21-p13 (Sulisalo et al., 1993b, 1994a). The pleiotrophic features of CHH include disproportionate short stature, hypoplastic (thin and sparse) hair, defective immunity, and, more rarely, Hirschsprung disease and defective erythrogenesis; an increased risk of malignancy also exists (Sulisalo et al., 1997).

Patient 1. A female with UPD9 (isodisomy) measured 83 cm (less than 1st centile), at 11 years of age. In addition to the severe, disproportionate dwarfism with skeletal dysplasia, she had had numerous bacterial and viral infections, severely depressed T-cell mediated immunity and beta-cell deficiency.

Analysis of many chromosome-9-specific polymorphic markers showed no paternal contribution and complete homozygosity for maternally derived alleles (Sulisalo et al., 1997). A large number of additional markers on other chromosomes confirmed paternity. In this case, it is possible that a lack of paternal 9 transmission at fertilization was compensated for by mitotic duplication of the normally transmitted maternal 9. Alternatively, early mitotic loss of a paternal 9 could have been associated with maternal chromosome 9 duplication caused by isochromatid nondisjunction. A third, less likely, possibility is that of maternal Me2 nondisjunction of an achiasmatic chromosome and fertilization with a nullisomic paternal gamete.

Patient 2. A female child with the skeletal anomalies of CHH and severe hair hypoplasia measured below 60 cm (less than 1st centile) at 2 years of age. She also had aortic valve stenosis and radiologic evidence of megacolon.

She was found to have maternal UPD9 (iso- and heterodisomy) (Sulisalo et al., 1997). All 13 informative markers for chromosome 9 were maternally derived. The majority were homozygous, but in the mid- and distal portions of 9q, two loci were heterozygous, indicating two recombinations in Me1. In this case, the most likely mechanism appeared to be maternal UPD9 resulting from Me2 nondisjunction coinciding with loss of the paternal chromosome, by either sperm nullisomy 9 or rescue of a trisomy 9. The CHH phenotype in these two cases is probably the result of a homozygosity of a recessive CHH allele. In fact Ridanpaa et al., 2001, identified a single nucleotide substitution in the RMRP gene in these two patients. This gene, which encodes the RNA component of an endoribonuclease, is responsible for CHH.

Case 5

A healthy 34-year-old female with a history of repeated spontaneous abortions was found to have isochromosomes i(9p) and i(9q). The origin of these isochromosomes was investigated using microsatellites from 9p and 9q. The patient was homozygous for maternally derived alleles for all the informative markers tested; no paternal contribution of chromosome 9 alleles was observed. Nonpaternity was excluded using additional markers. The patient thus had maternal UPD9 and no apparent clinical symptoms, which strongly indicated that there are no maternally imprinted genes on chromosome 9 (Anderlid et al., 1999).

Case 6

The twin female probands were born to nonconsanguineous parents. The mother was 46 years old (Tiranti et al., 1999). After the age of 8 months, both patients developed a rapidly progressive clinical syndrome characterized by failure to thrive, psychomotor regression, hypotonia, ophthalmoparesis, bilateral optic atrophy, and ataxia. At 18 months, both had mild lactic acidosis. Brain imaging studies showed symmetrical paramedian lesions in the mesencephalon and brain stem, which suggested the diagnosis of Leigh syndrome, an autosomal recessive disease. Both patients died of respiratory failure in the third year of life. Needle muscle biopsies at 24 months showed diffuse reduction of the histoenzymatic reaction of cytochrome c oxidase (COX). Biochemically, COX activity in muscle homogenates was markedly reduced: 12.1 nmol/min/mg in one patient and 3.6 in the other (mean normal values 68; SD 20). Sequence analysis of the nine exons of the SURF1 gene on chromosome 9q34 revealed homozygosity for a frameshift mutation. The mother was heterozygous for this mutation, which was absent from the paternal DNA. Nonpaternity was excluded by the use of numerous DNA polymorphic markers. Furthermore, the genotyping of many markers on chromosome 9 revealed maternal UPD9 with areas of isodisomy and heterodisomy probably due to cross-overs in the maternal meiosis (Tiranti et al., 1999).

Paternal UPD9

To our knowledge, no paternal UPD9 has been described yet (data up to December 2000).

CHROMOSOME 10 UPD

Maternal UPD10

Case 1

A male infant was delivered by C-section at 41 weeks' gestation to a 41-year-old mother. Polyhydramnios as of the 26th week subsided spontaneously toward the end of pregnancy (Jones et al., 1995). The child's birthweight was 3100 gm, Apgar scores were 7 and 9 at 1 and 5 minutes, and no anomalies or dysmorphic features were noted. At 8 months, both weight and length were at the 25th centile and head circumference was at the 80th centile; gross and fine motor skills and social demeanor were normal.

The pregnancy had come to attention for a 46,XY/47,XY+10 mosaicism in some cells of direct preparations of a 10-week chorionic villus sampling for maternal age, while 40 cells in longer-term cultures were all 46,XY. At amniocentesis, the karyotype was only 46,XY (16 cells). At birth, cord blood and cord tissue biopsy (20 cells) and amniotic membrane cultures (80 cells) only displayed the 46,XY type, but placental cultures showed trisomy 10 in 10 of 60 cells (Jones et al., 1995). Dinucleotide repeat polymorphic markers for chromosome 10 were studied (as well as markers for chromosomes 7 and 13, which confirmed paternity). The chromosome 10 markers on DNA extracted from cultured amniotic fluid cells, cord blood and tissue, fetal membranes and placenta showed UPD10, i.e., inheritance of only maternal alleles in heterozygosity at all three loci. There was lack of paternal alleles. The failure to identify paternally derived alleles in term placenta was assigned to selection for diploid cells during prolonged *in vitro* culture to obtain DNA extracts.

The authors (Jones et al., 1995) concluded that: (1) Maternal meiotic nondisjunction of chromosome 10 led to trisomy 10 in a conceptus that, by early loss of paternal 10, converted to maternal UPD10 in the fetus, and mosaic trisomic 10 confined to extraembryonic tissues. (2) Maternal UPD10 did not produce any major imprinting effect on intrauterine growth and development, not precluding more subtle effects at longer term. (3) Trophoblast cell mosaicism on CVS, without mesenchymatous anomalies on longer-term cultures, should be investigated for UPD.

Case 2

A male infant died at 37 days with a 47,XY, + 10/46,XY,UPD(10)mat mosaicism. He was a very malformed, severely retarded child with microcephaly, microphtalmia, cleft lip and palate, thoracic scoliosis, anal atresia, and heart and limb malformations (Hahnemann et al., 1996). The karyotype showed mosaicism trisomy 10/euploid in skin fibroblasts (25/30 mitoses) and in postmortem tissues (tendon, cartilage, pericardium) with trisomy 10 in 19/30, 30/30 and 26/30 mitoses, respectively. However, lymphocytes cultures from blood taken *in vivo* had only shown 46,XY (50/50 mitoses). Microsatellite marker analysis showed maternal meiotic origin of the extra chromosome 10 and maternal UPD10 in the euploid blood sample cultures (Hahnemann et al., 1996).

This is an example of an incomplete, probably belated trisomy "rescue" and it is apparent that trisomy 10 must have played the essential role in this patient's severe phenotype.

Paternal UPD10

A case of partial paternal UPD10 in a patient with multiple congenital anomalies was presented in a meeting (Kousseff et al., 1992). However, on subsequent analysis there was no evidence of UPD10 (Kousseff; personal communication).

CHROMOSOME 11 UPD

Maternal UPD11

No cases of whole or segmental maternal UPD11 have yet been described (December 2000). It is likely that maternal UPD11 is lethal. This will be further discussed in Chapter 8.

Paternal UPD11

There are three cases described of paternal UPD11 involving the entire chromosome (Grundy et al., 1991; Webb et al., 1995; Dutly et al., 1998). All other cases are segmental for 11p15.5, as a result of somatic recombination (Henry et al., 1991). The resulting Beckwith-Wiedemann syndrome will be discussed in Chapter 8 and these cases are therefore not included in this chapter.

Case 1

A male newborn was delivered at 42 weeks' gestation by C-section for transverse lie (Grundy et al., 1991; Saracco et al., 1988). He weighed 3750 gm and had normal Apgar scores. Disseminated over the upper back, neck, and scalp were some 20 red-violeceous papular and nodular lesions of 3–10 mm, which proved to be metastases originating from an adrenocortical tumor. Imaging studies also suggested brain frontal and parietal metastases. No organomegaly was noted. At the age of 4 weeks, the baby underwent excision of the well encapsulated adrenal tumor that contained large polymorphic cells with granular eosinophilic cytoplasm. Regression of cutaneous nodules began in the second month and all had spontaneously resolved by 4 months. Right hemihypertrophy was noticed at 2 months of age. One year after diagnosis, the patient showed no evidence of tumors and was developing normally. The karyotype was normal. At a later time, samples of blood and adrenal tumor were examined in search of tumor specific alterations detectable by RFLPs (Grundy et al., 1991). Thirteen probes, mapping to 11p and 11q, showed that the child's DNA was homozygous at all polymorphic sites. There was only paternally inherited loci at FSHB [11p13], PTH, and INS (11p15) genes. It was concluded that the proband had paternal UPD11 (isodisomy). The parental origin of the 11q markers had not been determined, but the fact that all tested markers were homozygous strongly favors the UPD11 for the entire chromosome. Other loci on chromosomes 1, 3, 7, 12, and 17 showed biparental inheritance. The authors considered the hypothesis that homozygosity for a recessive paternal allele may result in the patient's phenotype. It is also likely that the phenotype is due to the effect of imprinted loci on 11p. The child did not have other signs of the BWS aside from hemi-hypertrophy, and the adrenocortical and renal tumors that he later developed.

Case 2

This was the first pregnancy of 35-year-old parents, characterized by IUGR, high alpha-fetoprotein, low HCG, and oligohydramnios, ending in IUD at about 20 weeks. There was a 3- to 4-week delay in fetal growth, and the placental growth was severely delayed with a weight roughly 1/4 to 1/5 of what would be the normal at 19 weeks. The fetus had penile hypospadias and inverted rotation of small and large bowel (Webb et al., 1995).

Direct chromosome preparations of placental biopsy showed a 47,XY,+11 karyotype and long-term cultures a 46,XY/47,XY,+11 karyotype. After delivery, interphase cells from skin and several placental sites, studied with a chromosome

11 alpha-satellite repeat probe by FISH, revealed trisomy in about 75% of the placental cells and disomic signals in 85% of skin cells; only 7/280 showing three signals. Molecular analyses using several markers from 11p and one on 11q24.5 revealed that in fetal tissues there were no maternal alleles at 11p13, 11p15.5, and 11q24.5; homozygosity for only paternally derived alleles was observed, indicating paternal UPD11 (isodisomy) at these sites. In the placenta, the 11p15.5 locus showed three alleles, one of which was of maternal derivation. The authors concluded that these results were compatible with a rescue of a paternally originated trisomy 11 conceptus; the maternal chromosome 11 was eliminated in somatic cells, resulting in paternal UPD11. The pericentromeric area of chromosome 11 was heterodisomic. This is compatible with paternal Me1 nondisjunction.

Case 3

About 20% of patients with sporadic WBS have segmental paternal UPD of 11p. Mitotic recombination at the 11p region has been suggested to be responsible for the somatic mosaicism in these patients. These cases will be discussed in Chapter 8. In one study, however, of seven sporadic WBS patients, there was one case with mosaic paternal uniparental disomy for the whole chromosome 11 (Dutly et al., 1998), as evidenced by the study of 18 polymorphic markers covering from 11p15.5 to 11q25. Apparently, the clinical findings for this patient did not differ from data reported for other WBS patients. This female patient was from Turkey and was born at 33 weeks' gestation with a birthweight of 2250 gm when her mother was 31 and the father 34 years old. She was examined at 9 months of age and displayed growth asymmetry, omphalocele, umbilical hernia, hepatomegaly, macroglossia, ear lobe grooves, naevus flammeus, and hypoglycemia. This case makes it unlikely that the proximal short arm and the long arm of chromosome 11 contain imprinted genes with a phenotype recognizable prenatally or in infancy, and lends some support to the hypothesis that nonmosaic UPD11 is prenatally lethal.

CHROMOSOME 12 UPD

No cases of maternal or paternal UPD12 have yet been described (December 2000). Even in cases of mosaicism for chromosome 12, there is no formal proof of uniparental disomy in the euploid cells (Bischoff et al., 1995; Meck et al., 1994).

CHROMOSOME 13 UPD

Maternal UPD13

Case 1

A pregnant (10-week) 23-year-old woman of normal phenotype had a chromosome study because of her mother's reproductive history: eight miscarriages and two liveborns, including a brother with multiple congenital anomalies who died at 3 days (Stallard et al., 1995). The proband and her mother shared the karyotype 45,XX,t(13q13q). When the proposita inherited the t(13q13q), she received two copies of 13q from her mother. Moreover, she and her mother shared the same homozygous pattern of alleles from seven highly polymorphic microsatellite repeats localized along 13q. No evidence of paternal markers from 13 was detected, although biparental inheritance was demonstrated with DNA markers from chromosomes 2 and 17. Therefore, cytogenetic and molecular findings indicated that the proposita's chromosomal complement included maternal UPD13. The proposita's normal phenotype suggested that no maternally imprinted genes map to 13q. It was likely that the paternal chromosome 13 was either lost shortly after fertilization (trisomy rescue) or there was a fertilization with a nullisomic 13 paternal gamete (complementation). Unfortunately, the proband had a miscarriage at 12 weeks of gestation and no fetal analyses were reported.

Case 2

This case resulted from the maternal contribution of a t(13;13) to a male offspring; both mother and child were phenotypically normal. Again, the indication for a chromosome study was the history of five first-trimester spontaneous abortions in the mother (Slater et al., 1994). The 5-year-old boy, whose height, weight, and head circumference were in the 50th to 75th centiles, was healthy and had average intelligence, as did his mother. Both their karyotypes showed 45 chromosomes with a t(13;13)(p11.2;p11.2) balanced rearrangement (the mother's translocation had, in turn, been inherited from her father as described in the next case). FISH, using probe alpha R1 as well as a chromosome 13 paint, showed the translocation chromosome to be dicentric and entirely made up of chromosome 13.

Molecular studies of three dinucleotide repeat polymorphisms showed mother and child to be identically homozygous at these loci; there was no contribution from the father's alleles. Analysis of polymorphic markers in other chromosomes excluded nonpaternity.

The authors (Slater et al., 1994) concluded that: (1) The translocation appeared to have produced a true dicentric isochromosome 13, since the highly polymorphic markers were homozygous. (2) This isochromosome could have resulted from fusion, at breakpoint 13p11.2, of both isochromatids of a chromosome 13, before Me2 anaphase. (3) The lack of paternal 13 contribution could have been either the result of a nullisomic sperm or subsequent zygotic loss of the paternal chromosome 13. (4) Human chromosome 13 is syntenic (homologous) to mouse 14, which apparently does not contain imprinted genes. (5) Cases of maternal UPD13 are rarely recognized, perhaps owing to limited ascertainment due to the lack of deleterious phenotypic effects.

Paternal UPD13

Case 1

This case also belongs to the same family as the previous case. Although the maternal grandfather of the proband in the previous case was deceased, it was implied from molecular analysis that he transmitted the t(13;13) rearrangement to his daughter, the latter being the mother of case 2 above (Slater et al., 1994). When looking for the transmission of two informative chromosome 13 dinucleotide polymorphisms from the maternal grandmother in case 2 above, it was observed that there was no such contribution from the grandmaternal alleles. "By inference (this) t(13;13) must have been transmitted by the grandfather, either from a pre-existing constitutional or a de novo translocation" (Slater et al., 1995). This, in turn, suggested to the authors that there were no paternal or maternal imprinted genes on human chromosome 13.

The amazing fact is that, apparently twice in two generations, the phenomenon of gamete complementation or trisomy rescue occurred, accounting for two most improbable healthy live births (a daughter and grandson), proving that nontraditional inheritance such as UPD can work "miracles" (more adventurous minds might also infer from this that, to some extent, an active mechanism of trisomy rescue might also exist!).

Cases 2–3

Two cases of paternal UPD 13 involving isochromosomes 13 were recently described (Berend et al., 1999). Both cases were referred for UPD studies due to the formation of a *de novo* rea(13q13q). Case 3 was complicated by the segregation of a familial rob(13q14q) of maternal origin. Both propositi were phenotypically normal at the time of examination. Polymorphic marker analysis in case 2 showed the distribution of alleles of markers along chromosome 13 to display complete isodisomy, consistent with an isochromosome. This rearrangement could have occurred either meiotically, without recombination, or mitotically. A likely mechanism for UPD13 in this case is monosomy rescue, through postzygotic formation of the isochromosome. In case three, the distribution of proximal alleles indicated an isochromosome, but recombination was evident. Thus, this isochromosome must have formed prior to or during MeI. A likely mechanism for UPD13 in this case is gametic complementation, since the mother carries a rob(13q14q) and is at risk of producing aneuploid gametes. However, trisomy rescue of a trisomy 13 conceptus cannot be completely excluded. Given that both cases were phenotypically normal, these data further support the notion that paternal UPD13 does not have an adverse phenotypic outcome and, thus, does not show an apparent imprinting effect (Berend et al., 1999).

Case 5

A UPD13q (isodisomy) in a balanced karyotype, 45,XY,−13,−13, +i(13)(q10), was found in cultured amniocytes studied because of advanced maternal age (Jarvela et al., 1998). The isochromosome 13 was monocentric and *de novo* as both parents had normal chromosomes. Fetal blood was studied to exclude trisomy 13 mosaicism. All (100) lymphocytes studied had the same karyotype with i(13)(q10) as the amniocytes. To determine the origin of the isochromosome, six microsatellite markers from 13q were analyzed: D13S175, D13S166, D13S162, AC224, COLA61, and D13S122. The results indicated that the i(13)(q10) was of paternal origin and isodisomic. A healthy child was born at week 40 of pregnancy, supporting the hypothesis that there are no paternally imprinted genes on chromosome 13q. Analysis of extraembryonal tissue (four samples studied) revealed the same balanced karyotype with the i(13)(q10)pat chromosome. The authors speculated that the origin of the isochromosome 13 could be a transverse centromere cleavage of an achiasmatic 13, at the paternal Me2 or at an early mitosis (Jarvela et al., 1998).

Case 6

A case of paternal isodisomy 13 in a phenotypically normal girl was described. Prenatal diagnosis had shown a 46,XX,-13,der(13;13) karyotype in chorionic villi and a 45,XX,der(13;13) karyotype in amniocytes and fetal blood. Molecular studies demonstrated that the *de novo* der(13;13) was an isochromosome 13 of paternal origin (Soler et al., 2000). This observation supports the lack of imprinting effects on chromosome 13 and trisomy rescue as a mechanism leading to uniparental disomy in cases involving isochromosomes.

CHROMOSOME 14 UPD

Maternal UPD14

Case 1

A boy, aged 17 at the time of the report, had been born prematurely at 32 weeks by C-section with a weight and head size in the 25th and 10th percentiles, respectively (Figure 2, left, in Chapter 5). Several problems developed over the years: (1) Initial hydrocephalus, with spontaneous arrest, was later shown to be associated with third and lateral ventricle enlargement. A somewhat delayed early motility was noted, later accompanied by fine hand tremor, ongoing poor coordination, and signs of mild right-sided hemiplegia. Eyes, ears, vision, and hearing were normal; speech acquisition had been early and intelligence was within normal limits. (2) Scoliosis, noted at 1 year, and bloc vertebrae at T5-T6 required surgical management at the age of 12 years. (3) Orchidopexy for undescended testis took place at 4 years, but puberty had an early onset, at age 10, with normal pubic hair but persistence of small testes. (4) Growth was delayed, since at age 15, his height was at the 3rd centile (154.7 cm) and hand-bone epiphyses were closed.

At age 17, he had a prominent forehead and supraorbital ridges, accentuating the discrepancy between head size and height. The philtrum was short and the mouth downturned, with a high-arched palate and bifid uvula. Hands and feet were slender, with 5th finger clinodactyly. Scoliosis was mild and truncal obesity contrasted with relatively short arms and legs. Nipples were widely spaced. Secondary sexual hair was developed, genitalia were relatively small with testicular volume below the 3rd centile. Levels of testosterone, thyroxine, TSH, and LH-FSH were normal.

Both the proband and his mother had 45 chromosomes and a Robertsonian translocation t(13;14) (Temple et al., 1991). Fluorescent cytogenetic polymorphisms suggested a paternally derived free chromosome 13 and a maternally derived free chromosome 14, along with the translocation. Molecular analyses were performed for seven polymorphisms on chromosome 13 and six on chromosome 14. There was paternal allele contribution for chromosome 14 alleles, while biparental inheritance of chromosome 13 alleles was noted. The authors concluded that there was inheritance of both the free and translocated chromosome 14 from the mother, without paternal contribution (maternal UPD14) (Temple et al., 1991). The proband was homozygous for all six polymorphisms on chromosome 14, including those at two distal loci for which the mother was heterozygous. This reduction to homozygosity suggested homologous pairing and at least one recombination event involving the distal site. Markers elsewhere confirmed paternity.

The authors also suggested that some clinical features might not be related to maternal UPD14, i.e., short stature was perhaps related to scoliosis and premature puberty to hydrocephalus (although the testes had remained small). It was however clear that, at least, maternal UPD14 was compatible with normal intellectual development. The authors had ascertained this case while screening persons with familial Robertsonian translocations for UPD14, to evaluate the role of translocations as a predisposing factor. They noted that human chromosome 14 has homologies with mouse 14 (not imprinted) and mouse 12, the imprinting status of which was not yet clarified.

Case 2

This case was ascertained at 20 years of age because of the association of two unrelated disorders: rod monochromacy and growth retardation of prenatal onset (Pentao et al., 1992). Rod monochromacy, or compete congenital achromatopsy resulting from retinal cone impairment (OMIM 603096), is a rare autosomal recessive eye disorder causing total absence of color discrimination, reduced central vision, severe light discomfort, and infantile nystagmus. There are generally no other associated illnesses with this disorder. The patient was born at term to a 24-year-old mother and a 26-year-old father, with a weight of 2,216 gm, below the 5th centile. Her health problems affected different systems: (1) The ocular symptoms and signs were pathognomonic of rod monochromacy and led her to a state school for the visually impaired. Recurrent otitis media also resulted in conductive hearing loss. (2) Growth and development: At 9 months, body weight, fronto-occipital circumference, and height were − 1.5, − 1.2, and − 2.7 SD, respectively. She sat unassisted at 1 year of age. Motor retardation, hypotonia, and head lag and kyphosis were noted. Speech was delayed, and a pediatric neurology consult concluded "diffuse static encephalopathy." Between 12 and 14 years, she scored 70–80 on two different IQ evaluations. At age 20, her height was still below the 5th centile. (3) Genital development: At age $8\frac{1}{2}$, early pubertal changes (breast enlargement, axilliary hair) were noted. (4) At age 20, she was pleasant and cooperative, was of small stature (144.4 cm), and moderately obese (74.4 kg), with a normal head size. Both hands and feet were below the 5th percentile. She married at age 19 and had three spontaneous first-trimester miscarriages within a year. Cytogenetic analysis revealed a 45,XX,rob(14;14). The karyotype of the mother was normal.

The nature and origin of the t(14;14) rearrangement were characterized with nine highly polymorphic VNTR markers and two dinucleotide repeat polymorphic sequences mapping to chromosome 14. The patient was homozygous for all markers tested, having received two copies of a single maternal allele, as verified by densitometric measurements, with no paternal allelic contributions. These data are consistent with maternal UPD14 (isodisomy) and suggested that the apparent translocation was, in fact, an isochromosome derived from one maternal chromosome. The authors (Pentao et al., 1992) considered that the eye phenotype was due to homozygosity for a recessive maternal allele. Since case 1 (Temple et al., 1991) also displayed short stature and premature puberty, this suggested that the clinical phenotype resulted from an imprinting of maternal chromosome 14 genes affecting growth and sexual maturation.

A general recommendation from the study of this case was that in patients with 2 etiologically unassociated developmental problems, UPD studies should be carried out.

Case 3

This female patient was referred to a genetics clinic for short stature and mild dysmorphic features. She was the product of a 32- to 33-week gestation with a birthweight of 1759 gm. Delivery was induced because of Rh incompatibility and rising antibody titers. The neonatal period was marked by hypotonia, feeding difficulties, jaundice, and temperature instability (Antonarakis et al., 1993). The clinical phenotype subsequently included: (1) *Central nervous system*: Early signs of hydrocephalus led to a CT scan at 4 months, showing enlargement of the lateral ventricles; hydrocephalus arrested spontaneously and subsequent head size was within normal limits. She exhibited mild developmental delay with normal gross motor and sensory examination, normal reflexes and gait. At 9 years of age, her Wechsler intelligence scale tests ranged from 80–95, with a full scale IQ of 86. (2) *Growth and development*: The clinical course was characterized by short stature (below 5% at 9 years), thoracolumbar scoliosis, hyperextensible joints, recurrent otitis media, and hypercholesterolemia (which her father also had).

X-rays confirmed the thoraco-lumbar scoliosis and revealed bilateral absence of the patellae. Laboratory testing was normal for thyroid function, somatomedin C, growth hormone stimulation tests, and IGF levels. (3) *Sexual development*: Early onset of puberty was noted, with normal external genitalia and Tanner stage 2 breast and pubic hair development at 9 years. LH and FSH levels were suggestive of central onset puberty.

On physical examination at age 9, weight was 31 kg (50–75th centile) and height 110.5 cm (average height of a 5.4-year-old girl). Head circumference was at the 50th percentile, the ears were rather small, the palate high and narrow with crowded teeth. Chest circumference was increased and there was thoraco-lumbar scoliosis. The hands were below the 3rd centile, both the total length and palmar length being small; the feet were also small.

Chromosomal banding studies showed a *de novo* apparently balanced Robertsonian translocation, t(13;14). For DNA analyses, seven dinucleotide repeat polymorphisms, of which four proved to be informative, allowed characterization of chromosomes 14 and two more served to trace the origin of chromosomes 13. These revealed that one chromosome 13 came from each parent, but that the mother alone had contributed the chromosomes 14, suggesting maternal UPD14.

Longer film exposures, however, with the radiolabeled probes revealed a paternally derived allele for each marker studied, thus uncovering the presence of a mosaicism. This was confirmed by extended chromosome studies of 100 metaphases each from blood and fibroblasts. It was then found that 5% of lymphocyte metaphases showed trisomy 14, consisting of two free chromosomes 14 and a third in the translocated form, t(13;14). In the cells with UPD14, there was isodisomy for the proximal part of chromosome 14 and heterodisomy for the distal part. Studies with other markers (i.e., chromosome 21) confirmed paternity.

Several conclusions were drawn from this case (Antonarakis et al., 1993): (1) The *de novo* t(13;14) and disomy 14 had occurred during maternal meiosis, which was consistent with the observation of recombination events for chromosome 14. (2) The conceptus, which had effectively been trisomic for chromosome 14, underwent early loss of the paternal chromosome 14, and a mosaicism for trisomy 14 could be found in blood cells, but not the fibroblasts. (3) As in the earlier cases of maternal UPD14, a translocation between 13 and 14 was observed that must have favored a maternal 14 meiotic nondisjunction. (4) Comparison of the first two reports of maternal UPD14 and this case led the authors to underline the common phenotypic features, such as short stature (3/3), arrested hydrocephalus (2/3), small hands and feet (3/3), and early to precocious puberty (3/3), which might have resulted from maternally imprinted genes that map to human chromosome 14. A role for a trisomy 14 residual effect was not ruled out, however. A potential novel syndrome appeared to emerge from these three cases. (5) The authors proposed that on prenatal diagnosis of a *de novo* Robertsonian translocation, analysis of polymorphisms should be performed with DNA markers for both acrocentrics involved in the centric fusion, to attempt detection of a UPD-associated condition. (6) Mouse chromosomes 12 and 14 have homologies with human chromosome 14, without, at present, the knowledge of imprinted genes in these mouse chromosomes.

Case 4

This little girl was born at 38 weeks to a 30-year-old mother and 32-year-old father, weighing only 2000 gm. She had developed, early in life, a communicating hydrocephalus that had later been stabilized (Healey et al., 1994). At $4\frac{1}{2}$, height and weight were below the 3rd centile and head circumference at the 75th centile. She had mild to moderate developmental delay. Other findings included sparse scalp hair, frontal bossing, short philtrum, high palate, and small hands (below 3rd centile) with hyperextensible joints. There was a right single palmar crease and seven ulnar loops at the fingertips. She had suffered recurrent otitis media, and had asthma and hypercholesterolemia.

Cytogenetic studies revealed a *de novo* translocation t(13q14q); no mosaicism could be detected from a study of 200 blood cells. As in the previous case, chromosome 13 markers identified a paternal and maternal chromosome. Several DNA polymorphisms for chromosome 14 (two fully informative) tested in lymphocytes, hair follicles, and buccal mucosa cells showed both chromosomes 14 to be only maternally derived. Heterodisomy was shown for all except the most distal markers; other microsatellites confirmed paternity. The mechanism suggested for the *de novo* translocation was the same as in case 3 (Antonarakis et al., 1993). This study (Healey et al., 1994) also emphasized the similarity of clinical features shown by the four probands diagnosed with centric fusion between 13 and 14 and maternal UPD14 with isodisomy or heterodisomy.

Case 5

Then a report appeared of an essentially normal 30-year-old woman, G4Ab4, ascertained for miscarriages, occurring at 6–8 weeks' gestation (Papenhausen et al., 1995). She enjoyed good health and normal intelligence; her height was 167.6 cm and weight 62 kg. Menstrual periods began at age 13. There were three normal siblings.

Cytogenetic studies revealed a 45,XX,t(14q14q) karyotype with a single centromere on the der(14). DNA polymorphisms indicated a single allele of maternal origin for 2 polymorphic loci, and the authors concluded that there was maternal UPD14 (isodisomy). The t(14;14) was probably *de novo* although the father was deceased and could not be studied. If considered as an i(14q), the variant, in this case, was likely to represent monosomy rescue by maternal 14 duplication, compensating for a paternal 14 nullisomy. Thus one might conclude from this case that maternal UPD14 per se is harmless (except for homozygosity for recessive alleles). Since only two polymorphic loci on chromosome 14 were examined and the father was unavailable for study, there is a possibility of biparental inheritance in this case (Robinson and Langlois, 1996).

Case 6

A newborn girl carried a *de novo* balanced translocation t(13;14) detected at prenatal diagnosis (Coviello et al., 1996). She was initially evaluated because of severe hypotonia, hyperextensible joints, minor facial dysmorphism including asymmetry, and frontal bossing. At the age of 15 months, this child also showed severe scoliosis and motor delay, but apparently normal intelligence. Several microsatellite markers showed both chromosomes 14 to be maternally derived (UPD14), while nonpaternity was excluded. The maternal 14 pair was mainly heterodisomic.

The similarities of this case to the previous ones suggested to the authors the imprinting of maternal chromosome 14, again contributing to the definition of a possible maternal UPD14 syndrome, and a role for centric fusion in the occurrence of maternally disomic 14 gametes, causing a trisomic condition subsequently reducible to maternal UPD14 by paternal 14 loss. These same authors later reported a study of 10 subjects carrying t(13;14) Robertsonian translocations, four of whom were detected on prenatal diagnosis. In this group of 10 predisposed individuals, only the above reported case proved to have UPD14.

Case 7

Cytogenetic and molecular investigation of an 11-year-old boy with precocious puberty and motor developmental delay revealed a 45,XY,t(14q14q) or i(14q) karyotype with no paternal chromosome 14 contribution (Fig. 2, right, Chapter 5) (Tomkins et al., 1996) VNTR analysis of loci on four other chromosomes excluded nonpaternity with greater than 99% confidence. Results of polymorphisms at 10 loci along the entire length of chromosome 14 were consistent with homozygosity at all loci, suggesting that the chromosomal rearrangement was a maternal isochromosome for 14q. As the proband's father had a balanced Robertsonian translocation, t(13q14q), the authors suggested that the origin of the maternal UPD14 was fertilization by a nullisomy 14 sperm with formation of the isochromosome in the early embryo. Also, the proband has several clinical features in common with previously reported liveborn cases of maternal UPD14: hypotonia and motor developmental delay, mild dysmorphic facial features, low birthweight and growth abnormalities, and, more specifically, precocious puberty among the four cases old enough to be assessed. The emergence of a syndrome associated with maternal UPD14 suggests the possibility of genomic imprinting of regions of chromosome 14, especially a gene involved in the onset of puberty.

Case 8

Chorionic villus sampling, in a 40-year-old, had revealed a trisomy 14/euploid mosaicism, with 20% trisomic cells in cultured villi and 100% such cells in direct samples. Amniocentesis was then performed and a normal 46,XX karyotype was found in 70 cells examined (Morichon-Delvallez et al., 1994). The parents chose to continue the pregnancy and, per recent communication with the authors, it appears that the child is normal aside from growth retardation. There was no isochromosome or translocation in this case, and the trisomy 14 was confined to the placenta. The study of DNA polymorphic markers evenly distributed along chromosome 14 indicated maternal UPD14 (heterodisomy). The mechanism for this case was a trisomy 14 rescue as described above.

Case 9

Another case of fetal mosaicism 46,XX/47,XX+14 was diagnosed at amniocentesis (Sirchia et al., 1994). Molecular analysis of five polymorphic loci of the short tandem repeat type was performed. Markers D14S43 and D14S49 showed the presence of maternal UPD14 in the apparently normal cell line. The distribution of the markers analyzed along the chromosome suggests maternal heterodisomy with a large isodisomic segment in the telomeric region, possibly caused by meiotic crossing-over. The pregnancy was terminated.

Case 10

An infant female was studied and reported in an abstract form with *de novo* t(13,14), cleft palate, retrognathism, and "other problems" (Barton et al., 1996). Chromosome 14 polymorphic markers showed maternal UPD14 with isodisomy. In addition, a more sensitive analysis of the polymorphic markers revealed a minor paternal chromosome 14 contribution (<10%), suggesting the infant to be mosaic for trisomy 14 and maternal UPD14. This mosaicism was subsequently confirmed by cytogenetic analyses. As in case 3, the authors proposed the formation of a *de novo* maternal rob t(13;14) giving rise to a conceptus trisomy 14, with the subsequent loss of the paternal chromosome 14 resulting in maternal UPD14.

Case 11

In another abstract, a 12-year-old boy of normal intelligence, with an early history of hypotonia, failure to thrive, developmental delay, and hypogonadism, followed by precocious puberty and obesity was reported to carry a familial balanced, t(13;14) Robertsonian translocation (Linck et al., 1996). He had inherited his single "free" chromosome 14 from his mother and preliminary results from the analysis of DNA polymorphisms demonstrated maternal UPD14 (heterodisomy). The phenotype in this case included short stature, small hands and feet, mild developmental delay, hypotonia, premature puberty, hypogonadism, scoliosis, and mild dysmorphic features.

Case 12

In an abstract, a 41-year-old mother was reported who had an amniocentesis at 17 weeks. All amniocytes were 46,XX. At delivery, the infant was noted to have significant growth restriction, minor dysmorphic features (crumpled ear, simian crease, anteriorly placed anus), hypotonia, and poor suck-swallow coordination. Cytogenetic analysis of blood showed a normal 46,XX karyotype, whereas placental tissue revealed 47,XX,+14 in every colony studied. Molecular analyses of the newborn's blood using chromosome 14 polymoprhic markers showed maternal UPD14 and isodisomy for all markers (Walgenbach et al., 1997). The placental DNA showed maternal and paternal alleles consistent with a trisomy rescue mechanism for the UPD14. Nonpaternity was excluded with many other polymorphic markers. This case also highlights the importance of studying placental tissue in pregnancies with unexplained fetal growth restriction in the presence of normal karyotype at amniocentesis.

Case 13

A 26-year-old female was examined because of short stature (Splitt and Goodship, 1997). She had low birthweight (2 kg at term, <3rd percentile), her menarche begun at age 8 years, and she had a history of migraine from age 6. Her height was 132 cm; she had very small hands and feet, cubitus valgus, mild lumbar scoliosis, and mild obesity. Head cicumference was 51 cm (<3rd percentile) and brain CT did not show any evidence of ventricular dilatation. The karyotype showed a Robertsonian t(14;14) translocation. Analysis of DNA polymorphisms revealed maternal UPD14 (isodisomy). No evidence for mosaicism involving trisomy 14 was found.

Case 14

A male child was examined at 2 years 9 months because of motor and mental retardation. Birthweight at 36 weeks' gestation was 1900 gm (<3rd percentile). At the time of the examination microcephaly was noted (HC 44.5 cm), as well as hypotonic facies with large and protruding ears, depressed nasal bridge, narrow and upturning nose, constantly open mouth, irregular position of the upper and lower incisors, right undescended testis, transverse palmar crease on the right hand, small hands (length 9.5, <3rd percentile), small feet (foot length 14 cm, ~10th percentile). Developmentally, he corresponded to an infant of about 20 months. The karyotype revealed a translocation t(14q;14q). DNA polymorphisms analyses showed no paternal inheritance for multiple chromosome 14 markers. There was maternal UPD14, and all the markers were homozygous (isodisomy). Paternity was confirmed using other polymorphisms (Robinson et al., 1994).

Case 15

A 13-year-old female was evaluated for short stature (−4 SD). Her past medical history was remarkable for low birthweight and premature puberty. She also had generalized hypotonia, nail hypoplasia, hyperextensible joints, and small feet. She had normal intelligence. Cytogenetic analysis of blood lymphocytes revealed a translocation t(13;14). Her mother and maternal uncle were also carriers of the same balanced translocation. DNA analysis using microsatellite polymorphic markers showed maternal UPD14 (heterodisomy) (Desilets et al., 1997).

Case 16

A case of a Japanese infant with an isochromosome 14 [i(14q)] and intrauterine growth retardation (IUGR) was also reported (Miyoshi et al., 1998). The infant was one of triplets comprising a boy (the patient) and two karyotypically normal girls. He was delivered at 29 gestational weeks with a weight of 634 gm (the other two members of the triplet were 1110 gm at birth), length of 31 cm, and HC of 25.6 cm. At 9 months, his length was 59.3 cm, weight 4210 gm, and HC 41 cm. He was hypotonic, with small hands and feet, short neck, and patent ductus arteriosus. Analysis of polymorphic markers along the entire length of chromosome 14 was consistent with maternal UPD14 with alternating regions of isodisomy and heterodisomy. Thus, the derivative chromosome 14 had arisen through a translocation between maternal homologous chromosomes 14 after at least two crossing-over events at Me1.

Case 17

A case of maternal UPD14 detected prenatally, in a pregnancy with mosaicism for trisomy 14 observed in both a chorionic villus sample (CVS) and in amniocytes, was reported (Ralph et al., 1999). Analysis of polymorphic microsatellites showed that the fetus was isodisomic for one of the mother's chromosomes 14 and that recombination had introduced a mid-long-arm region of heterodisomy. The fetus, which died in utero at 18 weeks, showed no apparent pathological features.

Case 18

To determine the frequency of UPD14 among carriers of balanced Robertsonian translocations involving chromosome 14, 17 families [12 with t(13;14), 4 with t(14;21), and 1 with t(14;22)] were screened using polymorphic markers for each of the chromosomes involved. Eleven of these cases were from a prenatal diagnosis program. Maternal UPD14 was found in a 3-year-old male with a maternally inherited t(13;14), ascertained because of short stature, hypotonia, dysmorphic features, and developmental delay (Harrison et al., 1998).

Case 19

The occurrence of maternal UPD 14 in a 25-year-old woman with a normal karyotype and normal intelligence but low birthweight, short stature (143 cm; <3rd percentile), small hands, and early onset of puberty (at age 8 years 10 months) was reported (Fokstuen et al., 1999). She was born at term with a birthweight of 1960 gm, length of 47 cm, and HC of 31.5 cm (all <10th percentile). The authors commented that the comparison of her phenotype with those of 15 previously described liveborn patients with maternal UPD14 gives further evidence for an imprinted gene region on chromosome 14 and highlights the necessity to consider this cause in children with intrauterine growth retardation and early onset of puberty associated with acceleration of skeletal maturation.

Case 20

A child with clinical features of "maternal UPD14 syndrome" was reported who exhibited maternal UPD14 confined to a specific interstitial segment of chromosome 14 (Martin et al., 1999). The proband presented at the age of 3 years 6 months for evaluation of developmental delay. Her parents reported no problems at birth except hypotonia. No birth measurements were available. On physical examination, her height was at the 25th centile, weight at the 10th centile, and head circumference above the 97th centile. She had a normal head MRI. On formal developmental testing, there was mild global developmental delay. The karyotype was normal. Analysis of DNA polymorphisms showed maternal UPD14 for markers D14S277 and D14S77. However, marker D14S66 that is proximal to the previous loci, and D14S1028, D14S61, and D14S67 that are distal, were biparentally inherited. The authors interpreted these findings as evidence for segmental maternal UPD14 for the region 14q23-14q24.2. This region may define the shortest chromosomal region associated with the matUPD14 phenotype. The authors also hypothesized that the segmental UPD14 could have occurred after early postzygotic recombination events between the paternal and one of the two maternal chromosomes in a trisomy 14 zygote. This was followed by mitotic nondisjunction of the recombinant derivative chromosome containing the paternally derived interstitial segment (Martin et al., 1999).

The Phenotype of Maternal UPD14 Syndrome

The description of the above cases and the comparison of the phenotypes of the different probands allow the initial definition of the phenotype of the maternal UPD14 syndrome (Ledbetter and Engel, 1995). A tabular presentation of the clinical data from these patients has been done recently (Fokstuen et al., 1999). Table 4 in Chapter 5 lists the clinical data from the above cases.

Paternal UPD14

Case 1

This female patient was born with a length and weight at the 10th centile. Her multiple congenital anomalies included bilateral subarachnoid hygromas (head circumference at the 75th centile), short webbed neck, small thoracic cage with marked angulation of the ribs and restrictive lung disease, and facial dysmorphism (blepharophimosis, small ears, anteverted nostrils, and protruding philtrum) (Wang et al., 1991).

At the age of 9 years, she was severely retarded, and had a seizure disorder, coarse facial features, frontal bossing, prominent jaws (both maxilla and mandible), and severe kyphoscoliosis. Her karyotype was 45,XX,t(13q14q) inherited from her father. The mother was also the carrier of a reciprocal translocation coincidentally involving a chromosome 14: 46,XX,t(1;14)(q32;q32).

The free chromosome 14 of the proband was also probably derived from the father, as it showed a similar nucleolus organizer region (for both stalk and satellites) on QFQ banded analysis. This was confirmed by analysis of DNA polymorphic markers (paternal UPD14) that also indicated heterodisomy for chromosome 14. The fortuitous occurrence of a translocation involving chromosome 14 in both parents was reminiscent of the experimental mouse situation, whereby breeding of genitors heterozygous for similar balanced translocations produces a relatively high yield of complemented offspring. Consequently, it appeared likely that a 3:1 segregation in the child's mother produced a nullisomy 14 ovum fertilized by a sperm with both a translocated and a free 14, as the result of a 2:1 segregation. Other explanations, including trisomy 14 and subsequent mitotic loss of a maternal chromosome 14, were also possibilities, but careful search for a trisomy 14 mosaicism remained negative. The authors (Wang et al., 1991) attributed the child's phenotypic anomalies to imprinted genes that map to paternal chromosome 14.

Case 2

In case 2 (see Figure 4, Chapter 5), a little girl, with a *de novo* t(14q14q), was born to a 13-year-old mother and 15-year-old father (Papenhausen et al., 1995). The pregnancy had been complicated by progressive polyhydramnios and eclampsia. At delivery (32–34 weeks), birthweight was 1840 gm, birth length was 41 cm, and Apgar scores 6 and 8 at 1 and 5 minutes. Respiratory distress necessitated assisted ventilation. Tracheostomy was performed at 2 months, for laryngomalacia and gastrostomy for recurrent aspiration. At birth, the infant was remarkable for a hairy forehead, puckered lips, retrognathia, ventral wall hernia, and hypotonia along with mild contractures of hips, ankles, and fingers. At 1 year of age, cognitive and motor skills were a few months behind, and at 20 months, the patient was at or below the 5th centile for weight and length.

Prenatal analyses had already documented a *de novo* 45,XX, t(14q14q) karyotype, the derivative chromosome 14 showing a single centromere. A chromosome 14-specific DNA polymorphism indicated paternal UPD14 (isodisomy). There were phenotypic differences between this patient and case 1 of paternal UPD14 (Wang et al., 1991); striking laryngotracheomalacia, ventral wall hernia, and mild contractures were present in case 2, whereas severe mental retardation, seizures, marked rib and spine anomalies were present in case 1. There was no recognizable mosaicism for trisomy 14 in either case. The authors discussed that it is likely that there are paternally imprinted genes on chromosome 14.

Case 3

The 25-year-old mother with gestational diabetes was seen at 30 weeks of gestation for polyhydramnios (Walter et al., 1996). The fetal limb length was below the 5th centile, other measurements being normal for age except for a small thorax and unusually positioned hands. Polyhydramnios increased and spontaneous labor occurred at 34 weeks with vaginal delivery. Apgar were 4 and 5 in a limp and inactive infant, requiring respiratory assistance. Length was below the 3rd centile, head was large with marked scalp edema. Palpebral fissures were short, ears small (3rd centile) and mildly dysplastic; the neck was short and broad, the narrow thorax with prominent sternum was bell-shaped with upslanted ribs on X-rays. Nipples were hypoplastic. There was atrial septal defect and patent ductus arteriosus. There was also diastasis recti, unpalpable testes, and normal phallus. Fingers were long (97th centile) with proximal, interplalangeal joint mild contractures, adducted thumbs, short palms, and flexed wrists. Mild contractures were also present at elbows and knees with limbs at the 3rd centile. On X-ray, the fibulae were long and the proximal tibial epiphyses small. Hypoxia and apneic episodes occurred after tracheal extubation and esophageal intubation was needed for feeding. CT and MRI showed no brain anomalies but for prominence of the sylvian fissure.

At 4 months, he was below the 5th centile for weight (3.5 kg), length (49 cm), and head circumference (36.5 cm). The child had brachycephaly, high arched palate, small ears, short broad nose and long philtrum, small thorax, short limbs, and contractures. He was alert and responsive, smiled and cooed, rolling from back to side and seizing objects. At 6 months, congestive heart failure developed with obstructive hypertrophic cardiomyopathy. Cytogenetic results showed a *de novo* Robertsonian translocation t(14;14), which proved to be dicentric with C-banding, and FISH analysis with an alpha-satellite probe, 45,XY,idic(14)(p11).

Analysis of 19 highly polymorphic dinucleotide repeats spanning chromosome 14 showed paternal UPD14 (isodisomy). Of the potential mechanisms producing this dicentric chromosome 14, the authors envisaged as the most likely the mechanism that a monosomic chromosome 14 conceptus had been rescued by an isodicentric chromosome derived from the singly inherited paternal 14. All other mechanisms should have led to some degree of mosaicism for a “duplication trisomy” 14, which was not seen on either lymphocyte or fibroblast cultures. The authors also listed several characteristics in common with the other paternal UPD14 cases: polyhydramnios, digit contractures, small thorax, abnormal ribs, low birthweight, reduced birth length (below 10th centile), the requirement for gastrostomy, and small ears. As in other cases of balanced Robertsonian translocations associated with phenotypic abnormalities, the authors indicated that molecular analyses are necessary to rule out the possibility of uniparental disomy.

Case 4

The female proposita was born, at 29 weeks, to 21- and 20-year-old parents, after a pregnancy complicated by polyhydramnios. Birthweight, length, and head circumference were 1960 gm, 40 cm, and 29 cm respectively (Cotter et al., 1997). There was a short neck and narrow thorax, depressed nasal bridge, small ears, protruding philtrum, and short palpebral fissures. There was also hypotonia, poor suck and assisted respiration with intubation after tracheostomy, and the need for oro-gastric tubing. When the baby died, at 6 months, a skeletal survey showed shortening of the long bones, microcrania, ossification defect of the cranial base, a bell-shaped, short thorax with thin ribs, elongated clavicles, and coxa valga (see Figure 3 in Chapter 5).

Chromosome analysis showed a Robertsonian translocation with a karyotype 45,XX,t(13;14). Parental karyotypes revealed that both parents were carriers of Robertsonian translocations; the father had a t(13;14) and the mother t(14;21). Molecular studies with polymorphic markers showed paternal UPD14 and absence of maternally derived alleles. Paternal heterodisomy was documented at D14S52 and D14S68 and isodisomy for D14S81 and D14S12 markers. Eleven other markers were uninformative but consistent with either paternal isodisomy or heterodisomy. Biparental inheritance was confirmed for markers on chromosomes 13 and 21.

Comparing this case to the three earlier ones, the authors reached the conclusion that patients with paternal UPD14 share several findings such as polyhydramnios, characteristic facial anomalies, small thorax, abnormal ribs, and short limbs. Paternal UPD14 could then be considered the cause of an identifiable genetic syndrome, although some variability might result from homozygosity of recessive alleles, undetected mosaicism for trisomy 14, and/or other extraneous factors (e.g., extreme prematurity). Clearly, more cases are needed for the comprehensive definition of this emerging syndrome.

Paternal UPD14 in case 4 was thought to result from MeI paternal nondisjunction with a 2/1 adjacent segregation causing disomy 14 in a gamete. As in another case of paternal UPD14 (Wang et al., 1991), both parents happened to have a translocation involving chromosome 14, which, by increasing the propensity to missegregation, would favor gamete complementation. As the mother had a t(14;21), it is quite likely that she produced an ovum with nullisomy 14 (by 1:2 adjacent segregation) that was fertilized by a sperm with the t(13:14) and disomy for chromosome 14 (Cotter et al., 1997).

It is possible that both maternally and paternally derived genes on chromosome 14 are imprinted in humans, eventually from a single domain with differential expression for each sex.

CHROMOSOME 15 UPD

Maternal UPD15

It was already known that interstitial deletion of the proximal region of paternal chromosome 15 at 15q11-q13, would cause the Prader-Willi syndrome (PWS). In 1989, maternal UPD15, another mechanism resulting in the absence of paternal contribution of chromosome 15 genes, including region 15q11-q13 was also found to be associated to PWS. Since then it became clear that genes in this region needed to be inherited from each parent for normal human development to proceed (Nicholls et al., 1989). The two first clear demonstrations of this remarkable fact, one with a familial rob t(13;15) and the other with a pair of free maternal chromosomes 15, began a long string of such cases in which maternal UPD15 causes PWS (Nicholls et al., 1989).

Case 1

The patient had typical PWS, with growth delay and hypotonia in infancy, short stature, hyperphagia, obesity, small hands and feet, hypogonadism, and mild mental retardation (Nicholls et al., 1989). A familial balanced Robertsonian translocation t(13;15) similar to that observed in the proband was found in the mother and unaffected maternal relatives. The difference was that the proband had inherited his second, free chromosome 15 also from his mother, while all other healthy siblings inherited it from their respective fathers. In this family and that of the next case, RFLP analyses were performed at nine loci on proximal 15q. In the PWS proband, two maternal alleles but no paternal allele at the locus were detected by probe 3.21 (which is absent in PWS patients with deletions). One of the two alleles of this marker was shared by the proband, his mother and half-sister carrying the translocation, and was therefore a marker for the t(13;15) chromosome; the other allele was also detected in mother and proband, but was absent in the half-sister and could thus be traced to the free chromosome 15 also shared by mother and proband but not by other translocation carriers. The finding was confirmed with additional markers such as by CAW-1, a multiallelic DNA marker, distal to the region of deletion associated with PWS. The segregation pattern, while showing only maternal alleles, also confirmed an absence of paternal alleles. Nonpaternity was ruled out by analysis of the hypervariable region of the alpha-globin locus on chromosome 16. The authors, from these data, concluded that there was maternal UPD15 because only maternally derived alleles were present in the DNA of the proband. Although the full extent of UPD15 was not determined for the whole length of the chromosome, it spread well beyond the proximal 15q area which is deleted in other cases (Nicholls et al., 1989).

Case 2

The patient with PWS had two intact, free chromosomes 15, with only maternal alleles, as shown by polymorphisms 34 and 1R39d; dosage analysis demonstrated the absence of deletion at these loci (Nicholls et al., 1989). The long-range structure of 15q11-q13 was analyzed by pulsed-field electrophoresis, showing that both the above probes detected the same large maternal 2500-kb NotI fragment in this patient. Nonpaternity was also excluded.

In addition to the above two cases, the authors referred to preliminary data on four more PWS patients, strongly suggestive of a maternal UPD15 etiology.

It was thus "well shown that the association of PWS with maternal UPD for regions 15q11-13 implicates a role for genomic imprinting in the etiology of the PWS" and that "... normal human development requires input from both parents, whereas the absence of a paternal contribution to region 15q11-q13, whether by paternal deletion or maternal uniparental disomy ... would result in PWS." It was also proposed that, by reason of symmetry, paternal UPD15 might additionally result in the Angelman syndrome as does maternal deletion 15q11-q13 (Nicholls et al., 1989).

Well over 120 cases of PWS in the medical literature have been associated with maternal UPD15 (Robinson et al., 1996); of this large group, a few cases have been selected for individual mention, as they illustrate some points of interest in the PWS. The same approach is also adopted for presenting selected cases of Angelman syndrome associated with paternal UPD15.

Case 3

This child was hypotonic at birth. Chromosome studies had failed to reveal the deletion of 15q11-q13, and typical features of PWS developed over time (Cassidy et al., 1992). The mother, a gravida 3 para 3, 43 years old at conception, was treated successfully for hypothyroidism. CVS using direct harvest and cultured cells revealed a uniform 47,XX,+15 karyotype in about 40 cells, but amniocentesis from two cultures with a total of over 150 cells showed 46,XX. The pregnancy proceeded well until preeclampsia and premature labor developed, accompanied by decreased fetal movements, leading to a C-section at 35 weeks' gestation. Birthweight was 2200 gm, length 43 cm, and head circumference 31 cm. Hypotonia and a weak cry were noted with respiratory distress, episodes of apnea, poor sucking, and feeding problems. Facial features included a narrow face, small palpebral fissures, mild micrognathia, and hypotonia. Feeding problems gradually improved although bottle feeding had to be supplemented by gavage for the first 6 months. At 14 months, length was at the 20th centile, weight at 5–10th centile, and head circumference at the 40th centile. Developmental milestones were appropriate for 11 months and hypotonia persisted. Hands and feet were small. On karyotypic studies, the two chromosomes 15 showed different polymorphisms, one with a short satellite stalk, the other without a stalk. Molecular studies revealed maternal UPD15 for markers in distal 15q, with the absence of paternally derived alleles for polymorphic locus pMS620; polymorphic markers in the PWS region were uninformative. No allelic indications of a trisomic lineage were noted, even after overexposure of the Southern blots. Although these studies failed to show molecular evidence of pericentromeric heterozygosity, the noted cytogenetic heterozygosity strongly suggested MeI nondisjunction of a maternal pair and rescue from trisomy 15 in the early embryo by loss of the paternal chromosome 15.

This case, which is among the first of well-documented "trisomy rescue" mechanisms (Cassidy et al., 1992), also illustrated the need for molecular studies to rule out maternal UPD15 as a cause of PWS syndrome derived from trisomy-15-cases, in order to provide fetal diagnosis and appropriate genetic counseling.

Case 4

As in case 3, CVS for advanced maternal age led to the detection of placental euploid/trisomy mosaicism for chromosome 15, all cells being trisomic in short-term preparations and 50% trisomic in cultured villi. Amniotic fluid analysis, however, indicated a uniform 46,XY pattern and the pregnancy was carried to term (Purvis-Smith et al., 1992). The baby was extremely hypotonic and later displayed delayed developmental milestones. At 2 years, the weight was above the 97th percentile and PWS was diagnosed. Molecular analysis of the parental origin of the child's chromosomes 15, using 3 polymorphic microsatellites, indicated that there were only maternally inherited alleles. Of the two informative loci, one was homozygous and the other heterozygous, proving heterodisomy. Pericentromeric short-arm heteromorphisms were also distinctive, indicating maternal Me1 nondisjunction. Paternity was confirmed by using additional polymorphic markers. The correction of a trisomy by loss of the singly inherited paternal chromosome 15 appeared most likely, with maternal UPD15 present in the fetal cells, while extraembryonic tissues (CVS) contained trisomy 15 cells. The need for molecular investigation during the course of such prenatal diagnoses was again strongly advocated.

Case 5

Bloom syndrome (BS) is an autosomal recessive disorder characterized by increases in the frequency of sister-chromatid exchange and in the incidence of malignancy. The gene responsible for BS maps to chromosome 15q. A patient with features of both BS and Prader-Willi syndrome (PWS) was described and molecular analysis showed maternal UPD15 (Woodage et al., 1994). Meiotic recombination between the two maternal chromosomes 15 had resulted in heterodisomy for proximal 15q and isodisomy for distal 15q. The authors concluded that in this individual BS was probably due to homozygosity for a gene that is telomeric to D15S95 (15q25), rather than to genetic imprinting, the mechanism responsible for the development of PWS. This report was one of the first that applied analysis of UPD to the regional localization of a disease gene (Woodage et al., 1994). The gene was cloned a year later (Ellis et al., 1995) and the mapping prediction was correct.

Advanced Maternal Age and Maternal UPD15

Since one of the most common causes of maternal UPD15 is trisomy 15 rescue, the possibility of advanced maternal age, similar to that of trisomy 21 (Antonarakis et al., 1993), has been examined in these cases. Nondisjoined chromosomes 15 from 115 cases of maternal UPD15 (ascertained through Prader-Willi syndrome) and 13 cases of trisomy of maternal origin were densely genotyped for microsatellite loci spanning chromosome 15q. Of these 128 cases, a total of 97 meiosis 1 (Me1) errors, 19 meiosis 2 (Me2) errors, and 12 mitotic errors were identified (Robinson et al., 1998). The genetic length of chromosome 15 created from the Me1 errors was 101 cM, as compared with a maternal length of 137 cM based on control families. It was estimated that 21% of tetrads leading to Me1 nondisjunction were achiasmate, which may account for most or all of the reduction in recombination. Similar to the situation in trisomy, the mean age of mothers of cases involving Me1 errors of chromosome 15 was increased (34.9 versus 28 years in the general population). Interestingly, the mean maternal age in Me1 errors that showed no transitions from heterodisomy to isodisomy (no detectable cross-over) was also increased over the general population but was lower (32.7 years) than cases showing one or more observable recombinations (36.3 years) ($P < 0.003$).

Paternal UPD15

The first evidence of paternal UPD15 was reported 2 years after the prediction reported in (Nicholls et al., 1989) and was observed in two patients with Angelman syndrome (AS). The transmission of parental chromosome 15 was examined in 32 AS patients from 26 families who had shown no detectable deletion of chromosome 15 by either cytogenetic or molecular investigation (Malcolm et al., 1991).

Case 1

This baby girl weighed 3550 gm at term. Severe feeding difficulties occurred with micrognathia and a cleft soft palate. Motor milestones were delayed, with neither walking nor speech development at 4 years and intermittent seizures with characteristic EEG since 2 years of age. Head circumference was at the 10th centile, and the facies typical of AS (Malcolm et al., 1991).

Parental origin of the chromosomes 15 was assessed with two multiallele polymorphisms cMS620 at the distal end of 15q and CMW1 at 15q11-q13. In this child, both alleles originated from the father, and no maternally derived allele was observed. There was therefore paternal UPD15 with heterodisomy.

Case 2

This proband was a 6-year-old boy, born after a normal pregnancy and delivery at average weight (Malcolm et al., 1991). At 9 months, microcephaly was noted and motor milestones appeared delayed. AS was diagnosed at 2 years (facial appearance, happy disposition, jerky movements, and typical EEG). Strabismus was corrected, speech did not develop, ataxic and stiff-legged gait was present at 3 years. At 6 years, microcephaly was marked (3rd centile), height and weight were at the 25th and 10th centile. On molecular analysis, the patient appeared homozygous for paternally derived alleles and no maternal contribution was found for either CMV-1 or cMS60 polymorphisms, a pattern indicative of paternal UPD15 (isodisomy) as in case 1. High resolution chromosomal banding at the 850 band level did not indicate deletion of the critical interstitial segment (Malcolm et al., 1991).

The authors concluded that, in both of these cases, a maternal contribution at the AS locus is essential for normal development and that two intact paternal chromosomes 15 are no substitute, again underscoring the importance of parent-of-origin imprinting of certain genes. It was also thought unlikely that chromosome 15 contained imprinted loci outside the area ordinarily found deleted in both AS and PWS, and the view was expressed that loci imprinted in both syndromes are not homologous but closely linked, and mapped within the same deleted area, paternal in PWS and maternal in AS (Malcolm et al., 1991).

Case 3

The patient was born at 42 weeks' gestation to a 33-year-old mother and 26-year-old father. Weight and length were 3950 gm and 49.5 cm, respectively (Nicholls et al., 1992). Delayed development was suspected by age 9–10 months. Language delay was obvious by 15 months. Extremities were hypotonic and deep tendon reflexes present, with downward plantar reflexes. He could sit with support but remained unable to bear weight or take steps. Head circumference was in the 10th centile, length and weight in the 50th centile. At age 26 months, a composite development score (Bailey scale) was that of an 11-month-old. At 33 months, microcephaly was present (< 5th centile), while height and weight were in the 75th centile. Face was round with mid-facial hypoplasia, macrostomia, thick lips, and tongue protrusion. The child was of a happy disposition. Gait was unsteady. There were no seizures, but the EEG showed a background of 3–5 Hz or slower theta waves of 30–40 mV. Cranial magnetic resonance imaging was unremarkable. Angelman syndrome was then considered a likely diagnosis.

A normal 46,XY karyotype was found. A number of DNA polymorphic markers for 15q11-q13 were genotyped, including 34, 3-21 IR4-3R, IR10-1, 189-1, and CMW1 as well as marker MS620 that maps in the 5qter telomeric area. Allele copy numbers were studied by densitometry for loci at 15q11-q13. It was found that the DNA of the patient contained only paternally derived alleles for chromosome 15 markers (paternal UPD15), without contribution from the mother. Cytogenetically, this child also had received a pair of the only distinctly satellited paternal chromosome 15 (centromeric isodisomy). For all markers studied throughout chromosome 15, there was homozygosity for the paternally derived alleles (isodisomy). This was thus regarded as the duplication of a paternal chromosome 15. This was the only AS patient out of 10 without a proximal 15q deletion that was found to have paternal UPD15 (Nicholls et al., 1992). An additional 27 AS patients in the author's series were found with such proximal 15q deletions. The authors remarked that UPD15 is found much more rarely in AS than PWS, approaching only 4% in this series (Nicholls et al., 1992).

Case 4

This boy was born at term to a 31-year-old mother. The clomiphene assisted pregnancy was complicated in the third trimester by hypertension and edema (Freeman et al., 1993). Birthweight was 4100 gm. Motor delay became obvious at about 9 months. Depressed central motor tone was noted at 19 months, at which time the child was unable to stand, walk, or talk. At 35 months, weight was in the 95th centile, length in the 10–25th centile, and head circumference in the upper range (75–90th centile). He was blond, with hypertelorism, blue eyes, and mild prognathism. He had begun walking with a jerky, wide-based gait but still did not speak. He had outbursts of mixed emotions beginning with a laugh to end in a sound of anxiety or distress. There was no history of seizures but an EEG consistent with that reported in children with AS.

Cytogenetic studies showed a karyotype 45,XY,t(15q15q); the translocation chromosome was monocentric. Results for DNA polymorphisms at four chromosome 15 loci showed only a paternal contribution, and for three of these loci, there was homozygosity for one allele that was present in heterozygosity in his father. There was therefore paternal UPD15 with homozygosity for paternal alleles at all loci tested, including one located within the AS critical region, which made it likely that the chromosome 15 rearrangement was an isochromosome (Freeman et al., 1993).

The authors proposed that the i(15q) chromosome was more likely to be a compensation for maternal nullisomy than a postzygotic loss of the maternal chromosome with selection of the balanced cell line containing only the paternal i(15q). It was also remarked that this child did not display microcephaly, macrostomia, seizures, and several of the other features usually observed in AS (Freeman et al., 1993). The following two cases are described to illustrate some phenotypic peculiarities of paternal UPD15 cases, as compared to those caused by deletions.

Case 5

The 2970-gm infant was the product of a 43-year-old mother and 45-year-old father (Bottani et al., 1994) (see Figure 2, Chapter 7). While very quiet and rarely crying in the first months, sleep disturbances occurred between 9 and 21 months. She rolled over and crawled at 11 months, spoke her first words at 18 months, walked independently at $2\frac{1}{2}$ years, and was day toilet-trained at 7 years. Transient mild ataxia when walking later disappeared to the point that she could slide down gentle hills on skis. She could never speak more than three to five words, but could communicate with gestures and her understanding far exceeded her language communication. A short period of self-aggressive behavior ended with the second year and mouthing and drooling disappeared after 4 years. Outbursts of laughter started at the end of the first year, occurring even at night, by 5 years. Tonic-clonic seizures began at $4\frac{1}{2}$ but only occurred a few times. The diagnosis of AS was suspected only at age $7\frac{1}{2}$, on the basis of the EEG pattern (Bottani et al., 1994); there was also hyperactivity and severe mental retardation. There was no ataxia and gait was slightly wide-based.

The karyotype was 46,XX and no partial deletion of chromosome 15 was detected at a resolution of about 600 bands. On DNA studies, the patient showed no inheritance of a maternal allele for marker D15S12, which maps at the critical region, nor at D15S24, D15S108, and D15S86, mapping distal to the PWS/AS region on 15q. There was reduction to homozygosity of all chromosome 15 markers heterozygous in the father (6 out of 11), compatible with paternal UPD15 (isodisomy) (Bottani et al., 1994).

Case 6

The girl was the product of a normal pregnancy with vertex delivery at term, weighing 3200 g for a length of 48 cm (Bottani et al., 1994). The mother was 31 and the father 43. Excessive crying and tongue thrusting began before 2 months and breast feeding was difficult. Inguinal and umbilical hernias were operated on at 10 weeks. Frequent upper respiratory and urinary infections occurred, at which time, jerky arm movements accompanied high fever. Outbursts of laughter were heard around the age of 1 year. Psychomotor development was delayed, as she sat alone and transferred objects at 9 months and learned to walk at 28 months. The gait was initially wide-based, with arms typically held flexed. Balance improved considerably after the third year. The child could ride a bicycle and slide on skis. First words were uttered around 4 years, but speech, the most delayed feature in her impaired development, has not gone beyond four to five single words. Disturbed sleep was a major problem until age 10 and appetite has always been excessive. No genuine seizure was ever observed, but the EEG was abnormal.

Examination at 10 years included the following: hyperactivity, severe mental retardation, bouts of laughter, absent speech, brachycephaly with occipital grooves, black hair, brown eyes, macrostomia with intermittent tongue thrusting, pointed chin without prognathism and no ataxia.

Karyotype was 46,XX, with no detectable deletion of proximal chromosome 15. There was no maternal inheritance of chromosome 15 alleles at four loci and homozygosity for all paternal alleles was noted; the diagnosis therefore of paternal UPD15 (isodisomy) was established.

The authors noted that both these patients had some milder features than usually seen in AS, such as a less typical facies, mild epilepsy in both, nonsignificant ataxia, and prepubertal obesity (Bottani et al., 1994). Epilepsy was also mild in some other AS patients with paternal UPD15. The reason for this was unclear, but the hypothesis of "leaky" expression of the AS gene from the paternal chromosome was considered. The potential role of increased paternal age in the etiology of these cases was also raised, although paternal isodisomy for chromosome 15 might be best explained as a compensatory mechanism for maternal 15 meiotic nondisjunction, resulting in nullisomy 15.

Case 7

This male proband was born to healthy parents aged 29 and 32 with a birthweight of 3.2 kg. He was irritable, experienced poor sleep and motor delay, and walked at 19 months. At 4 years, he displayed ataxia, jerky movements, and absent speech. He combined happy sociable behavior with some autistic trends (Ramsden et al., 1996).

Facial dysmorphic features were minor, including prominent chin and protruding tongue. He had no seizures and appeared normally pigmented. The karyotype revealed a *de novo* t(15;15) translocation. FISH analysis with probes 15A (D15S11) and 15B (GABRB3) within 15q11-q13 showed no deletion. No maternal allele contribution was found at marker GABRA5 and PW71B markers. Both chromosomes 15 in the centric fusion were therefore of paternal origin (paternal UPD15).

The authors remarked that the phenotype of paternal UPD15 cases of AS differ somewhat from those cases with partial deletion 15q; the former usually exhibit a higher birthweight, lesser delay in motor milestones, reduced tendency to seizures, and no hypopigmentation, as they retain both alleles at the non-imprinted locus of type II oculocutaneous albinism, OCA2, when compared to deleted cases that are hemizygous for these alleles (Ramsden et al., 1996).

Case 8

A case with 45,XX,t(15q;15q) was described in which molecular analysis with numerous polymorphisms on chromosome 15 showed paternal UPD15. The patient at the last examination at 9 years of age had several characteristics of AS, including developmental delay, speech impairment, ataxic gait, severe mental retardation, apparent happy demeanor with frequent laughter, frequent drooling, and macrostomia (Fridman et al., 1998). However, the diagnosis of AS was delayed due to a late onset of seizures, a milder clinical course, and the presence of unusual findings such as hyperphagia, obesity, behavioral problems when food intake was refused, thick saliva, and skin picking. The height was on the 90th percentile (142 cm) and weight greater than the 98th percentile (47 kg). The authors suggested that some patients with clinical features of AS have hyperphagia and obesity with overgrowth; this unusual AS phenotype may be associated with t(15;15).

Case 9

The boy was born from 25-year-old parents at 40 weeks' gestation by forceps, with a birthweight of 3460 gm (50th centile), length of 51 cm (25th centile), and head circumference of 34 cm (25th centile) (Gillessen-Kaesbach et al., 1995). He was referred at 13 months for lack of speech, short stature, and microcephaly (height and head circumference being −1 and −2 SD, respectively).

The child was friendly, blond with a triangular face and a small pointed chin. The mouth was large, the upper incisors widely spaced, the tongue not protruding. Psychomotor delay was moderate; he sat unassisted at 9 months but did not walk independently at 13 months. The EEG showed theta/delta rhythm and spike/wave complexes. The diagnosis of AS was considered at 23 months, on the basis of microbrachycephaly, delayed psychomotor development, and absent speech. The height was 85 cm (−1 SD) and head circumference 46 cm (−2 SD). Walking with a broad ataxic gait was achieved at 30 months. His behavior was friendly but there were no outbursts of laughter. He did not speak but communicated with gestures.

The karyotype was 46,XY, with no detectable deletion at the proximal 15q. DNA polymorphic studies showed hemi- or homozyosity for paternal alleles and absence of maternal alleles at D15S18, D15S10, and D15SS12. Hybridization using the SNRPN probes showed two copies and the methylation status was of paternal type only. Once again, this was a case with minor epileptic status and relatively good gestual communication. The facies was less typical. The authors suggested that the relative scarcity of paternal UPD15 cases in AS might stem from diagnostic oversight since milder cases may escape diagnosis. Therefore, methylation testing (Gillessen-Kaesbach et al., 1995) was recommended in patients with unexplained absence of speech, microcephaly, developmental delay, and EEG abnormalities (Gillessen-Kaesbach et al., 1995).

To summarize, paternal UPD15 cases with AS tend to have a more usual facies, less ataxia and seizure activity, with more appropriate familial pigmentation leading to diagnostic delay or oversight.

CHROMOSOME 16 UPD

It has been observed for several years that there were discrepancies of chromosomal constitution between extraembryonic (mostly placental) and embryo-fetal tissues; this is the so-called "confined" placenta mosaicism (CPM), in which there is apparent confinement of various mosaic patterns to the placenta and the chromosomal abnormality is not seen in the fetus (Kalousek and Dill, 1983; Kalousek et al., 1987). The significance of this phenomenon became more clear with the introduction of chorionic villus sampling [also called "choriocentesis" (Brambati et al., 1987; Engel, 1984)] as a tool for prenatal diagnosis. Considering the high prevalence of trisomy 16 in human spontaneous abortions, it was no surprise that most cases of maternal UPD16 appeared to be the product of trisomy 16 rescue.

Maternal UPD16

Case 1

Ultrasonography of a primigravida pregnancy documented a dramatic fetal growth deceleration to below the 5th centile between the 13–16th and the 23rd week of gestation (Bennett et al., 1992). Chorionic villus biopsy and rapid karyotyping revealed trisomy 16. At elective termination, a 350-gm morphologically normal fetus (weight below 3rd centile) was delivered. Placental trophoblast cultures showed euploid/trisomy 16 mosaicism in 35 and 65 cells, respectively. Five fetal tissues were karyotypically normal. Two informative chromosome 16 polymorphisms revealed only maternal alleles in fetal tissues, in contrast to three alleles (two maternal and one paternal) at each tested locus in the placenta, confirming fetal maternal UPD16 with apparently confined placental mosaicism.

Case 2

A phenotypically normal girl was born at 35 weeks' gestation to a 36-year-old mother, whose CVS had revealed confined placental abnormality for trisomy 16, in both direct and cultured samples. Amniotic fluid cultures (30 cells) and cord blood (50 cells) had normal chromosome counts (Dworniczak et al., 1992). Birthweight was not reported, but severe intrauterine growth retardation had been noted. Apgar scores were excellent. Surgery was performed at 3 months for incarcerated inguinal hernia. Neurologic examination showed adequate milestones and, at $3\frac{1}{2}$ months, physical development was reported as normal.

Three highly polymorphic microsatellites on chromosome 16 revealed two maternal and one paternal alleles per locus in the placenta, and maternal UPD16 was demonstrated in the child. Nonpaternity was excluded using polymorphic markers for four different chromosomes. Aside from having demonstrated maternal UPD16 by presumptive early loss of paternal chromosome 16 in the conceptus, the authors (Dworniczak et al., 1992) indicated that, from this limited observation, UPD16 might be compatible with normal development and therefore there may not be imprinted genes on chromosome 16. They considered that IUGR might have resulted either from the UPD16 or from placental malfunction due to trisomy 16 confined placental anomaly.

Cases 3 to 6

The next four cases of maternal UPD16 were found in a collaborative study of nine cases with trisomy 16 prenatally diagnosed at CVS. All but one had complete placental aneuploidy, i.e., with 100% trisomy 16 in cells examined on direct preparation or long-term cultures or both; the only mosaic case had 70% aneuploid cells in CV cultures (Kalousek et al., 1993). There were, however, no trisomy 16 cells on amniocentesis, in these nine cases.

The features of these four cases with maternal UPD16 were as follows: (i) Three of four had IUGR; (ii) the 4th case experienced unexplained intrauterine death at around 16 weeks' gestation; (iii) all four cases also had 100% aneuploidy in term placentae, in cells of cultured villi, and/or chorionic plate; (iv) one had imperforate anus and another had a two vessel cord; (v) one case was aborted at 25 weeks' gestation because of maternal hypertension.

The features of four further cases with a fetal biparental pair 16 were as follows: (i) Two of the four also had IUGR; (ii) these same two also presented 100% aneuploid cells in cultured villi or cultured chorionic plate cells of their term placentae; (iii) in one newborn, there was marked asymmetric IUGR with normal cranial growth; in another, there was mild glandular hypospadias. In two cases of biparental 16, and one whose chromosome 16 origin could not be determined, there were few or no trisomic cells in villus or cultured chorionic plate at term. These infants had normal birthweight.

In spite of the small numbers, it appeared that normal birthweight correlated with low-level mosaicism or absence of the extraembryonic trisomy for chromosome 16 at delivery. The authors (Kalousek et al., 1993) concluded, "... Whether the growth retardation and fetal demise observed in ... cases are related to the presence of the trisomic cell line in the placenta or to fetal UPD, remains to be determined."

Case 7

In this peculiar case, maternal UPD16 (heterodisomy) and residual trisomy 16 cell line were found in the fibroblasts of the proband but not in blood (Lindor et al., 1993). Born to a 28-year-old mother at 35 weeks with Apgar scores of 6 and 8, the female infant was below the 3rd centile for weight (1680 gm), while length was 41 cm and HC 30 cm (10th centile). Hypotonia was initially present and cerebral ultrasonography showed grade I intraventricular hemorrhage. At 11 months, neurologic development and motor milestones were normal and she walked at 14 months. Social and communication skills appeared normal. On physical examination, dolichocephaly and widely spaced eyes had been noted at birth. Ears were asymmetrically placed, with poorly folded upper helices and angulated antihelices and a left preauricular pit. Toward the end of the first year, height was at the 10th centile, weight 25th centile. The right zygomatic arch was underdeveloped, hands and feet were disproportionately small (below 3rd centile), with hypoplasia of the proximal and middle phalanges of the 5th finger. Toenail hypoplasia was also noted.

On cytogenetic analysis, there were colonies from distinct primary amniotic fluid cultures that showed euploid/trisomy 16 mosaicism. Following delivery, blood was found to be uniformely diploid (in over 100 cells), but trisomy 16 mosaicism was present in skin fibroblasts (6/30). Cultures of term placenta were uniformly trisomic (30/30 cells). The study of three highly polymorphic loci for chromosome 16, on blood and fibroblast DNA, clearly revealed inheritance of two different maternally derived alleles (maternal UPD16, heterodisomy). This example again documents very small birth size (less than 3rd percentile), in the presence of trisomy 16 in term placenta. The phenotypic contribution of the mosaicism for trisomy 16 in some tissues is unknown. In additional cases of mosaic trisomy 16, there is extensive tissue variation in the proportion and distribution of trisomic cells (Kalousek, 1994); Devi et al., 1993).

Case 8

A CVS at $9\frac{1}{2}$ weeks in a 35-year-old woman showed 47,XX,+16 in all cells and in early amniocentesis, 2 weeks later, a 46,XX pattern (Sutcliffe et al., 1993). Intrauterine growth retardation was present at delivery of a normal female with a right club foot. Apgar scores were 6 and 7. Of five term-placenta biopsies, all showed a 47,XX,+16 in 162 metaphases, except for three cells from the maternal surface biopsy and one from the fetal surface. Parental origin of chromosome 16, from CVS and cord blood, was determined with polymorphisms D16S7 and

D16S85; there was maternal UPD16 (isodisomy) at both loci. A third paternal allele was visible only in the placenta.

The authors concluded that their finding underlined a Me2 error causing isodisomy. The child was said to be developmentally normal at 9 months, but measurements were not given (Sutcliffe et al., 1993).

Cases 9–10

Case 9 came to the authors' attention in the work-up of a 25-year-old primagravida. She had high levels of maternal serum alphafetoprotein (MSAFP) with spontaneous rupture of the membranes at 18 weeks' gestation; IUGR was observed from 21 weeks' gestation, ultrasound measurements having fallen below the 5th centile (Vaughan et al., 1994). CVS revealed confined placental mosaicism with trisomy 16, in direct and long-term cultures. After elective termination of the pregnancy, the fetus weighed 350 gm (less than 5th centile for 24 weeks' gestation), and had an imperforate anus, large immature ears, and a single unilateral transverse palmar crease; cord and membranes were normal. Villus stromal cultures on delivery placenta confirmed trisomy 16 in about 70% of 50 cells. Four different fetal tissues revealed a normal female karyotype. DNA from placenta and fetal lung was studied using polymorphic markers on chromosome 16. This analysis showed two maternal alleles in the fetus (maternal UPD16) and three alleles per locus, including a paternal one, in the placenta.

Case 10 was detected in a 33-year-old primagravida with markedly elevated hCG at 16 weeks' gestation and showed fetal growth impairment, and ultrasound measurements falling below the 3rd centile (Vaughan et al., 1994). Amniocentesis revealed a 46,XY karyotype. At 28 weeks, no fetal growth had occurred for the last 3 weeks; abnormal Doppler velocities and cardiotocograph recordings led to an emergency C-section. The male newborn weighed 520 gm and had low Apgar scores; an imperforate anus, and a two-vessel cord; he died at 8 days due to persistent hyaline membrane formation. The placenta was microinfarcted with perivillous fibrin deposition.

The study of polymorphic markers for fetal DNA showed only the maternal chromosome 16 alleles in the blood; there was, however, a paternal allele contribution in the placenta DNA. Placental cultures and karyotyping showed about 30% trisomy 16 cells out of 29 metaphases. It is of interest that in both cases, there was abnormal MSAFP or increased hCG; in addition, anal imperforation was noted in both.

This group of investigators later screened for uniparental disomy for 12 candidate chromosomes in 35 babies with idiopathic intrauterine growth retardation (< 5th percentile). They only found two cases of maternal UPD16 (cases 9 and 10). They concluded that UPD for the chromosomes tested does not explain the etiology of the majority of cases of intrauterine growth retardation (< 5th percentile). Maternal UPD16 accounted for 5% of this cohort. Structural chromosomal abnormalities accounted for 11% (Moore et al., 1997).

Case 11

Ultrasound examination suggested a 3-week growth delay in a 16-week fetus, in the course of a pregnancy in which maternal serum screening had revealed a relatively high risk of trisomy 21 (O'Riordan et al., 1996). CVS documented uniform trisomy 16 in both direct and cultured samples. In amniotic fluid cultures, however, a normal 46,XY karyotype was observed. Three polymorphic loci were examined, in search of UPD in the fetus and the newborn. There were only maternally derived alleles in the fetus (maternal UPD16; heterodisomy). After a C-section delivery, birthweight was 1790 gm (10th centile for 33-week gestation), and a diagnosis of IUGR was made. On day 4, a cardiac anomaly was diagnosed that consisted of a secundum atrial septal defect with left to right shunt, a small perimembranous ventricular septal defect with minor left-right shunt and a dilated right ventricle. Pulmonary hypertension was also present. The authors remarked that, including their case, half the reported maternal UPD16 infants had documented structural anomalies (O'Riordan et al., 1996).

Other cases

Additional cases of maternal UPD16 have been subsequently reported. These include the following.

(i) A mosaic trisomy 16 case diagnosed by amniocentesis; following elective termination, this case demonstrated a trisomy 16 cell line in fetal skin (4%) and placental tissue (64%). Molecular studies on the disomic cell line indicated a maternal UPD16 with heterodisomy (Garber et al., 1994).

(ii) A growth-retarded infant with congenital heart disease. Nonmosaic trisomy 16 was detected at mid-trimester chorionic villus sampling, performed because biochemical screening indicated an increased Down's syndrome risk. Karyotype analysis of the placenta, after delivery, showed a 50% mosaic trisomy 16. The infant had an atrioventricular (A-V) canal defect, scoliosis, and several minor dysmorphic features (Whiteford et al., 1995).

(iii) Cytogenetic analysis of chorionic villus in this case obtained at 10 weeks' gestation for advanced maternal age showed trisomy 16. At 15 weeks, amniocentesis demonstrated low-level mosaicism 47,XY,+16[1]/46,XY[25]. Decreased fetal growth was noted in the last 2 months of pregnancy and the infant was small for gestational age at birth. There was maternal UPD16 (heterodisomy) in a peripheral blood sample from the child. Although short stature remained a concern at age 4 years, there were no major cognitive effects (Schneider et al., 1996) (see Figure 6, Chapter 5).

(iv) This case presented initially at prenatal diagnosis with a karyotype of 47, XX+16 on chorionic villus sampling at 11 weeks' gestation. Follow-up amniocentesis showed a normal female karyotype. At birth, the child was healthy, but had intrauterine growth retardation. She had unilateral talipes equinovarus and unilateral renal agenesis. Her growth had improved to within the normal range by age 3 years.

On examination, she had epicanthic folds, and a flat midface. A maternal UPD16 was demonstrated in the DNA of this proband (Woo et al., 1997).

(v) A 2-year-old, short, microcephalic, and developmentally retarded boy with multiple minor anomalies, hypospadias, and a dysplastic right kidney was described. Maternal age at delivery was 41 years. His karyotype showed two cell lines, one apparently normal, the other with a 1p+ chromosome. The segment attached to 1p was from chromosome 16, and molecular investigations disclosed maternal UPD16 (heterodisomy) except for the segment (16)(pter → p13.1), for which there was mosaicism between trisomy 16 and UPD16. This case was formed by a complicated mechanism. The authors concluded that the phenotype might be due to underlying factors such as UPD16 or undetected mosaicism in addition to the more obvious implications of the chromosome rearrangement itself (e.g., partial trisomy) (Schinzel et al., 1997).

(vi) A family was studied in which the mother had mosaicism for a balanced reciprocal translocation between chromosomes 10 and 16, which was associated with a break in chromosome 16 centromere alpha-satellite DNA 46,XX,t(10;16)(q11.2;q11.1) [29]/46,XX[25]. The same translocation was present in all cells in her son who was found prenatally to have trisomy 16 mosaicism 46,XY,t(10;16)(q11.2;q11.1)mat[22]/47,idem, + 16[4]. Trisomy 16 cells were subsequently determined to be confined to the placenta. DNA polymorphism analyses demonstrated maternal UPD16 in the diploid child. The child, at age 7 months, had minor facial peculiarities (Wang et al., 1998).

(vii) The physical and psychomotor development of three maternal UPD16 patients, aged 1.5, 3.2, and 4.5 years, was reported (Exeler et al., 1996). No catch-up growth in these cases with severe initial intrauterine retardation was observed, as postnatal length and weight acquisitions proceeded along the 3rd centile. One proband had congenital hypothyroidism, while a cardiac septal defect has been diagnosed in the two others. Psychomotor development was within normal limits in two and slightly retarded in one.

A study from the Netherlands attempted to estimate the frequency of maternal UPD16 in all cases (of a 4-year period) of mosaic or nonmosaic trisomy detected in chorionic villus in semidirect preparations and suspected to be confined to the placenta. Fluorescent *in situ* hybridization (FISH) on uncultured amniotic fluid cells to differentiate between generalized mosaicism and CPM was performed. Twenty-nine cases of CPM were found and the incidence of UPD was studied in 23 of these cases. Normal biparental chromosome contributions were found in 22 cases. In one case, there was a maternal UPD16 (heterodisomy) (Van Opstal et al., 1998). The authors concluded that maternal UPD16 appeared to be a rare phenomenon in the cases of CPM that they encountered in 3958 consecutively investigated CV samples.

A similar study was performed in 101 cases of confined placental mosaicism (CPM) involving autosomal trisomy. The origin of the trisomic cell line was determined in 54 cases (from 51 pregnancies), 47 of which were also analyzed for the presence of uniparental disomy (UPD) in the disomic cell line. An additional 47 cases were analyzed for parental origin in the disomic cell line only. Fetal maternal UPD was found in 17 of 94 informative CPM cases, involving UPD2 (1 case), UPD7 (1 case), UPD22 (2 cases), and UPD16 (13 cases) (Robinson et al., 1997). Concerning chromosome 16, there were also 15 cases with biparental inheritance in the fetus following the presumed trisomy 16 rescue. The placental trisomy was of meiotic origin in all cases associated with fetal UPD. Abnormal pregnancy outcome (usually IUGR) correlated with meiotic origin ($P = 0.0003$), the presence of fetal UPD ($P = 4 \times 10^{-7}$), and the level of trisomy in trophoblast ($P = 3 \times 10^{-7}$) but not with the level of trisomy in CVS or term chorion. The outcome of the CPM trisomy 16 cases was of interest: In 11 out of 13 cases with UPD16, there was IUGR, IUD, or another abnormality; in contrast, only 5 out of 13 cases with biparental chromosomes 16 were associated with these problems ($p = 0.02$). The authors concluded that imprinting may exist for chromosome 16 genes, but that the effect is limited to the placental tissues and in utero growth. In their view, "... it is also possible that the UPD16 cell line simply does not outcompete the trisomy cell line as well as a normal biparental disomic 16 cell line does ..." (Robinson et al., 1997).

Five cases with maternal UPD16 and CPM trisomy 16 were analyzed for common regions of isodisomy using polymorphic markers distributed along the length of chromosome 16. In each case, the aberration was consistent with a maternal meiosis 1 error. Complete isodisomy was not detected in any of the patients. Interestingly, the patient with the greater region of isodisomy was the most severely affected. The authors concluded that because there were no common regions of isodisomy in any of the patients, it is likely that imprinted genes, rather than recessive mutations, may play a role in the shared phenotypes (Abu-Amero et al., 1999).

This is certainly not an exhaustive list of all the maternal UPD16 cases reported, but an effort was made to include the majority of the cases including those that add substantially to our current understanding of the emerging clinical impression of this condition.

Considering these initial set of data, one has the impression that chromosome 16 contains maternally imprinted gene(s) that affect overall pre- and postnatal body size development as well as cell growth and differentiation of some tissues and organs.

Paternal UPD16

Case 1

There is a reported case of a 20-week-old fetus with hydrops fetalis and only Bart's hemoglobin (Ngo et al., 1993) that was homozygous for alpha-thalassemia-1. The father was heterozygous for alpha-thalassemia-1 and the mother homozygous normal. DNA analysis with a 5′HVR polymorphisms in the alpha-globin gene on chromosome 16 showed fetal inheritance of two paternal copies only and no contribution of maternal alleles.

Case 2

A 19-year-old presented for prenatal cytogenetic analysis at 21 weeks' gestation following the results of serum screening (AFP 3.7 MoM, hCG 8.0 MoM, uE3 0.9 MoM). The karyotype of cultured amniotic fluid cells revealed trisomy 16 mosaicism in one culture and 46,XX in 25 metaphases of a second culture. Repeated cytogenetic analysis at 23 weeks failed to detect trisomy 16 in 78 cells from three amniotic fluid cultures. Analysis of six polymorphic loci from chromosome 16 showed paternal UPD16 (isodisomy) in the DNA of amniotic fluid cells. The pregnancy was continued and at 35 weeks a female with a birthweight of 1790 gm was born. Cytogenetic and DNA analysis from the newborn's blood failed to detect mosaicism for trisomy 16 and confirmed the paternal UPD16 (Bartels et al., 1999). No abnormalities were detected in the newborn at the first year of age. The authors concluded that paternal UPD16 seems to be associated with normal development.

Case 3

A case of prenatal detection and follow-up of isodisomic paternal UPD16 in a child with normal development was reported (Kohlhase et al., 2000). The authors concluded that isodisomic pat UPD16 is associated with a normal outcome if no recessive mutation is reduced to homozygosity.

CHROMOSOME 17 UPD

Maternal UPD17

Case 1

Mosaicism for trisomy 17 in amniocyte cultures is a rare finding, and postnatal cases are also very rare. Three newly detected prenatal cases of trisomy 17 mosaicism identified in cultured amniotic fluid were studied. The second of these cases was associated with a normal pregnancy outcome and postnatal development, and only euploid cells were found in peripheral blood after birth. The mosaic trisomy 17 was found in two different cultures of amniotic fluid cells (6 of 21 clones). Umbilical cord blood and lymphocytes from the newborn showed only normal karyotype. Maternal UPD17 (isodisomy) consequent to a Me2 error and loss of a chromosome 17 homologue was detected in blood postnatally by using six polymorphic DNA markers. The baby boy was born after 38 weeks of gestation with a birthweight of 3030 gm. His growth and psychomotor development were normal at the age of 2 years. The authors concluded that the absence of phenotypic anomalies in the child with maternal UPD17 suggests that chromosome 17 is not likely to be subject to imprinting in maternal gametes (Genuardi et al., 1999).

Paternal UPD17

Case 1

One case of limited somatically acquired paternal interstitial UPD17q has been described, as a mosaic condition, in a 10-month-old male with *de novo* neurofibromatosis 1 (NF1), complicated by myeloid leukemia (Stephens et al., 1996). In contrast to other tissues, polymorphic markers in blood and marrow cells documented the absence of maternal alleles on a 17q11.2 segment encompassing loci UT172, NF1, DS806, DS787, and GH. Loci flanking this segment were heterozygous and biparental. Demonstration that segmental paternal UPD17 disomy, rather than maternal interstitial deletion, was causing such a loss of heterozygosity (LOH) was derived from FISH analysis and PCR gene dosage assays on blood, bone marrow, and lymphoblastoid cell lines from the patient and blood from both parents. These data showed that the NF1 *de novo* mutation had occurred in the paternal allele that had become homozygous in the malignant marrow clone, supporting evidence that NF1 is likely to be a tumor suppressor in myeloid cells (Stephens et al., 1996).

CHROMOSOME 18 UPD

Maternal UPD18

To our knowledge, no maternal UPD18 has been described yet (data up to December 2000).

Paternal UPD18

To our knowledge, no paternal UPD18 has been described yet (data up to December 2000).

CHROMOSOME 19 UPD

Maternal UPD19

To our knowledge, no maternal UPD19 has been described yet (data up to December 2000).

Paternal UPD19

To our knowledge, no paternal UPD19 has been described yet (data up to December 2000).

CHROMOSOME 20 UPD

Maternal UPD20

Case 1

Maternal UPD20 was observed in a 4-year-old boy with severe pre- and postnatal growth retardation [body height: 85 cm (< 3rd centile), head circumference: 48 cm (< 3rd centile)], a few minor facial findings, and with apparent hyperactivity. His intelligence was within the normal range for his age. Karyotype analysis revealed two cell lines, one apparently normal with 46,XY, the other with a marker (47,XY,+mar). Microdissection and reverse chromosome painting, as well as PCR analysis, revealed that the marker was from chromosome 20 and contained only the centromere and pericentromeric segments. Microsatellite analysis of 25 chromosome 20 loci disclosed maternal UPD20 for all 16 informative markers. Heterodisomy was evident for seven loci of the short-arm segment 20p11.2-pter. Isodisomy was found at five loci; three of them mapping to the proximal 20p11.2 segment and two to 20q (Chudoba et al., 1999).

Paternal UPD20

Case 1

The patient had multiple congenital abnormalities, including absent left ear, a small right ear remnant, microcephaly, congenital heart malformation, and Hirschsprung disease. The karyotype, in blood and marrow, revealed a 45,XY,−20,−20,+ter rea(20;20)(p13;p13). In skin fibroblasts, 8 of 100 cells showed trisomy 20 (46, XY,−20,+ter rea(20;20)(p13;p13). Analysis of informative DNA polymorphisms on blood cells revealed paternal UPD20 (isodisomy) (Spinner et al., 1994). No conclusion was drawn as to the possible correlation of developmental defects with aneuploidy, hypothetical paternally imprinted genes or putative reduction of alleles to homozygosity (Spinner et al., 1994).

CHROMOSOME 21 UPD

Maternal UPD21

Case 1

An early such case was reported in 1987 (Creau-Goldberg et al., 1987). Trisomy 21 in a newborn with a t(21q21q) was traced to the same but balanced translocation in the mother. She, in turn, had parents and numerous siblings with normal karyotypes. Analysis of several RFLPs for chromosome 21 showed that the woman's two copies of chromosome 21 in the t(21q21q) were derived solely from her mother (maternal UPD21). The authors (Creau-Goldberg et al., 1987) as an explanatory mechanism favored fertilization of a disomic 21 gamete by a nullisomy 21 spermatocyte, an event considered far from improbable, given the documentation of such a sperm type in serial studies of gametes. The formation of the t(21q21q) was considered a possible result of a misdivision. This case suggested that there are no maternally imprinted genes on chromosome 21.

Case 2

One case of growth failure, microcephaly, facial dysmorphism, muscular hypertonia, and severe psychomotor retardation was described with cytogenetic mosaicism in lymphocytes and skin fibroblasts of deletion of chromosome 21 and monosomy 21. At a later age, the lymphocyte karyotype changed almost completely to 46,XX, but the mosaicism persisted in the fibroblasts. DNA polymorphism analysis indicated that the 46,XX lymphocytes contained two identical chromosomes 21 from the mother (maternal UPD21, isodisomy). Nonpaternity was excluded. The isodisomy was probably the result of duplication of a chromosome in mitosis after the loss of the homologous abnormal chromosome ("compensatory isodisomy"). The phenotypes were attributed to the cells with monosomy 21 (Bartsch et al., 1994; Petersen et al., 1992).

Case 3

A phenotypically normal child with maternal UPD21 due to a *de novo* der(21;21)(q10;q10) was described (Rogan et al., 1999). The karyotype was done after amniocentesis in a 38-year-old woman referred for prenatal testing. The parental cytogenetic analysis did not detect the der(21;21). This karyotype was also confirmed in the newborn. At the age of 1 year, the child was clinically and developmentally normal. This suggested that chromosome 21 is not imprinted in the maternal germ line.

Other cases

A systematic search for uniparental disomy in tissues from 23 cases of early embryonic failure was carried out, using variable number tandem repeat (VNTR) analysis and PCR amplification of polymorphic short-sequence repeats. Two cases of maternal UPD21 (heterodisomy) were identified. One case occurred in conjunction with trisomy for chromosomes 7 and 9, but in the other case maternal UPD21 heterodisomy was the only chromosomal abnormality found (Henderson et al., 1994). The authors therefore postulated that there may be developmentally important genes on human chromosome 21 which are imprinted such that both parental copies are essential for normal embryogenesis.

Acute leukemia in Down's syndrome (DS) is often associated with additional changes in the number or structure of chromosome 21. One patient was described that developed acute lymphocytic leukemia (type L1); disomy for chromosome 21 was evident in all blast cells examined. Loss of the paternal chromosome in the leukemic clone produced maternal UPD21 with isodisomy over a 25-cM interval (Rogan et al., 1995).

Paternal UPD21

Case 1

The 40-year-old healthy male proband was ascertained following the birth of an offspring affected with Down's syndrome (Blouin et al., 1993). He had been the 3400-gm healthy product of a normal pregnancy, mother and father being 29 and 35 at the time of his birth. Puberty occurred at 14 years; at 40, his height was 172 cm, weight 60 kg, and head circumference 56.5 cm. He was found to have a t(21q21q), which proved to be a dup(21q) that led to the trisomy 21 offspring.

Chromosomes 21, in this man and his parents, were studied from leukocyte DNA for 17 highly informative short-sequence repeat polymorphisms, mapping from the pericentromeric area to terminal 21q. For the nine informative probes, there was no maternal contribution to the proband's genotype, which displayed homozygosity for paternal alleles.

It was concluded that this man's chromosome 21 pair was made up of an i(21q) and that the paternal UPD21 (isodisomy), associated with a normal phenotype, suggests a lack of imprinted genes on paternal chromosome 21 (Blouin et al., 1993).

Case 2

Another case of growth failure, microcephaly, facial dysmorphism, muscular hypertonia, and severe psychomotor retardation was described with mosaicism in lymphocytes and skin fibroblasts for a ring chromosome 21 and partial monosomy 21. At a later age, the lymphocyte karyotype changed almost completely to 46,XX, but the r(21) remained in fibroblasts. DNA polymorphism analysis indicated that the 46,XX lymphocytes contained two identical chromosomes 21 inherited from the father (paternal UPD21, isodisomy) (Bartsch et al., 1994; Petersen et al., 1992). The isodisomy was the result of duplication of a chromosome in mitosis after the loss of the r(21) ("compensatory isodisomy"). The phenotypes were attributed to the cells with partial monosomy 21.

Case 3

This case is described as case 5 in (Robinson et al., 1994). She was a phenotypically normal female with a *de novo* t(21q21q)pat, born to a 31-year-old mother and 35-year-old father, the last of eight healthy children. Paternal UPD21 was established after analysis of DNA polymorphic markers. This case also supports the hypothesis that there are no paternally imprinted genes on chromosome 21.

Other cases

Already in 1984, Niikawa and Kajii had made two important observations, based on the study of fluorescent chromosome 21 heteromorphisms in mosaic Down's syndrome: (1) that UPD occurred in the disomic line of these euploid/trisomy 21 mosaics and, consequently, (2) that the diploid line is derived from the putative loss of the singly inherited parental chromosome 21. Such cell lines, therefore, were an early illustration of what was later called "trisomy rescue" (Niikawa and Kajii, 1984). This was also one of the initial accounts giving practical consideration to the then speculative concept of UPD (Engel, 1980).

CHROMOSOME 22 UPD

Maternal UPD22

Case 1

The proband and his wife came to medical attention because of five miscarriages occurring at 6 and 14 weeks of pregnancy (Schinzel et al., 1994). He was a 25-year-old male, born to 17-year-old parents from Eastern Anatolia. Athough no birth measurements were available, his height of 177 cm was well within the range of that of his father and three brothers. Intelligence was reported to be normal, as was the rest of the phenotype. He was found to be the *de novo* balanced carrier of a t(22q22q) karyotype and his wife had a normal karyotype. Analysis of polymorphic microsatellite markers for chromosome 22 in blood DNA of both parents and the proband showed only the presence of maternal alleles in homozygosity (maternal UPD22 with isodisomy) and the absence of paternally derived alleles. Paternity was confirmed for alleles at loci on chromosomes other than 22. The authors concluded that this complete isodisomy for a maternal chromosome 22, as a result of isochromosome formation, is associated with normal phenotype and that there are therefore no maternally imprinted loci on chromosome 22 (Schinzel et al., 1994).

Cases 2–3

Two unrelated women, one with five miscarriages, the other with 10, each gave birth to a normal daughter, who, in turn, had six and seven spontaneous abortions respectively and no liveborn children. One, 27 years old, measured 155 cm and was of normal phenotype and intelligence (Palmer et al., 1980). The other, a 26-year-old, was also normal and healthy (Kirkels et al., 1980). Both had homologous translocations of chromosome 22, the former 45,XX,t(22qter->cen->22qter), the latter a 45,XX,t(22)(p13;q11). In both cases, the mothers showed the same karyotype, proving transmission of the homologous 22 translocation from balanced carriers to phenotypically normal daughters. The authors (Palmer et al., 1980; Kirkels et al., 1980) of each study spelled out the more plausible explanation of such an observation: namely, postfertilization rescue of trisomy 22 conceptus by loss of the paternal chromosome 22 or the fertilization of a translocation disomy 22 ovum by a nullisomy 22 sperm. Molecular studies were unavailable at the time, excluding further genomic investigation of these "familial" homologous inter-22 centric fusions.

Case 4

A case of generalized mosaicism for trisomy 22 was reported. At chorionic villus sampling (CVS) in the 37th week of pregnancy, a 47,XX,+22 karyotype was detected in all cells. The indication for CVS was intrauterine growth retardation (IUGR) and a ventricular septal defect (VSD) in ultrasound examination. In cultured cells from amniotic fluid taken simultaneously, only two out of 10 clones were trisomic. At term, a growth-retarded girl with mild dysmorphic features was born. Lymphocytes showed a normal 46,XX[50] karyotype; analysis of DNA polymorphisms revealed that both chromosomes 22 were maternal in origin (maternal UPD22). Investigation of the placenta postdelivery using fluorescence *in situ* hybridization showed a low presence of trisomy 22 cells in only one out of 14 biopsies. In cultured fibroblasts of skin tissue, a mosaic 47,XX,+22[7]/46,XX[25] was observed (de Pater et al., 1997).

Case 5

A child with developmental delay and multiple anomalies was identified that on cytogenetic analysis showed an abnormal chromosome complement of 47,XX,+der(22)t(11;22)(q23; q11) in all 50 cells analyzed. FISH analysis showed a pattern consistent with a reciprocal translocation of 11q23 and 22q11, respectively. Parental karyotypes were normal. RFLP analysis of locus D22S43, which maps above the t(11;22) breakpoint, showed that the der(22) was paternal in origin and indicated that the normal chromosomes 22 were the probable result of maternal UPD22 (heterodisomy). RFLP analysis of locus D22S94, which maps below the t(11;22) breakpoint, also suggested that both normal chromosomes 22 of the child represented the two maternal homologues. Nonpaternity was excluded through the analysis of many other nonchromosome 22 polymorphisms. In this case, the abnormal karyotype probably resulted from a *de novo* translocation in the paternal germline after unbalanced adjacent 1 segregation and maternal nondisjunction of chromosome 22 in meiosis I (Dawson et al., 1996).

In the study mentioned above (see maternal UPD16) of 101 cases of confined placental mosaicism (CPM) involving autosomal trisomy, the origin of the trisomic cell line was determined in 54 cases (from 51 pregnancies). Forty-seven of these cases were also analyzed for the presence of uniparental disomy (UPD) in the disomic cell line. Fetal maternal UPD22 was found in 2 of 94 informative CPM cases (Robinson et al., 1997). In one case, the meiotic error was assigned to Me1 and the other to Me2. In both cases, the newborns have intrauterine growth retardation assigned to the mosaic trisomy 22.

Paternal UPD22

Case 1

A t(22q22q) was identified in a normal male as a result of an evaluation of a couple with three abortions (Miny et al., 1995). Their 10-year-old phenotypically normal child also carried the paternal translocation. Investigation of a microsatellite locus showed an exclusively paternal contribution and no maternal alleles in the child's DNA, an observation providing evidence that paternal UPD22 is compatible with a normal phenotype.

Trisomy 22/euploid mosaicism has occasionally been detected in fibroblasts, but not in blood cultures. In females, the phenotype often resembles that of Turner syndrome. In an early case (Wertelecki et al., 1986), a potential paternal UPD22 in the disomic line of such a mosaic individual was observed using fluorescent heteromorphisms. This isodisomy was hypothesized to be a potential factor in the variable expression of the clinical phenotypes.

X CHROMOSOME UPD

Maternal UPDX

Case 1

The proband, a female, was affected with Duchene muscular dystrophy (DMD) as a result of homozygous deletion of exon 50 of the dystrophin gene (Quan et al., 1997). Her karyotype appeared normal and a UPDX (isodisomy) was observed that explained the phenotype. She was 6 years old, had unilateral calf pain after increased activity, and a tendency to toe-walking. Of note were mild hypotonia, and weakness of neck flexion, the proximal musculature of all limbs, and foot dorsiflexion. Elevated serum CK levels, muscle biopsy, and immunohistochemistry confirmed the diagnosis.

PCR amplified X-linked microsatellites within the dystrophin gene, as well as others spanning the whole length of the X chromosome were analyzed. Of 12 markers outside the dystrophin gene, 11 were informative for parental origin: In no instance was a paternal allele inherited by the proband, who was homozygous for a maternal allele at all tested loci, including 10 for which the mother was heterozygous. These results indicated that the proband's X chromosomes were both identical and maternally derived, confirming also that homozygosity for the mutant dystrophin allele was the consequence of an UPDX with isodisomy over the whole length of the X chromosome.

The authors (Quan et al., 1997) proposed that the most likely explanation for the aberrant transmission of two identical copies of the maternal X was its mitotic duplication to make up for paternal X or Y loss from the fertilizing sperm or early conception. This case was the first demonstration that UPD is an additional mechanism (aside from a skewed pattern of X inactivation or plain X monosomy) for the manifestation of X-linked recessive disorders in females. It is of interest that except for a severe form of DMD, the proband did not display any associated phenotypic abnormalities suggesting that there were no maternally imprinted genes on the X chromosome.

Case 2

In this interesting case, the severe mental retardation and multiple congenital abnormalities in a female with a mosaic 45,X/46,X,del(X)(q21.3-qter)/ 46X,r(X) karyotype are associated with overexpression of the genes within Xpter to Xq21.31 in many of her cells. Her normal X, ring X, and deleted linear X chromosomes originated from the same maternal X chromosome, and all were transcriptionally active. None expressed X inactive specific transcript (XIST), although the locus and region of the putative X inactivation center (XIC) were present on both normal and linear deleted X chromosomes. This was the first report of a functional maternal X isodisomy with the normal X and its deleted derivatives escaping inactivation. In addition, these results showed the authors that: (1) *cis* inactivation does not invariably occur in human females with two X chromosomes, even when the XIC region is present on both of them; (2) provided evidence for a critical time prior to the visible onset of X inactivation in the embryo when decisions about X inactivation are made; and (3) support the hypothesis that the X chromosome counting mechanism involves chromosomal imprinting, occurs prior to the onset of random inactivation, and is required for subsequent inactivation of the chromosome.

The proband was much more severely affected than a typical Turner syndrome patient. At 11 years of age, she was a nonverbal, nonambulatory person with severe mental retardation and microcephaly, plus malformations such as partial corpus callosum agenesis, seizures, cranial and facial asymmetry, scoliosis, camptodactyly, joint contractures (elbows, knees, fingers), rocker-bottom feet, and linear pigmentary dysplasia (Baschko lines). Of Turner syndrome characteristics, she had the short stature (less than 2nd centile), acral neonatal lymphedema, short neck, mild webbing, low hairline, shieldlike chest, inverted nipples, hypoplastic labia, and hypogenitalism. She died around 11 years of age of cardiorespiratory and liver failure, having, for years, suffered from severe pulmonary problems (Migeon et al., 1996).

Case 3

The parental origin and X inactivation status of X-derived marker [mar(X)] or ring X [r(X)] chromosomes in six Turner syndrome patients were analyzed (Yorifuji et al., 1998). Two of these patients had mental retardation of unknown cause in addition to the usual Turner syndrome phenotype. By FISH analysis, the mar(X)/r(X) chromosomes of all patients retained the X centromere and the XIST locus at Xq13.2. By polymorphic marker analysis, both patients with mental retardation were shown to have UPDX; one had maternal UPDX while the other had paternal UPDX. The other four patients had both a maternal and paternal contribution of X chromosomes. By RT-PCR analysis and the androgen receptor assay, it was shown that in one of these mentally retarded patients, the XIST on the mar(X) was not transcribed and consequently the mar(X) was not inactivated, leading to functional disomy X. In the other patient, the XIST was transcribed but the r(X) appeared to be active by the androgen receptor assay.

These results suggest that UPDX may not be uncommon in mentally retarded patients with Turner syndrome. Functional disomy X seems to be the cause of mental retardation in these patients, although the underlying molecular basis could be diverse. In addition, even without unusual dysmorphic features, Turner syndrome patients with unexplained mental retardation need to be investigated for possible mosaicism including such mar(X)/r(X) chromosomes (Yorifuji et al., 1998).

Three normal females were reported to have maternal UPDX (Avivi et al., 1992). They were found among the children (84 males, 33 females) of 36 unrelated couples. A full account of this study has not yet been published.

Paternal UPDX

Case 1

The 13 $\frac{1}{2}$-year-old girl with growth retardation, mild gonadal dysfunction, and a low frequency of lymphocytes with 45,X had, from polymorphic marker studies, homozygosity for paternal X alleles in diploid cells and an absence of X-linked demonstrable maternal alleles in all cells (Schinzel et al., 1993). She was born at 40 weeks' gestation to a mother and father (24 and 33 years of age, respectively) who were themselves of normal height. Thelarche and pubarche occurred during the course of the 11th year. At $13\frac{1}{2}$, she measured 135.5 cm (less than the 3rd centile) and weight was 43 kg (25–50th centile), while head circumference was in the 75–90th centile. Bone age was chronologically normal and predicted final height was 140 cm. The only other Turner stigmata were short neck, broad chest, and cubitus valgus. Ovaries and uterus were of normal size on ultrasound examination. Mental development and school achievement were normal and menarche occurred toward the end of the 13th year. Genetic studies showed that from roughly $11\frac{1}{2}$ to $13\frac{1}{2}$ years, the proportion of 45,X in blood cultures declined from 13–8% to none (in 50 cells). Various hormonal studies showed normal results except for the GnRH stimulation test, suggesting mild ovarian insufficiency. The clinical picture, basically one of growth retardation, remains etiologically unsolved, since no other tissues could be studied for 45,X cell line mosaicism, so that the role of paternal disomy remains uncertain in this case.

Case 2

One additional case of paternal UPDX was described by (Yorifuji et al., 1998). The authors analyzed the parental origin and X inactivation status of X-derived marker [mar(X)) or ring X (r(X)] chromosomes in six Turner syndrome patients. Two of these patients had mental retardation of unknown cause in addition to the usual Turner syndrome phenotype. By polymorphic marker analysis, both patients with mental retardation were shown to have UPDX (one paternal and one maternal), whereas the four others had both a maternal and paternal contribution of X chromosomes. These results suggest that uniparental disomy X may not be uncommon in mentally retarded patients with Turner syndrome.

UPD(XY)

This is the 47th possible uniparental pair, which can only be heterodisomic and of paternal origin.

A boy with hemophilia A was investigated using molecular and cytogenetic techniques because his father and his uncle also suffered from hemophilia A. In addition, his mother, who had no family history of hemophilia A, had normal coagulation assays. The cytogenetic analysis was normal. More than 15 X-linked DNA markers were used to determine whether the disease was due to a *de novo* mutation or transmitted by the father (Vidaud et al., 1989). The data indicated that the propositus inherited the X chromosome from his father. The analysis of DNA markers from chromosomes 2, 7, 11, 16, 17, and 18 showed that the parental UPDXY was restricted to the sex chromosomes. The authors favored the occurrence of gamete complementation, an XY sperm having fertilized a presumably X-nullisomy ovum (Vidaud et al., 1989)

REFERENCES

Abramowicz, M. J., Andrien, M., Dupont, E., et al. Isodisomy of chromosome 6 in a newborn with methylmalonic acidemia and agenesis of pancreatic beta cells causing diabetes mellitus. *J Clin Invest* 94:418–421, 1994.

Abu-Amero, S. N., Ali, Z., Abu-Amero, K. K., Stanier, P. and Moore, G. E. An analysis of common isodisomic regions in five mUPD 16 probands. *J Med Genet* 36:204–207, 1999.

Anderlid, B. M., Bjorck, E. J. and Blennow, E. Maternal isodisomy of chromosome 9 with no impact on the phenotype in a woman with isochromosomes i(9p) and i(9q). *Cytogenetics and Cell Genetict* 85:137, 1999. (Abstract)

Antonarakis, S. E., Avramopoulos, D., Blouin, J. L., Talbot, C. C. J. and Schinzel, A. A. Mitotic errors in somatic cells cause trisomy 21 in about 4. 5% of cases and are not associated with advanced maternal age [see comments]. *Nat Genet* 3:146–150, 1993.

Antonarakis, S. E., Blouin, J. L., Maher, J., Avramopoulos, D., Thomas, G. and Talbot, C. C. J. Maternal uniparental disomy for human chromosome 14, due to loss of a chromosome 14 from somatic cells with t(13;14) trisomy 14. *Am J Hum Genet* 52:1145–1152, 1993.

Avivi, L., Korenstein, A., Braier-Goldstein, O., Goldman, B. and Ravia, Y. Uniparental disomy of sex chromosomes in man. *Am J Hum Genet* 51:A33,1992. (Abstract)

Bartels, I., Janssen, B., Waldenauer, K. and Kohlhase, J. First case with confirmed paternal uniparental disomy 16 (UPD16). *Med Gen* 11:179, 1999. (Abstract)

Barton, D. E., McQuaid, S., Stallings, R., Griffin, E. and Geraghty, M. Further evidence for an emerging maternal uniparental disomy chromosome 14 syndrome: Analysis of a phenotypically abnormal *de novo* Robertsonian translocation t(13;14) carrier. *Am J Hum Genet* 59:A125, 1996. (Abstract)

Bartsch, O., Petersen, M. B., Stuhlmann, I., et al. "Compensatory" uniparental disomy of chromosome 21 in two cases. *J Med Genet* 31:534–540, 1994.

Beaudet, A. L., Perciaccante, R. G. and Cutting, G. R. Homozygous nonsense mutation causing cystic fibrosis with uniparental disomy [letter]. *Am J Hum Genet* 48:1213, 1991.

Beldjord, C., Henry, I., Bennani, C., Vanhaeke, D. and Labie, D. Uniparental disomy: a novel mechanism for thalassemia major [letter]. *Blood* 80:287–289, 1992.

Benlian, P., Foubert, L., Gagne, et al. Complete paternal isodisomy for chromosome 8 unmasked by lipoprotein lipase deficiency. *Am J Hum Genet* 59:431–436, 1996.

Bennett, P., Vaughan, J., Henderson, D., Loughna, S. and Moore, G. Association between confined placental trisomy, fetal uniparental disomy, and early intrauterine growth retardation [letter]. *Lancet* 340:1284–1285, 1992.

Berend, S. A., Feldman, G. L., McCaskill, C., Czarnecki, P., Van Dyke, D. L. and Shaffer, L. G. Investigation of two cases of paternal disomy 13 suggests timing of isochromosome formation and mechanisms leading to uniparental disomy. *Am J Med Genet* 82:275–281, 1999.

Bernasconi, F., Karaguzel, A., Celep, F., et al. Normal phenotype with maternal isodisomy in a female with two isochromosomes: i(2p) and i(2q) [see comments]. *Am J Hum Genet* 59:1114–1118, 1996.

Betz, A., Turleau, C. and de Grouchy, J. [Heterozygosity and homozygosity for a pericentric inversion of human chromosome 3]. *Ann Genet* 17:79–80, 1974.

Bischoff, F. Z., Zenger-Hain, J., Moses, D., Van Dyke, D. L. and Shaffer, L. G. Mosaicism for trisomy 12: four cases with varying outcomes. *Prenat Diagn* 15:1017–1026, 1995.

Bittencourt, M. C., Morris, M. A., Chabod, J., et al. Fortuitous detection of uniparental isodisomy of chromosome 6. *J Med Genet* 34:77–78, 1997.

Blouin, J. L., Avramopoulos, D., Pangalos, C. and Antonarakis, S. E. Normal phenotype with paternal uniparental isodisomy for chromosome 21. *Am J Hum Genet* 53:1074–1078, 1993.

Bottani, A., Robinson, W. P., DeLozier-Blanchet, C. D., et al. Angelman syndrome due to paternal uniparental disomy of chromosome 15: a milder phenotype? [see comments]. *Am J Med Genet* 51:35–40, 1994.

Brambati, B., Oldrini, A., Ferrazzi, E. and Lanzani, A. Chorionic villus sampling: an analysis of the obstetric experience of 1,000 cases. *Prenat Diagn* 7:157–169, 1987.

Brzustowicz, L. M., Allitto, B. A., Matseoane, D., et al. Paternal isodisomy for chromosome 5 in a child with spinal muscular atrophy. *Am J Hum Genet* 54:482–488, 1994.

Carpenter, N. J., Say, B. and Barber, N. D. A homozygote for pericentric inversion of chromosome 4. *J Med Genet* 19:469–471, 1982.

Cassidy, S. B., Lai, L. W., Erickson, R. P., et al. Trisomy 15 with loss of the paternal 15 as a cause of Prader-Willi syndrome due to maternal disomy. *Am J Hum Genet* 51:701–708, 1992.

Chavez, B., Valdez, E. and Vilchis, F. Uniparental disomy in steroid 5alpha-reductase 2 deficiency. *J Clin Endocrinol Metab* 85:3147–3150, 2000.

Chen, H., Young, R., Mu, X., et al. Uniparental isodisomy resulting from 46,XX,i(1p),i(1q) in a woman with short stature, ptosis, micro/retrognathia, myopathy, deafness, and sterility. *Am J Med Genet* 82:215–218, 1999.

Christian, S. L., Rich, B. H., Loebl, C., et al. Significance of genetic testing for paternal uniparental disomy of chromosome 6 in neonatal diabetes mellitus [see comments]. *J Pediatr* 134:42–46, 1999.

Chudoba, I., Franke, Y., Senger, G., et al. Maternal UPD 20 in a hyperactive child with severe growth retardation [In Process Citation]. *Eur J Hum Genet* 7:533–540, 1999.

Cogliati, F., Russo, S., Macchi, M., et al. Uniparental disomy and PEG1/MEST gene in Silver-Russell syndrome. *Eur J Hum Genet* 6:156, 1998. (Abstract)

Cotter, P. D., Kaffe, S., McCurdy, L. D., Jhaveri, M., Willner, J. P. and Hirschhorn, K. Paternal uniparental disomy for chromosome 14: a case report and review. *Am J Med Genet* 70:74–79, 1997.

Coviello, D. A., Panucci, E., Mantero, M. M., et al. Maternal uniparental disomy for chromosome 14. *Acta Genet Med Gemellol (Roma)* 45:169–172, 1996.

Creau-Goldberg, N., Gegonne, A., Delabar, J., et al. Maternal origin of a *de novo* balanced t(21q21q) identified by ets-2 polymorphism. *Hum Genet* 76:396–398, 1987.

Das, S., Lese, C. M., Song, M., et al. Partial paternal uniparental disomy of chromosome 6 in an infant with neonatal diabetes, macroglossia, and craniofacial abnormalities. *Am J Hum Genet* 67:2000.

Dawson, A. J., Mears, A. J., Chudley, A. E., Bech-Hansen, T. and McDermid, H. Der(22)t(11;22) resulting from a paternal *de novo* translocation, adjacent 1 segregation, and maternal heterodisomy of chromosome 22. *J Med Genet* 33:952–956, 1996.

De Marchi, N., Antonarakis, S. E. and Jackson, L. No uniparental disomy for chromosome 3 in Brachmann-De Lange syndrome [letter]. *Am J Med Genet* 49:133–135, 1994.

de Pater, J. M., Schuring-Blom, G. H., van den Bogaard, R., et al. Maternal uniparental disomy for chromosome 22 in a child with generalized mosaicism for trisomy 22. *Prenat Diagn* 17:81–86, 1997.

Desilets, V. A., Yong, S. L., Kalousek, D. K., Pantzar, T. J., Kwong, L. C. and Langlois S. Maternal uniparental disomy for chromosome 14. *Am J Hum Genet* 61:A122, 1997. (Abstract)

Devi, A. S., Velinov, M., Kamath, M. V., et al. Variable clinical expression of mosaic trisomy 16 in the newborn infant [see comments]. *Am J Med Genet* 47:294–298, 1993.

Dufourcq-Lagelouse, R., Lambert, N., Duval, M., et al. Chediak-Higashi syndrome associated with maternal uniparental isodisomy of. *Eur J Hum Genet* 7:633–637, 1999.

Dupond JM, Cuisset L, LeTessier D, et al. Cytogenetic and molecular studies in 32 Silver-Russel syndrome patients. *Eur J Hum Genet* 6:97, 1998. (Abstract)

Dutly, F., Baumer, A., Kayserili, H., et al. Seven cases of Wiedmann-Beckwith syndrome, including the first reported case of mosaic paternal isodisomy along the whole chromosome 11. *Am J Med Genet* 79:347–353, 1998.

Dworniczak, B., Koppers, B., Kurlemann, G., Holzgreve, W., Horst, J. and Miny, P. Uniparental disomy with normal phenotype [letter]. *Lancet* 340:1285 1992.

Eggerding, F. A., Schonberg, S. A., Chehab, F. F., Norton, M. E., Cox, V. A. and Epstein, C. J. Uniparental isodisomy for paternal 7p and maternal 7q in a child with growth retardation. *Am J Hum Genet* 55:253–265, 1994.

Eggermann, T., Wollmann, H. A., Kuner, R., et al. Molecular studies in 37 Silver-Russell syndrome patients: frequency and etiology of uniparental disomy. *Hum Genet* 100:415–419, 1997.

Ellis, N. A., Groden, J., Ye, T. Z., et al. The Bloom's syndrome gene product is homologous to RecQ helicases. *Cell* 83:655–666, 1995.

Engel, E. A new genetic concept: uniparental disomy and its potential effect, isodisomy. *Am J Med Genet* 6:137–143, 1980.

Engel, E. La choricentese; un diagnostic prenatal a la dizieme semaine par biopsie et analyse des villosites chorioniques. *Medicine et Hygiene* 1984.

Exeler, R., Dworniczak, B., Tercanli, S., et al. Physical and psychomotor development of three children with uniparental disomy 16. *Am J Hum Genet* 59:A92, 1996. (Abstract)

Field, L. L., Tobias, R., Robinson, W. P., Paisey, R. and Bain, S. Maternal uniparental disomy of chromosome 1 with no apparent phenotypic effects [letter] [see comments]. *Am J Hum Genet* 63:1216–1220, 1998.

Fokstuen, S., Ginsburg, C., Zachmann, M. and Schinzel, A. Maternal uniparental disomy 14 as a cause of intrauterine growth retardation and early onset of puberty. *J Pediatr* 134:689–695, 1999.

Freeman, S. B., May, K. M., Pettay, D., Fernhoff, P. M. and Hassold, T. J. Paternal uniparental disomy in a child with a balanced 15;15 translocation and Angelman syndrome. *Am J Med Genet* 45:625–630, 1993.

Fridman, C., Varela, M. C., Nicholls, R. D. and Koiffmann, C. P. Unusual clinical features in an Angelman syndrome patient with uniparental disomy due to a translocation 15q15q. *Clin Genet* 54:303–308, 1998.

Garber, A., Carlson, D., Schreck, R., et al. Prenatal diagnosis and dysmorphic findings in mosaic trisomy 16. *Prenat Diagn* 14:257–266, 1994.

Gardner, R. J., Mungall, A. J., Dunham, I., et al. Localisation of a gene for transient neonatal diabetes mellitus to an 18. 72 cR3000 (approximately 5. 4 Mb) interval on chromosome 6q. *J Med Genet* 36:192–196, 1999.

Gardner, R. J., Robinson, D. O., Lamont, L., Shield, J. P. and Temple, I. K. Paternal uniparental disomy of chromosome 6 and transient neonatal diabetes mellitus. *Clin Genet* 54:522–525, 1998.

Gelb, B. D., Willner, J. P., Dunn, T. M., et al. Paternal uniparental disomy for chromosome 1 revealed by molecular analysis of a patient with pycnodysostosis. *Am J Hum Genet* 62:848–854, 1998.

Genuardi, M., Tozzi, C., Pomponi, M. G., et al. Mosaic trisomy 17 in amniocytes: phenotypic outcome, tissue distribution, and uniparental disomy studies. *Eur J Hum Genet* 7:421–426, 1999.

Gillessen-Kaesbach, G., Albrecht, B., Passarge, E. and Horsthemke, B. Further patient with Angelman syndrome due to paternal disomy of chromosome 15 and a milder phenotype [letter; comment]. *Am J Med Genet* 56:328–329, 1995.

Gillessen-Kaesbach, G., Gross, S., Kaya-Westerloh, S., Passarge, E. and Horsthemke, B. DNA methylation based testing of 450 patients suspected of having Prader-Willi syndrome. *J Med Genet* 32:88–92, 1995.

Grundy, P., Telzerow, P., Paterson, M. C., et al. Chromosome 11 uniparental isodisomy predisposing to embryonal neoplasms [letter]. *Lancet* 338:1079–1080, 1991.

Hahnemann, J. M., Nir, M., Engel, U., Friberg, M., Bugge, M. and Brondum-Nielsen, K. Trisomy 10 mosaicicm and maternal uniparental disomy in a lifeborn infant with severe congenital malformations. *Eur J Hum Genet* 4:156, 1996. (Abstract)

Hansen, W. F., Bernard, L. E., Langlois, S., et al. Maternal uniparental disomy of chromosome 2 and confined placental mosaicism for trisomy 2 in a fetus with intrauterine growth restriction, hypospadias, and oligohydramnios. *Prenat Diagn* 17:443–450, 1997.

Harrison, K., Eisenger, K., Anyane-Yeboa, K. and Brown, S. Maternal uniparental disomy of chromosome 2 in a baby with trisomy 2 mosaicism in amniotic fluid culture. *Am J Med Genet* 58:147–151, 1995.

Harrison, K. J., Allingham-Hawkins, D. J., Hummel, J., et al. Risk of uniparental disomy in Robertsonian translocation carriers:identification of UPD14 in a small cohort. *Am J Hum Genet* 63:A11, 1998. (Abstract)

Healey, S., Powell, F., Battersby, M., Chenevix-Trench, G. and McGill, J. Distinct phenotype in maternal uniparental disomy of chromosome 14. *Am J Med Genet* 51:147–149, 1994.

Heide, E., Heide, K. G. and Rodewald, A. Maternal uniparental disomy (UPD) for chromosome 2 discovered by exclusion of paternity. *Am J Med Genet* 92:260–263, 2000.

Henderson, D. J., Sherman, L. S., Loughna, S. C., Bennett, P. R. and Moore, G. E. Early embryonic failure associated with uniparental disomy for human chromosome 21. *Hum Mol Genet* 3:1373–1376, 1994.

Henry, I., Bonaiti-Pellie, C., Chehensse, V., et al. Uniparental paternal disomy in a genetic cancer-predisposing syndrome [see comments]. *Nature* 351:665–667, 1991.

Hoglund, P., Holmberg, C., de la Chapelle, A. and Kere, J. Paternal isodisomy for chromosome 7 is compatible with normal growth and development in a patient with congenital chloride diarrhea. *Am J Hum Genet* 55:747–752, 1994.

Hubbard, V. S., Davis, P. B., di Sant'Agnese, P. A., Gorden, P. and Schwartz, R. H. Isolated growth hormone deficiency and cystic fibrosis: a report of two cases. *Am J Dis Child* 134:317–319, 1980.

James, R. S., Temple, I. K., Dennis, N. R. and Crolla, J. A. A search for uniparental disomy in carriers of supernumerary marker chromosomes. *Eur J Hum Genet* 3:21–26, 1995.

Jarvela, I., Savukoski, M., Ammala, P. and von Koskull, H. Prenatally detected paternal uniparental chromosome 13 isodisomy. *Prenat Diagn* 18:1169–1173, 1998.

Jones, C., Booth, C., Rita, D., et al. Identification of a case of maternal uniparental disomy of chromosome 10 associated with confined placental mosaicism. *Prenat Diagn* 15:843–848, 1995.

Kalousek, D. K. Variable clinical expression of mosaic trisomy 16 in the newborn infant [letter; comment]. *Am J Med Genet* 52:115–116, 1994.

Kalousek, D. K. and Dill, F. J. Chromosomal mosaicism confined to the placenta in human conceptions. *Science* 221:665–667, 1983.

Kalousek, D. K., Dill, F. J., Pantzar, T., McGillivray, B. C., Yong, S. L. and Wilson, R. D. Confined chorionic mosaicism in prenatal diagnosis. *Hum Genet* 77:163–167, 1987.

Kalousek, D. K., Langlois, S., Barrett, I., et al. Uniparental disomy for chromosome 16 in humans. *Am J Hum Genet* 52:8–16, 1993.

Kalousek, D. K., Langlois, S., Robinson, W. P., et al. Trisomy 7 CVS mosaicism: pregnancy outcome, placental and DNA analysis in 14 cases. *Am J Med Genet* 65:348–352, 1996.

Karanjawala, Z. E., Kaariainen, H., Ghosh, S., et al. Complete maternal isodisomy of chromosome 8 in an individual with an early-onset ileal carcinoid tumor [In Process Citation]. *Am J Med Genet* 93:207–210, 2000.

Kirkels, V. G., Hustinx, T. W. and Scheres, J. M. Habitual abortion and translocation (22q;22q): unexpected transmission from a mother to her phenotypically normal daughter. *Clin Genet* 18:456–461, 1980.

Kohlhase, J., Janssen, B., Weidenauer, K., Harms, K. and Bartels, I. First confirmed case with paternal uniparental disomy of chromosome 16. *Am J Med Genet* 91:190–191, 2000.

Kotzot, D., Schmitt, S., Bernasconi, F., et al. Uniparental disomy 7 in Silver-Russell syndrome and primordial growth retardation. *Hum Mol Genet* 4:583–587, 1995.

Kousseff, G., Gallardo, L. A. and Miller, O. T. Unusual clinical presentation associated with uniparental disomy of chromosome 10 in a child presymptomatic for multiple endocrine neoplasia type 2A. *Am J Hum Genet* 51:A219, 1992. (Abstract)

Langlois, S., Yong, S. L., Wilson, R. D., Kwong, L. C. and Kalousek, D. K. Prenatal and postnatal growth failure associated with maternal heterodisomy for chromosome 7. *J Med Genet* 32:871–875, 1995.

Ledbetter, D. H. and Engel, E. Uniparental disomy in humans: development of an imprinting map and its implications for prenatal diagnosis. *Hum Mol Genet 4* Spec No:1757–1764, 1995.

Linck, L., McMilin, K., Popovich, B. and Magenis, R. E. Maternal uniparental disomy for chromosome 14. *Am J Hum Genet* 59:A124, 1996. (Abstract)

Lindenbaum, R. H., Woods, C. G., Norbury, C. G., Povey, S. and Rysiecki, G. An individual with maternal disomy of chromoosme 4 and iso4p, iso4q. *Am J Hum Genet* 49:285–285, 1991. (Abstract)

Lindor, N. M., Jalal, S. M., Thibodeau, S. N., Bonde, D., Sauser, K. L. and Karnes, P. S. Mosaic trisomy 16 in a thriving infant: maternal heterodisomy for chromosome 16. *Clin Genet* 44:185–189, 1993.

Lopez-Gutierrez, A. U., Riba, L., Ordonez-Sanchez, M. L., Ramirez-Jimenez, S., Cerrillo-Hinojosa, M. and Tusie-Luna, M. T. Uniparental disomy for chromosome 6 results in steroid 21-hydroxylase deficiency: evidence of different genetic mechanisms involved in the production of the disease. *J Med Genet* 35:1014–1019, 1998.

Malcolm, S., Clayton-Smith, J., Nichols, M., et al. Uniparental paternal disomy in Angelman's syndrome. *Lancet* 337:694–697, 1991.

Martin, R. A., Sabol, D. W. and Rogan, P. K. Maternal uniparental disomy of chromosome 14 confined to an interstitial segment (14q23-14q24. 2) [In Process Citation]. *J Med Genet* 36:633–636, 1999.

Meck, J. M., Kozma, C., Tchabo, J. G., King, J. C., Lencki, S. and Pinckert, T. L. Prenatal diagnosis of trisomy 12 mosaicism: physical and developmental follow-up. *Prenat Diagn* 14:878–883, 1994.

Migeon, B. R., Jeppesen, P., Torchia, B. S., et al. Lack of X inactivation associated with maternal X isodisomy: evidence for a counting mechanism prior to X inactivation during human embryogenesis. *Am J Hum Genet* 58:161–170, 1996.

Miny, P., Koppers, B., Bogadanova, N., Schulte-Vallentin, M., Horst, J. and Dworniczak, B. Paternal uniparental disomy 22. *Med Gen* 7:216, 1995. (Abstract)

Miura, Y., Hiura, M., Torigoe, K., et al. Complete paternal uniparental isodisomy for chromosome 1 revealed by mutation analyses of the TRKA (NTRK1) gene encoding a receptor tyrosine kinase for nerve growth factor in a patient with congenital insensitivity to pain with anhidrosis [In Process Citation]. *Hum Genet* 107:205–209, 2000.

Miyoshi, O., Hayashi, S., Fujimoto, M., Tomita, H., Sohda, M. and Niikawa, N. Maternal uniparental disomy for chromosome 14 in a boy with intrauterine growth retardation. *J Hum Genet* 43:138–142, 1998.

Monk, D., Wakeling, E. L., Proud, V., et al. Duplication of 7p11. 2-p13, including GRB10, in Silver-Russell syndrome. *Am J Hum Genet* 66:36–46, 2000.

Moore, G. E., Ali, Z., Khan, R. U., Blunt, S., Bennett, P. R. and Vaughan, J. I. The incidence of uniparental disomy associated with intrauterine growth retardation in a cohort of thirty-five severely affected babies. *Am J Obstet Gynecol* 176:294–299, 1997.

Morichon-Delvallez, N., Segues, B. , Pinson, M. P., et al. Maternal uniparental disomy for chromosome 14 by secondary non disjunction of a initial trisomy. *Am J Hum Genet* 55:A379, 1994. (Abstract)

Ngo, K. Y., Lee, J., Dixon, B., Liu, D. and Jones, O. W. Paternal uniparental isodisomy in a hydrops fetalis alpha-thalassemia fetus. *Am J Hum Genet* 53:A1207, 1993. (Abstract)

Nicholls, R. D., Knoll, J. H., Butler, M. G., Karam, S. and Lalande, M. Genetic imprinting suggested by maternal heterodisomy in nondeletion Prader-Willi syndrome. *Nature* 342:281–285, 1989.

Nicholls, R. D., Pai, G. S., Gottlieb, W. and Cantu, E. S. Paternal uniparental disomy of chromosome 15 in a child with Angelman syndrome. *Ann Neurol* 32:512–518, 1992.

Niikawa, N. and Kajii, T. The origin of mosaic Down syndrome: four cases with chromosome markers. *Am J Hum Genet* 36:123–130, 1984.

O'Riordan, S., Greenough, A., Moore, G. E., Bennett, P. and Nicolaides, K. H. Case report: uniparental disomy 16 in association with congenital heart disease. *Prenat Diagn* 16:963–965, 1996.

Palmer, C. G., Schwartz, S. and Hodes, M. E. Transmission of a balanced homologous t(22q;22q) translocation from mother to normal daughter. *Clin Genet* 17:418–422, 1980.

Palmer, S. L., Christian, S. L., Danney, M. M., Odom, M. W. and Ledbetter, D. H. Neonatal diabetes mellitus due to uniparental disomy of chromosome 6. *Am J Hum Genet* 63:116, 1998. (Abstract)

Pan, Y., McCaskill, C. D., Thompson, K. H., et al. Paternal isodisomy of chromosome 7 associated with complete situs inversus and immotile cilia [letter]. *Am J Hum Genet* 62:1551–1555, 1998.

Papenhausen, P. R., Mueller, O. T., Johnson, V. P., Sutcliffe, M., Diamond, T. M. and Kousseff, B. G. Uniparental isodisomy of chromosome 14 in two cases: an abnormal child and a normal adult [see comments]. *Am J Med Genet* 59:271–275, 1995.

Penaherrera MS, Bernard L, Wang M, Van Allen MI, Langlois S and Robinson WP. UPD 7 in individuals with Silver-Russel like phenotype. *Am J Hum Genet* 59:A101, 1996. (Abstract)

Pentao, L., Lewis, R. A., Ledbetter, D. H., Patel, P. I. and Lupski, J. R. Maternal uniparental isodisomy of chromosome 14: association with autosomal recessive rod monochromacy. *Am J Hum Genet* 50:690–699, 1992.

Petersen, M. B., Bartsch, O., Adelsberger, P. A., Mikkelsen, M., Schwinger, E. and Antonarakis, S. E. Uniparental isodisomy due to duplication of chromosome 21 occurring in somatic cells monosomic for chromosome 21. *Genomics* 13:269–274, 1992.

Piantanida, M., Dellavecchia, C., Floridia, G., et al. Ataxic gait and mental retardation with absence of the paternal chromosome 8 and an idic(8)(p23. 3): imprinting effect or nullisomy for distal 8p genes? *Hum Genet* 99:766–771, 1997.

Preece, M. A., Price, S. M., Davies, V., et al. Maternal uniparental disomy 7 in Silver-Russell syndrome. *J Med Genet* 34:6–9, 1997.

Pulkkinen, L., Bullrich, F., Czarnecki, P., Weiss, L. and Uitto, J. Maternal uniparental disomy of chromosome 1 with reduction to homozygosity of the LAMB3 locus in a patient with Herlitz junctional epidermolysis bullosa. *Am J Hum Genet* 61:611–619, 1997.

Purvis-Smith, S. G., Saville, T., Manass, S., et al. Uniparental disomy 15 resulting from "correction" of an initial trisomy 15 [letter]. *Am J Hum Genet* 50:1348–1350, 1992.

Quan, F., Janas, J., Toth-Fejel, S., Johnson, D. B., Wolford, J. K. and Popovich, B. W. Uniparental disomy of the entire X chromosome in a female with Duchenne muscular dystrophy. *Am J Hum Genet* 60:160–165, 1997.

Ralph, A., Scott, F., Tiernan, C., et al. Maternal uniparental isodisomy for chromosome 14 detected prenatally. *Prenat Diagn* 19:681–684, 1999.

Ramsden, S., Gaunt, L., Seres-Santamaria, A. and Clayton-Smith, J. A case of Angelman syndrome arising as a result of a *de novo* Robertsonian translocation. *Acta Genet Med Gemellol (Roma)* 45:255–261, 1996.

Ridanpaa, M., van Eenennaam, H., Pelin, K., Chadwick, R., Johnson, C., Yuan, B., van Venrooij, W., Pruijn, G., Salmela, R., Rockas, S., Makitie, O., Kaitila, I., de la Chapelle, A. Mutations in the RNA component of RNase MRP cause a pleiotropic human disease, cartilage-hair hypoplasia. *Cell* 104:195–203, 2001.

Robinson, W. P., Barrett, I. J., Bernard, L., et al. Meiotic origin of trisomy in confined placental mosaicism is correlated with presence of fetal uniparental disomy, high levels of trisomy in trophoblast, and increased risk of fetal intrauterine growth restriction. *Am J Hum Genet* 60:917–927, 1997.

Robinson, W. P., Bernasconi, F., Basaran, S., et al. A somatic origin of homologous Robertsonian translocations and isochromosomes. *Am J Hum Genet* 54:290–302, 1994.

Robinson, W. P., Kuchinka, B. D., Bernasconi, F., et al. Maternal meiosis I non-disjunction of chromosome 15: dependence of the maternal age effect on level of recombination. *Hum Mol Genet* 7:1011–1019, 1998.

Robinson, W. P. and Langlois, S. Phenotype of maternal UPD(14) [letter; comment] [see comments]. *Am J Med Genet* 66:891996.

Robinson, W. P., Langlois, S., Schuffenhauer, S., et al. Cytogenetic and age-dependent risk factors associated with uniparental disomy 15. *Prenat Diagn* 16:837–844, 1996.

Rogan, P. K., Close, P., Blouin, J. L., et al. Duplication and loss of chromosome 21 in two children with Down syndrome and acute leukemia. *Am J Med Genet* 59:174–181, 1995.

Rogan, P. K., Sabol, D. W. and Punnett, H. H. Maternal uniparental disomy of chromosome 21 in a normal child. *Am J Med Genet* 83:69–71, 1999.

Saracco, S., Abramowsky, C., Taylor, S., Silverman, R. A. and Berman, B. W. Spontaneously regressing adrenocortical carcinoma in a newborn. A case report with DNA ploidy analysis. *Cancer* 62:507–511, 1988.

Schinzel, A., Kotzot, D., Brecevic, L., et al. Trisomy first, translocation second, uniparental disomy and partial trisomy third: a new mechanism for complex chromosomal aneuploidy. *Eur J Hum Genet* 5:308–314, 1997.

Schinzel, A. A., Basaran, S., Bernasconi, F., Karaman, B., Yuksel-Apak, M. and Robinson, W. P. Maternal uniparental disomy 22 has no impact on the phenotype. *Am J Hum Genet* 54:21–24, 1994.

Schinzel, A. A., Robinson, W. P., Binkert, F., Torresani, T. and Werder, E. A. Exclusively paternal X chromosomes in a girl with short stature. *Hum Genet* 92:175–178, 1993.

Schneider, A. S., Bischoff, F. Z., McCaskill, C., Coady, M. L., Stopfer, J. E. and Shaffer, L. G. Comprehensive 4-year follow-up on a case of maternal heterodisomy for chromosome 16. *Am J Med Genet* 66:204–208, 1996.

Shaffer, L. G., McCaskill, C., Egli, C. A., Baker, J. C. and Johnston, K. M. Is there an abnormal phenotype associated with maternal isodisomy for chromosome 2 in the presence of two isochromosomes? [letter; comment]. *Am J Hum Genet* 61:461–462, 1997.

Shaffer, L. G., Overhauser, J., Jackson, L. G. and Ledbetter, D. H. Genetic syndromes and uniparental disomy: a study of 16 cases of Brachmann–de Lange syndrome. *Am J Med Genet* 47:383–386, 1993.

Shuman, C., Weksberg, R., Nedelcu, R., Northey, A. and Scherer, S. Chromosome 7 uniparental disomy in Russel-Silver syndrome. *Am J Hum Genet* 59:A284, 1996. (Abstract)

Sirchia, S. M., De Andreis, C., Pariani, S., et al. Chromosome 14 maternal uniparental disomy in the euploid cell line of a fetus with mosaic 46,XX/47,XX,+14 karyotype. *Hum Genet* 94:355–358, 1994.

Slater, H., Shaw, J. H., Bankier, A., Forrest, S. M. and Dawson, G. UPD 13: no indication of maternal or paternal imprinting of genes on chromosome 13 [letter; comment]. *J Med Genet* 32:4931995.

Slater, H., Shaw, J. H., Dawson, G., Bankier, A. and Forrest, S. M. Maternal uniparental disomy of chromosome 13 in a phenotypically normal child [see comments]. *J Med Genet* 31:644–646, 1994.

Soler, A., Margarit, E., Queralt, R., et al. Paternal isodisomy 13 in a normal newborn infant after trisomy rescue evidenced by prenatal diagnosis. *Am J Med Genet* 90:291–293, 2000.

Spence, J. E., Perciaccante, R. G., Greig, G. M., et al. Uniparental disomy as a mechanism for human genetic disease. *Am J Hum Genet* 42:217–226, 1988.

Spinner, N. B., Rand, E., Bucan, M., et al. paternal uniparental isodisomy for human chromosome 20 and absence of external ears. *Am J Hum Genet* 55:A118, 1994. (Abstract)

Splitt, M. P. and Goodship, J. A. Another case of maternal uniparental disomy chromosome 14 syndrome [letter]. *Am J Med Genet* 72:239–240, 1997.

Spotila, L. D., Sereda, L. and Prockop, D. J. Partial isodisomy for maternal chromosome 7 and short stature in an individual with a mutation at the COL1A2 locus. *Am J Hum Genet* 51:1396–1405, 1992.

Stallard, R., Krueger, S., James, R. S. and Schwartz, S. Uniparental isodisomy 13 in a normal female due to transmission of a maternal t(13q13q). *Am J Med Genet* 57:14–18, 1995.

Stephens, K., Weaver, D. D., Leppig, K., Side, L. E., Shannon, K. M. and Maruyama, K. Somatic confined uniparental disomy of the NF1 gene region in myeloid leukemic cells of an NF1 patient mimics a loss of heterozygosity due to deletion. *Am J Hum Genet* 59:A51996. (Abstract)

Sulisalo, T., Francomano, C. A., Sistonen, P., et al. High-resolution genetic mapping of the cartilage-hair hypoplasia (CHH) gene in Amish and Finnish families. *Genomics* 20:347–353, 1994a.

Sulisalo, T., Makitie, O., Sistonen, P., et al. Uniparental disomy in cartilage-hair hypoplasia. *Eur J Hum Genet* 5:35–42, 1997.

Sulisalo, T., Sistonen, P., Hastbacka, J., et al. Cartilage-hair hypoplasia gene assigned to chromosome 9 by linkage analysis. *Nat Genet* 3:338–341, 1993b.

Sutcliffe, M. J., Mueller, O. T., Gallardo, L. A., Papenhausen, P. R. and Tedesco, T. A. Maternal isodisomy 16 in a normal 46,XX following trisomic conception. *Am J Hum Genet* 53:A1464, 1993. (Abstract)

Takizawa, Y., Pulkkinen, L., Chao, S. C., et al. Mutation report: complete paternal uniparental isodisomy of chromosome 1: a novel mechanism for Herlitz junctional epidermolysis bullosa. *J Invest Dermatol* 115:307–311, 2000.

Takizawa, Y., Pulkkinen, L., Shimizu, H., et al. Maternal uniparental meroisodisomy in the LAMB3 region of chromosome 1 results in lethal junctional epidermolysis bullosa. *J Invest Dermatol* 110:828–831, 1998.

Temple, I. K., Cockwell, A., Hassold, T., Pettay, D. and Jacobs, P. Maternal uniparental disomy for chromosome 14. *J Med Genet* 28:511–514, 1991.

Temple, I. K., Gardner, R. J., Robinson, D. O., et al. Further evidence for an imprinted gene for neonatal diabetes localised to chromosome 6q22-q23. *Hum Mol Genet* 5:1117–1121, 1996.

Temple, I. K., James, R. S., Crolla, J. A., et al. An imprinted gene(s) for diabetes? [letter]. *Nat Genet* 9:110–112, 1995.

Tiranti, V., Lamantea, E., Uziel, G., et al. Leigh syndrome transmitted by uniparental disomy of chromosome 9. *J Med Genet* 36:927–928, 1999.

Tomkins, D. J., Roux, A. F., Waye, J., Freeman, V. C., Cox, D. W. and Whelan, D. T. Maternal uniparental isodisomy of human chromosome 14 associated with a paternal t(13q14q) and precocious puberty. *Eur J Hum Genet* 4:153–159, 1996.

Turleau, C., Viot, G., Burglen, L., et al. Uniparental disomy 8 due to pseudodicentric chromosome: possible mechanisms and consequnces. *Am J Hum Genet* 63:A1531998. (Abstract)

van den Berg-Loonen, E. M., Savelkoul, P., van Hooff, H., van Eede, P., Riesewijk, A. and Geraedts, J. Uniparental maternal disomy 6 in a renal transplant patient. *Hum Immunol* 45: 46–51, 1996.

Van Opstal, D., Van den Berg, C., Deelen, W. H., et al. Prospective prenatal investigations on potential uniparental disomy in cases of confined placental trisomy. *Prenat Diagn* 18:35–44, 1998.

Vaughan, J., Ali, Z., Bower, S., Bennett, P., Chard, T. and Moore, G. Human maternal uniparental disomy for chromosome 16 and fetal development. *Prenat Diagn* 14:751–756, 1994.

Vidaud, D., Vidaud, M., Plassa, F., Gazengel, C., Noel, B. and Goossens, M. Father to son transmission of hemophilia A due to uniparental disomy. *Am J Hum Genet* 59:A226, 1989. (Abstract)

Voss, R., Ben-Simon, E., Avital, A., et al. Isodisomy of chromosome 7 in a patient with cystic fibrosis: could uniparental disomy be common in humans? *Am J Hum Genet* 45:373–380, 1989.

Walgenbach, D. D., Yang, S. P., McCaskill, C., Shaffer, L. G. and Towner, D. R. Maternal uniparental disomy of chromosome 14 in an infant with growth restriction, minor dysmorphology, and confined placental mosaicism. *Am J Hum Genet* 61:A380, 1997. (Abstract)

Walter, C. A., Shaffer, L. G., Kaye, C. I., et al. Short-limb dwarfism and hypertrophic cardiomyopathy in a patient with paternal isodisomy 14: 45,XY,idic(14)(p11). *Am J Med Genet* 65:259–265, 1996.

Wang, J. C., Mamunes, P., Kou, S. Y., Schmidt, J., Mao, R. and Hsu, W. T. Centromeric DNA break in a 10;16 reciprocal translocation associated with trisomy 16 confined placental mosaicism and maternal uniparental disomy for chromosome 16. *Am J Med Genet* 80:418–422, 1998.

Wang, J. C., Passage, M. B., Yen, P. H., Shapiro, L. J. and Mohandas, T. K. Uniparental heterodisomy for chromosome 14 in a phenotypically abnormal familial balanced 13/14 Robertsonian translocation carrier. *Am J Hum Genet* 48:1069–1074, 1991.

Webb, A., Beard, J., Wright, C., Robson, S., Wolstenholme, J. and Goodship, J. A case of paternal uniparental disomy for chromosome 11. *Prenat Diagn* 15:773–777, 1995.

Webb, A. L., Sturgiss, S., Warwicker, P., Robson, S. C., Goodship, J. A. and Wolstenholme, J. Maternal uniparental disomy for chromosome 2 in association with confined placental mosaicism for trisomy 2 and severe intrauterine growth retardation. *Prenat Diagn* 16:958–962, 1996.

Welch, T. R., Beischel, L. S., Choi, E., Balakrishnan, K. and Bishof, N. A. Uniparental isodisomy 6 associated with deficiency of the fourth component of complement. *J Clin Invest* 86:675–678, 1990.

Wertelecki, W., Breg, W. R., Graham, J. M. J., Iinuma, K., Puck, S. M. and Sergovich, F. R. Trisomy 22 mosaicism syndrome and Ullrich-Turner stigmata. *Am J Med Genet* 23:739–749, 1986.

Whiteford, M. L., Coutts, J., al-Roomi, L., et al. Uniparental isodisomy for chromosome 16 in a growth-retarded infant with congenital heart disease. *Prenat Diagn* 15:579–584, 1995.

Whiteford, M. L., Narendra, A., White, M. P., et al. Paternal uniparental disomy for chromosome 6 causes transient neonatal diabetes. *J Med Genet* 34:167–168, 1997.

Wilkinson, T. A., James, R. S., Crolla, J. A., Cockwell, A. E., Campbell, P. L. and Temple, I. K. A case of maternal uniparental disomy of chromosome 9 in association with confined placental mosaicism for trisomy 9. *Prenat Diagn* 16:371–374, 1996.

Willatt, L. R., Davison, B. C., Goudie, D., et al. A male with trisomy 9 mosaicism and maternal uniparental disomy for chromosome 9 in the euploid cell line. *J Med Genet* 29:742–744, 1992.

Woo, V., Bridge, P. J. and Bamforth, J. S. Maternal uniparental heterodisomy for chromosome 16: case report. *Am J Med Genet* 70:387–390, 1997.

Woodage, T., Prasad, M., Dixon, J. W., et al. Bloom syndrome and maternal uniparental disomy for chromosome 15. *Am J Hum Genet* 55:74–80, 1994.

Yang, X. P., Inazu, A., Yagi, K., Kajinami, K., Koizumi, J. and Mabuchi, H. Abetalipoproteinemia caused by maternal isodisomy of chromosome 4q containing an intron 9 splice acceptor mutation in the microsomal triglyceride transfer protein gene. *Arterioscler Thromb Vasc Biol* 19:1950–1955, 1999.

Yorifuji, T., Muroi, J., Kawai, M., et al. Uniparental and functional X disomy in Turner syndrome patients with unexplained mental retardation and X derived marker chromosomes. *J Med Genet* 35:539–544, 1998.

Yoshihashi, H., Maeyama, K., Kosaki, R., et al. Imprinting of human GRB10 and its mutations in two patients with Russell-Silver syndrome. *Am J Hum Genet* 67:476–482, 2000.

Chapter 5

"Old" and "New" Syndromes with Uniparental Disomy

OLD SYNDROMES

These are syndromes that have already been clinically recognized before the discovery of the UPD in humans. These syndromes are as follows.

1. Neonatal Transient Diabetes Mellitus

Neonatal transient diabetes mellitus (TNDM), with an incidence in newborns of 1/400,000 to 1/500,000 (Fosel, 1995; Shield et al., 1997), is characterized by intrauterine growth retardation, failure to thrive, hyperglycemia, glucosuria, dehydration, polyuria, occasional ketonemia and ketonuria, lethargy, and fever. Occasional features include anemia, macroglossia (Figure 1), and umbilical hernia. The diabetic condition may be permanent, subside to recur later as type 2 diabetes (Shield et al., 1997), or vanish completely (Christian et al., 1999).

TNDM may arise from paternal UPD6 (Temple et al., 1995; Hermann and Soltesz, 1997), from duplication of paternal chromosome 6q22–23 (Gardner et al., 1999; Temple et al., 1996; Arthur et al., 1997), or from other, as yet undefined, causes. Paternal UPD6 may account for approximately 20% of patients with TNDM

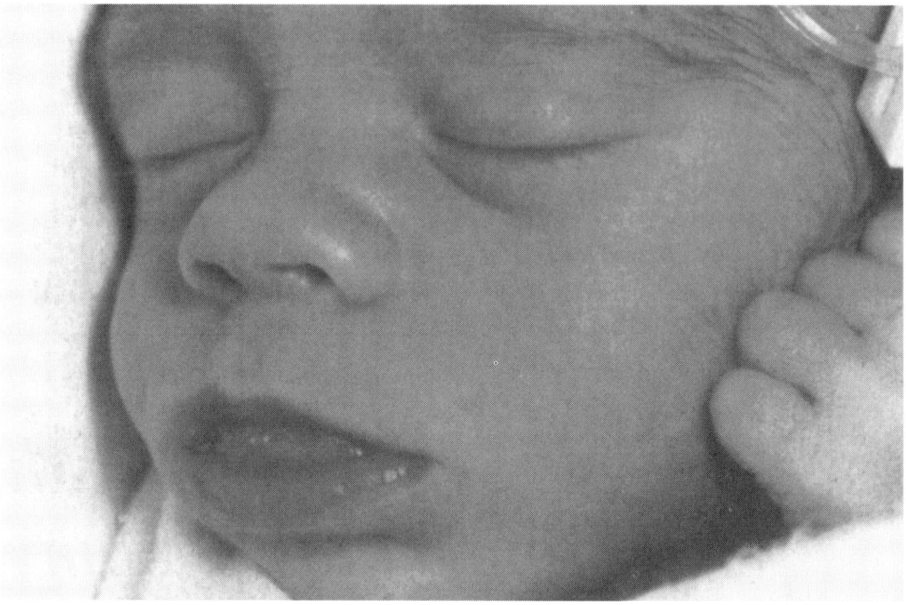

Figure 1 *The macroglossia observed in a patient with transient neonatal diabetes and paternal UPD6 (from Christian et al., 1999. Reproduced with permission).*

(Gardner et al., 1998). These individuals with TNDM and paternal UPD6 are likely to have macroglossia and transient diabetes. The condition is unlikely to recur in siblings (Christian et al., 1999).

It appears that TNDM is due to a duplication of the paternally inherited genes on chromosome 6 regardless of the presence or absence of maternally derived alleles for these genes. Using patients with partial duplication of the paternal chromosome 6, it was estimated that the imprinted gene for TNDM lies within an 18.72 cR3000 (approximately 5.4 Mb) interval on 6q24.1–q24.3 between markers D6S1699 and D6S1010 (Gardner et al., 1999).

The use of DNA polymorphic markers and the smallest region of overlap among patients with duplicated regions on 6q24 allowed the narrowing down of this region. The critical region that harbors the TNDM-related gene was further localized to a 300–400 kb DNA segment (Gardner et al., 2000). The partial sequence of this region resulted in the identification of several CpG sites that differ in their methylation status between the paternal and maternal alleles. Furthermore, two patients with TNDM have been identified without UPD6, nor paternal duplication of 6q24, in which the DNA methylation within the critical region is identical to that of patients with paternal UPD6. These results showed that TNDM is associated with a methylation change and identify a novel methylation imprint on chromosome 6 linked to TNDM.

2. The Russell-Silver Syndrome

Russell-Silver Syndrome (RSS) is characterized by pre- and postnatal growth retardation and a disproportionately large head, due to a broad and prominent forehead that contrasts with a small, narrow lower portion of the face, giving a triangular appearance. Delayed closure of the anterior fontanelle is characteristic, as are downturned corners of the mouth; hemi-hypotrophy of the face, trunk, and limbs; clinodactyly; brachymesophalangy of the fifth fingers; partial cutaneous syndactyly of the second and third toes; areas of hypo- or hyperpigmentation of the skin; diminished subcutaneous tissue; delayed bone maturation; and excessive sweating of the forehead (Taussig et al., 1973; Kotzot et al., 1995). Other features may include anteverted nostrils, low set ears, webbing of the neck, dislocation of elbows with difficult supination, single palmar creases, hyperextensibility of joints, shortening of upper limbs, cubitus valgus, slight ptosis, high-pitched squeaky voice, neonatal edema of extremities, hypertrichosis, and (congenitally) missing teeth. Although motor development may be delayed, mental development is generally normal (Kotzot et al., 1995). The most consistent feature is the primordial growth retardation (PGR), and most cases exhibit only some of the features listed above.

The syndrome is genetically heterogeneous. It has been associated with various chromosome rearrangements or deletions (Schinzel et al., 1994; Wahlstrom et al., 1993; Christensen and Nielsen, 1978; Rogan et al., 1996; Ayala-Madrigal et al., 1996; Midro et al., 1993; Tamura et al., 1993; Eggermann et al., 1998) and has occasionally shown various, apparently Mendelian, patterns of inheritance (Duncan et al., 1990; Al-Fifi et al., 1996). Most interestingly, in a subset of patients with RSS, less than 10% maternal uniparental disomy for chromosome 7 (UPD7) has been found (Table 1) (Eggermann et al., 1997; Kotzot et al., 1995; Ayala-Madrigal et al., 1996; Preece et al., 1997). It is not yet known if cases due to maternal UPD7 differ phenotypically from those of other etiologies, although no distinctive differential features are suggested by the preliminary analyses of the first few patients described. Cases associated with UPD7 strongly imply the presence of a maternally imprinted locus (or loci). It has been hypothesized that when maternal in origin, a mutation at such a locus would have no effect, whereas the transmission of a mutated paternal copy would cause the syndrome. It is possible that a simple paternal deletion of a

TABLE 1 Frequency of Maternal UPD7 Observed in Several Series of Russell-Silver Patients

No. of Cases	UPD(7)mat	References
25	3	Kotzot (1995)
14	1	Shuman (1996)
7	0	Ayala-Madrigal (1996)
33	2	Preece (1997)
65	3	Eggermann (1997)
144	9 (~6%)	

region of chromosome 7 containing a certain gene, with or without maternal UPD7, is sufficient for RSS.

Several genes that map on chromosome 7 have been proposed as candidates for potential involvement in RSS; these include an epidermal growth factor receptor (EGFR-7p12) (Wakeling et al., 1998) and IGFBP1 and IGFBP3 (Eggermann et al., 1999), PEG1/MEST at 7q32 (Kobayashi et al., 1997; Riesewijk et al., 1997, 1998; Cuisset et al., 1997) and GRB10/Meg1 (growth factor receptor binding protein 10) at 7q11.2–12 (Miyoshi et al., 1998).

3. Beckwith-Wiedemann Syndrome

The oncogenic nature of Beckwith-Wiedemann syndrome (BWS) seems to involve the dysregulation of both tumor-promoting and tumor-suppressor genes. Given the number of imprinted genes in the 11p15.5 region and the complexity of their interaction, the syndrome may originate from the combination of different alleles of the imprinted genes. The underlying molecular modifications of this richly imprinted domain, which leads to various combinations of the dysmorphic and neoplastic features characteristic of BWS, include:

1. Segmental (11p15.5) paternal UPD11 in mosaic form (Henry et al., 1991). A single case of a mosaic paternal UPD11 for the entire chromosome 11 has also been described in BWS (Dutly et al., 1998).
2. Trisomy 11p15 through paternal segmental duplication (Turleau et al., 1984).
3. Imprint relaxation through maternally transmitted breaks in the crucial domain (Rainier et al., 1993).
4. Mutations of essential genes such as p57Kip2 and KVLQT1 (Hatada et al., 1996; Lee et al., 1997).
5. Possibly imprinting mutations causing abnormal methylation of IGF2 and H19 (Reik et al., 1995; Hu et al., 1997).

Transient neonatal hypoglycemia, which may cause mental retardation if untreated, is a common feature of BWS (Engstrom et al., 1988). In approximately 6% of BWS patients, there is an increased risk of developing tumors and hemihyperplasia (Hoyme et al., 1998). Other variable defects, including visceromegaly, earlobe creases, nevus flammeus, and mid-face hypoplasia, are often observed (Pettenati et al., 1986).

The frequency of paternal UPD11 in the BWS is on the order of 20–30% (Henry et al., 1993). It is usually segmental, confined to the 11p15.5 region, and somatic in origin through chromatid interchange and thus present in a mosaic pattern. In rare cases, this mosaicism may involve one cell line with somatic UPD11 for an entire paternal pair (Dutly et al., 1998). Only a handful of cases with paternal UPD11 of meiotic origin have been described (Grundy et al., 1991; Webb et al., 1995).

The phenotypic presentation of BWS with mosaicism for paternal segmental UPD11 appears to be different from that of BWS without demonstrable UPD11. In one study (Slatter et al., 1994), nine cases of BWS with UPD11 were compared to 23 without UPD. The UPD11 cases appeared more likely to have hemihypertrophy (6/9 versus 1/23) and less likely to have exomphalos (0/9 versus 13/23). No differences were seen with respect to other features such as neoplasia, developmental delay, hypoglycemia, and nephromegaly. In another comparison between UPD11 and non-UPD cases of BWS, the incidence of tumors was higher in UPD11 cases (50 versus 8%), although the sample size was small (Henry et al., 1993). It would be tempting to speculate that mosaicism for normal cells and UPD11p cells accounts for asymmetrical tissue growth and hemihypertrophy (Slatter et al., 1994); however, the proband with paternal isodisomy UPD11 for the entire chromosome 11 resulting from a presumed meiotic origin was reported because of metastatic adrenal cortical carcinoma and right hemihypertrophy at birth or early infancy (Grundy et al., 1991; Saracco et al., 1988).

4. The Prader-Willi Syndrome

An excellent clinical summary of Prader-Willi syndrome (PWS) is provided in the literature (Holm et al., 1993; Cassidy, 1997). PWS is a multisystemic disorder with numerous manifestations of hypothalamic insufficiency. The neonatal and infantile period is marked by hypotonia, poor suck, feeding problems, and poor weight gain with a failure to thrive. In contrast, between 1 and 6 years of age, excessive weight gain with central obesity occurs, resulting from a voracious appetite. There is global developmental delay, mild to moderate mental retardation, and various learning problems. Characteristic cranio-facial features include dolichocephaly, narrow face or narrow forehead, almond-shaped eyes, small mouth with thin upper lip and downturned corners. Hypogonadism is marked, with hypoplasia of the genitals in both sexes and hypomaturity later on.

The minor diagnostic criteria include poor fetal movements, infantile lethargy, weak cry, and compulsive behavior with tantrums and outbursts. Also of note are sleep disturbances or apnea, short stature, hypopigmentation, small or narrow hands and feet, ocular anomalies, thick but scant saliva, speech articulation defects, and a tendency to skin “picking.” Other findings that often support the diagnosis include a high pain threshold, decreased vomiting, body temperature instability, kyphoscoliosis, osteoporosis, early adrenarche, unusual visual-spatial skills as seen from the ability to complete jigsaw puzzles, and normal neuromuscular abilities.

The study of the DNA from PWS patients and their families revealed four different classes of molecular lesions in the 15q11–q13 region (Nicholls et al., 1998; Jiang et al., 1998) as summarized in Table 2. These are: (1) patients with paternally inherited deletions of about 4 Mb in the chromosome 15q11–q13 region. The deletions are thought to occur via unequal crossing-over between copies of the HERC2 gene (see Chapter 7, also Ji et al., 1999) and account for 70% of PWS patients. The recurrence risk is extremely low. (2) The category of maternal UPD15 in which no paternally derived alleles are present accounts for approximately 25–30% of patients; the recurrence risk is negligible, unless there is a parental translocation that predisposes to UPD15. It is of interest that maternal UPD15 is

TABLE 2 Molecular Subclassification of PWS (according to Nicholls et al., 1998)

PWS Class	Chr15 Mechanism	Frequency	Methylation	Recurrence Risk
I	Paternal deletion 15q11–q13	70%	Abnormal	Extremely low
II	Maternal UPD15	25–30%	Abnormal	Extremely low
IIIa	Imprinting mutation with IC deletion	1–2%	Abnormal	Significant
IIIb	Imprinting mutation without IC deletion	1–2%	Abnormal	Low
IV	Balanced translocation involving 15q11–q13	<1%	?	Significant

much more common than paternal UPD15 which results in AS. This is probably due to the higher rates of maternal nondisjunction that results in trisomy 15 which is subsequently "corrected" to maternal UPD15 and loss of the paternally derived chromosome. (3) Another fraction of PWS patients, approximately 1–3%, shows "imprinting" mutations in which there is a maternal type of methylation on 15q11–q13 loci from the paternally inherited chromosome 15. In about half of these patients, there is a small deletion of the telomeric portion of the bipartite imprinting center (IC) at exon 1 of the SNRNP gene, which causes the inability to establish the normal methylation status (subcategory IIIa) (Ohta et al., 1999). The recurrence risk of these small deletions is high because most such cases are familial. The subcategory IIIb consists of cases with imprinting mutations but without any detectable deletion in the IC region. Their molecular mechanism is unknown, but their recurrence risk is low (Ohta et al., 1999). (4) There are a few rare PWS cases with a balanced translocation in 15q11–q13; the mechanism of PWS in these cases is unknown but it might involve a disruption of chromatin structure (Saitoh et al., 1997).

Phenotype-genotype correlations in PWS In order to detect phenotypic differences between PWS patients with different types of molecular lesions on chromosome 15q, analysis of a series of 167 cases, including 116 deletions and 51 nondeletion cases, was performed (Gillessen-Kaesbach et al., 1995). The diagnosis of the nondeletion cases was based on the methylation test. In the nondeletion group, molecular methods did not distinguish between maternal UPD15 and the rare cases of IC (imprinting center) mutations. No substantial (statistically proven) differences were found between the deletion and nondeletion cases. Weight and length of newborn PWS patients were significantly lower than those of healthy controls, but head circumferences were normal; however, male and female deletion cases had lower birthweights than nondeletion cases. The frequency of hypopigmentation was also different between the molecular classes, being present in about 50% of deletion cases (48 of 93) and only 25% (9 of 39) of nondeletion cases (Gillessen-Kaesbach et al., 1995).

There is an increase in maternal age in nondeletion cases as expected, since both trisomy rescue and gamete complementation follow maternal 15 meiotic nondisjunction. In one study of 128 cases with maternal UPD15, the mean maternal age was 34.9 versus 28 years in the general population (Robinson et al., 1998). Similarly in another study, the average maternal age was increased (34.4 years) among 17 PWS patients with maternal UPD15, as compared to 26.8 years in 37 PWS patients with deletions (Cassidy et al., 1997).

In another series of PWS patients, FISH for D15S11, SNRPN, or GABRB3 was used to detect chromosome 15 deletions and microsatellite analysis using at least six markers to detect maternal UPD15 (Cassidy et al., 1997). Statistically significant differences in the occurrence of PWS symptoms between 37 PWS patients with deletion of chromosome 15 and 17 PWS with maternal UPD15 cases were not observed, except for two features: The first was a trend to hypopigmentation, present in 50% of the cases with deletion compared to 15% with maternal UPD15. The second feature that may reflect the thorough and expert clinical dysmorphology assessment relates to facial appearance, which was judged to be somewhat less characteristic in UPD15 cases (typical facies in 10 of 17) than in deletions (33 of 36). The characteristic narrow frontal diameter, almond-shaped palpebral fissures, narrow nasal bridge, and downturned mouth were less often present in maternal UPD15 cases of PWS (Cassidy et al., 1997). Skin picking, high pain threshold, jigsaw puzzle skills, and articulation abnormalities were also more commonly reported in the deletion cases, although the differences were not statistically significant.

The age at diagnosis was also delayed in PWS patients with UPD15. In males, the mean age of diagnosis was 9 years in PWS with UPD15 versus 4 in PWS with deletions; in females, the ages were 9 versus 2, respectively, also suggesting a milder phenotype in maternal UPD15 versus deletion 15 PWS cases.

Significant differences were observed between females PWS with maternal UPD15 and those with deletions 15, particulary in the length of gavage feeding and a later onset of hyperphagia (Mitchell et al., 1996). These features suggest a milder presentation in the female UPD15 group. In both deletion 15 and UPD15 PWS patients, the length of time of gavage feeding was shorter in female than male babies (Cassidy et al., 1997).

A study of sleep anomalies, including onset of REM periods and excessive daytime sleepiness, did not reveal a consistent significant association with one or the other molecular type of PWS, although such disturbances were more common in patients with paternal deletions (Vgontzas et al., 1996).

The differences in the phenotypes of PWS patients with varying genomic alterations may result from factors such as: (i) preservation of the biallelic expression of nonimprinted gene pairs in UPD15 versus hemizygous expression in deletion (haploinsufficiency for nonimprinted loci); (ii) lack of possible interactions between regions of imprinted domains when deletions are present (LaSalle and Lalande, 1995, 1996); (iii) homozygosity for recessive alleles in the case of maternal isodisomy UPD15.

The molecular laboratory analysis of the 15q11–q13 region (methylation, FISH, and microsatellite analysis) of patients suspected to have PWS has contributed to the recognition of a phenotypic spectrum much broader than that with the consensus diagnostic criteria (Holm et al., 1993). Thus, among 50 cases, 35 with deletions of the 15q11–q13 region, 11 with maternal UPD15, and 4 with maternal methylation, nearly one-half had unusual findings. These consisted of normal height (seven cases), normal puberty (two cases), normal cognitive ability (five cases), and minimal speech acquisition (three cases). Macro- and microcephaly occurred in four and two cases, respectively, while other malformations such as gastroschisis, hypospadias, cataract, autism, and kidney stones were observed in single cases (Cassidy et al., 1997).

5. Angelman Syndrome

Angelman syndrome (AS) is a rare (approximately 1 in 15,000 births) neurobehavioral disorder that shows little specificity in the first year of life. There is, in general, a normal prenatal and birth history. Problems encountered in the first year include failure to thrive, feeding problems, seizures, and developmental delay, all already present by 6 months in some 75% of cases. Self-feeding is delayed and toilet training and walking are rarely achieved before 3 years, if at all; almost half of the patients are neither ambulatory nor toilet-trained. All patients are profoundly retarded and only a minority acquire minimal language; most use limited gestual communication and only one-half are able to understand speech to some extent. In general, a marked cheerfulness with frequent outbursts of laughter has been observed by 1 to 2 years of age. There is reduced need for sleep and hyperactive behavior; on the other hand, stubbornness and aggressive attitudes are uncommon. One-half of AS children have seizures of varying types, generally difficult to control but becoming milder with age. Affected children have a rigid jerky ataxia and microbrachycephaly. Macrognathia, macrosomia, protuding tongue, wide-spaced teeth, and drooling are common facial characteristics. Most have cutaneous ocular hypopigmentation. Height and weight are below the 2nd centile in about one-third and one-sixth of the patients, respectively. Hand and foot lengths are short in about 30% of cases. Visual defects such as strabismus, optic atrophy, and keratoconus are not rare, affecting some 40% of all individuals.

An extensive and authoritative clinical description of the syndrome can be found in Zori et al. 1992. To sum up, the major and most specific characteristics of the syndrome include microbrachycephaly, seizures with an abnormal EEG, stiff and jerky gait with flexed upper arms, severe language impairment, a dysmorphic face, and inappropriate laughter. Spontaneous, rhythmic, fast-bursting cortical myoclonus is also a prominent feature of AS (Guerrini et al., 1996). Clinical diagnosis is rarely made before the age of 1 year. Excellent recent reviews of the AS molecular etiology and clinical presentations were published (Lalande, 1996; Jiang et al., 1998, 1999; Nicholls et al., 1998).

The study of the DNA from AS patients and their families revealed six different classes of molecular lesions in the 15q11–q13 region, as summarized in Table 3 modified from (Jiang et al., 1999). These are:

1. Patients with maternally inherited deletions of about 4 Mb in the chromosome 15q11–q13 region. The deletions are thought to occur via unequal crossing-over between copies of the HERC2 gene (Ji et al., 1999) (see also Chapter 8) and account for 65–75% of AS patients. The recurrence risk is extremely low.

TABLE 3 Molecular Subclassification and Recurrence Risk for AS (according to Jiang et al., 1999)

AS Class	Chr15 Mechanism	Frequency	Methylation	Recurrence Risk
I	Maternal deletion 15q11–q13	65–70%	Abnormal	Extremely low
II	Complex rearrangement involving 15q11–q13	$<1\%$	Normal or abnormal	Significant
III	Paternal UPD15	3–5%	Abnormal	Extremely low
IIIa	Paternal UPD15 with predisposing parental translocation	$<1\%$	Abnormal	Significant
IVa	Imprinting mutation with IC deletion	3–5%	Abnormal	Significant
IVb	Imprinting mutation without IC deletion	3–5%	Abnormal	Low
V	UBE3A mutations	4–6%	Normal	Significant
VI	No detectable molecular abnormality	10–14%	Normal	Unknown, probably low

2. A closely related but rare category is that with unbalanced translocations or complex chromosomal rearrangements involving 15q11–q13 or inherited interstitial deletions; the recurrence risk is significant.
3. The category of paternal UPD15 in which no maternally derived alleles are present accounts for 3–5% of patients; the recurrence risk is negligible, unless there is a parental translocation that predisposes to UPD15 (subcategory IIIa).
4. Approximately 7–9% of AS patients show, "imprinting" mutations in which there is a paternal type of methylation on 15q11–q13 loci from the maternally inherited chromosome 15. In about half of these patients, there is a small deletion of the centromeric portion of the bipartite imprinting center (IC) near the SNRNP gene that causes the inability to establish the normal methylation status (subcategory IVa). The recurrence risk of these small deletions is high because most such cases are familial (Ohta et al., 1999). The subcategory IVb consists of cases with imprinting mutations but without any detectable deletion in the IC region. Their molecular mechanism is unknown, but their recurrence risk is low (Buiting et al., 1998).
5. In approximately 4–6% of AS cases, point mutations exist (mostly resulting in a truncated or nonfunctional protein) in the UBE3A gene that maps in the 15q11–q13 region and encodes the E6-AP ubiquitin-protein ligase (Malzac et al., 1998; Fang et al., 1999; Matsuura et al., 1997; Kishino et al., 1997). The mutations either occur de novo on the maternal chromosome or are inherited from a maternal allele. The same mutations inherited from the father are not associated with an abnormal phenotype. The recurrence risk in familial cases is 50%.
6. The last category includes those patients with the clinical diagnosis of AS but without a demonstrable abnormality in the 15q11–q13 region. Possibilities for this include incorrect diagnosis, undetected mutations in genes/sequences within 15q11–q13, mutations in genes outside 15q11–q13 that control the expression of UBE3A (Jiang et al., 1999).

Phenotype-genotype correlations in AS The phenotypes of AS with different molecular etiologies are not the same. After the initial proposal that the AS phenotype is milder in paternal UPD15 than in deletion 15 cases (Bottani et al., 1994), several other investigators have also made similar observations (Gillessen-Kaesbach et al., 1995; Smith et al., 1997). There is no consensus regarding a difference in the facial characteristics of paternal UPD15 versus deletion 15 AS patients (Bottani et al., 1994; Smith et al., 1997).

In the studies that compared the clinical features in AS patients with maternal deletion 15 versus the nondeletion cases, it was found that apart from the pigmentary abnormalities associated with deletion cases, nondeletion patients seem to have a milder phenotype (Moncla et al., 1999; Saitoh et al., 1994; Hou et al., 1997; Smith et al., 1996). In the largest most recent study (Moncla et al., 1999), the most significant phenotypic differences were: microcephaly (18/20 in maternal deletion 15 AS versus 7/20 in nondeletion AS; $p < 0.001$), ability to walk (5.2 years in maternal deletion 15 AS versus 2.6 years in nondeletion AS; $p < 0.001$), growth retardation (10/20 in maternal deletion 15 AS versus 2/20 in nondeletion AS; $p < 0.01$), seizures (20/20 in maternal deletion 15 AS versus 14/20 in nondeletion AS; $p < 0.02$), hypopigmentation (15/20 in maternal deletion 15 AS versus 0/20 in nondeletion AS; $p < 0.001$), obesity (3/20 in maternal deletion 15 AS versus 10/20 in nondeletion AS; $p < 0.01$). Facial dysmorphism also appeared milder in 13/20 nondeletion cases as compared to the 19/20 deletion AS patients.

There are also differences between paternal UPD15 AS cases versus the deletion patients (Smith et al., 1996, 1997; Saitoh et al., 1994). In a collection of 14 AS cases with paternal UPD15, microcephaly was present in 30%, growth retardation in 25%, seizures in 40%, and ataxia was either absent or mild in 40%; all these patients were able to walk.

In IC mutations (category IV), nine cases were compared to 9 age-matched AS deletion 15 patients (Burger et al., 1996). All had mental retardation, delayed motor development, and absent speech. However, hypopigmentation and microcephaly were present in only one of the IC mutation patients compared to seven of nine in the deletion group.

More recently, seven patients initially suspected of having PWS displayed an AS methylation pattern (Gillessen-Kaesbach et al., 1999). All these cases lacked the major features of AS and with one exception had normal birthweight and length; they also showed hypotonia, and all had obesity of early onset. Ataxia was absent, and epilepsy was noted in one. They were all mildly mentally retarded and in three cases there was language development. The mild phenotype may be explained by an incomplete imprinting defect or by cellular mosaicism.

In AS patients with point mutations in the UBE3A gene, ataxia, epilepsy, and microcephaly are either mild or absent (Fang et al., 1999; Moncla et al., 1999). In addition, there is intrafamilial variability of clinical manifestations in families with more than one affected individual. In the study (Moncla et al., 1999) of 14 AS patients with UBE3A mutations, microcephaly was present in 7/14 cases, ability to walk occurred at 2.8 years, growth retardation was present in 2/14, seizures in 11/14, obesity in 9/14.

Concerning specifically the microcephaly and epilepsy, there appear to be some frequency differences in the subcategories of patients with nondeletion 15 AS. In the cases collected from a number of studies of (Moncla et al., 1999; Tonk et al., 1996; Meijers-Heijboer et al., 1992; Freeman et al., 1993; Bottani et al., 1994; Prasad and Wagstaff, 1997; Gillessen-Kaesbach et al., 1995; Smith et al., 1997; Burger et al., 1996; Saitoh et al., 1997), microcephaly was present in 6/19 (32%) cases with paternal UPD15, 3/17 (18%) cases of imprinting mutations, and 14/22 (64%) cases with UBE3A mutations. For epilepsy, the numbers corresponding to these categories of molecular lesions were 8/17 (47%) for paternal UPD15, 11/17 (65%) for imprinting mutations, and 17/23 (74%) for UBE3A mutations. The epilepsy in all these categories is milder than in deletion 15 cases (Minassian et al., 1998).

NEW SYNDROMES

1. The Maternal UPD14 Syndrome

After the initial case (Temple et al., 1991), a total of 19 cases have been reported with maternal UPD14 (see Chapter 4). With the publication of the third (Antonarakis et al., 1993) and fourth (Healey et al., 1994) cases of maternal UPD 14, it became clear that the recurring signs and symptoms delineated a new syndrome. Indeed, the first three cases, a male (Temple et al., 1991) and two females (Pentao et al., 1992; Antonarakis et al., 1993), reported in their 17th, 20th, and 9th years had developed puberty at 10, 8.5, and 9 years, respectively; they were short, 154.7, 144.4, and 110.5 cm at 15, 20, and 9 years; they had kyphoscoliosis and unusually small hands and feet.

The fourth reported case, only 4.5 years old, although prepubertal, confirmed the presence of most of the signs (Healey et al., 1994), including arrested hydrocephalus, also noted in two of the three earlier cases and thus added convincing evidence to the existence of this "new" syndrome.

With the exception of one case of no apparent phenotype (Papenhausen et al., 1995), which was insufficiently documented at the molecular level (Robinson and Langlois, 1996), all other cases further illustrated this emerging syndrome (Coviello et al., 1996; Tomkins et al., 1996; Sirchia et al., 1994; Barton et al., 1996; Linck et al., 1996; Walgenbach et al., 1997; Splitt and Goodship, 1997; Robinson et al., 1994; Desilets et al., 1997; Miyoshi et al., 1998; Ralph et al., 1999; Harrison et al., 1998; Fokstuen et al., 1999; Martin et al., 1999). The birth is often premature and the birthweight low for gestational age. In about 30% of cases, the head grows rapidly in the postnatal period, owing to a hydrocephalic condition that, however, arrests spontaneously. A suggestive facies with prominent forehead (in association with arrested hydrocephalus) and supraorbital ridges, a short philtrum, and downturned mouth corners are present (Figure 2). There is hypotonia, hyperextensible joints, mild to moderate motor delay, small hands and feet, kyphoscoliosis (sometimes severe in degree), mild truncal obesity, sometimes with hypercholesterolemia. Intelligence is low-normal to normal. Otitis media occurs frequently. Table 4 (modified from Fokstuen et al., 1999) lists the phenotypic characteristics of the

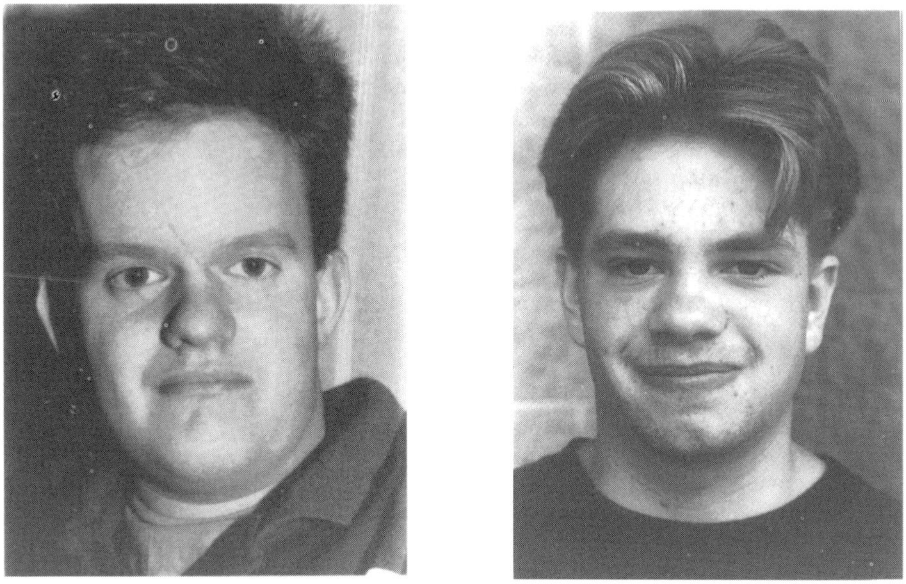

Figure 2 *Facial features of two males with the maternal UPD14 syndrome (left from Temple et al., 1991; right from Tomkins et al., 1996. Reproduced with permission).*

TABLE 4 Phenotypic and Cytogenetic Findings in the Maternal UPD14 Syndrome

	N	(%)
Maternal age	32.1 years	
Intrauterine growth retardation	9 of 11	82
Hypotonia	10 of 13	77
Hydrocephalus	4 of 13	31
Short stature	11 of 12	92
Developmental delay	9 of 13	70
Retarded intelligence	2 of 10	20
Small hands	10 of 11	91
Hyperextensible joints	5 of 9	56
Scoliosis	4 of 10	40
Early onset of puberty	8.7 years	
Advanced bone age	3 of 3	
Recurrent otitis media	5 of 9	56
Hypercholesterolemia	3 of 9	33
"Dysmorphic features"	6 of 13	46
Translocation t(13,14) *de novo*	5 of 18	28
Translocation t(13,14) familial	3 of 18	17
Translocation t(14,14) or isochromosome 14	6 of 18	33

maternal UPD14 syndrome. The presence of a recognizable phenotype strongly indicates that chromosome 14 harbors yet uncharacterized imprinted loci.

The karyotype of most of these cases includes a translocation t(13/14) in eight cases, either *de novo* (five cases), or maternally inherited (three cases); in six patients, an isochromosome 14q was found. In at least three cases, a mosaicism with a cell line of trisomy 14 of maternal origin was detected, and therefore the maternal UPD14 was likely to have originated from a mitotic loss of the paternal chromosome 14. The mosaicism for trisomy 14 may therefore contribute to the phenotype, although proven cases of somatic mosaicism for trisomy 14 display a distinct phenotype (Fujimoto et al., 1992; Cotter et al., 1997). It is of interest that the mean maternal age in maternal UPD14 is advanced (32.1 years).

2. A New Syndrome of Short Stature: The Paternal UPD14 Syndrome

The four known cases of paternal UPD14 are very similar in their presentation in that they allow a clear delineation of a new syndrome of short stature (Wang et al., 1991; Papenhausen et al., 1995; Walter et al., 1996; Cotter et al., 1997).

The maternal age was not increased. There was polyhydramnios (3/4) and premature vaginal delivery, induced or spontaneous between 29 to 34 weeks. At birth, the head was of normal size and appeared relatively large as compared to the reduced chest and limb sizes. Apgar scores were poor and the neonatal course rocky. Immediate respiratory distress required assisted ventilation and tracheal intubation or tracheostomy. This was probably due to a small narrow and bell-shaped thoracic cage, and thin, angulated upslanted ribs (Figure 3). The birthweight was low, the body length and limbs short. The abdomen was distended or showed ventral wall hernia or diastasis recti.

The face shows some striking features, namely, short palpebral fissures or blepharophimosis, unusually small ears, long philtrum, small mouth, sometimes with puckered lips (Figure 4). The neck is short, broad, and possibly webbed (Table 5). There were in some cases mild joint contractures of the hips, ankles, elbows, knees, and fingers. There was also hypotonia, wrist and finger anomalies, including "ulnar drift," contractures of proximal interphalangeal joints, adducted thumbs, and flexed wrists.

The evolution of the syndrome was characterized by pulmonary complications and feeding problems with hypoxia (even apnea), and recurrent food aspiration requiring tube feeding or gastrostomy. Severe growth retardation persisted, weight and length remained below the 5th centile, with persistant thoracic insufficiency and shortness of limbs. One child developed congestive heart failure with hypertrophic cardiomyopathy at 6 months (Walter et al., 1996), another died of pulmonary complications at 6 months (Cotter et al., 1997), a third, still alive at age 9 years, exhibited severe skeletal anomalies, particularly chest deformity and severe kyphoscoliosis and has had seizures and severe mental retardation (Wang et al., 1991). Another, at 1 year of age, showed the cognitive and motor skills of an 8- or 9-month-old. Investigation of the skeleton is recommended. Where described, the bones were

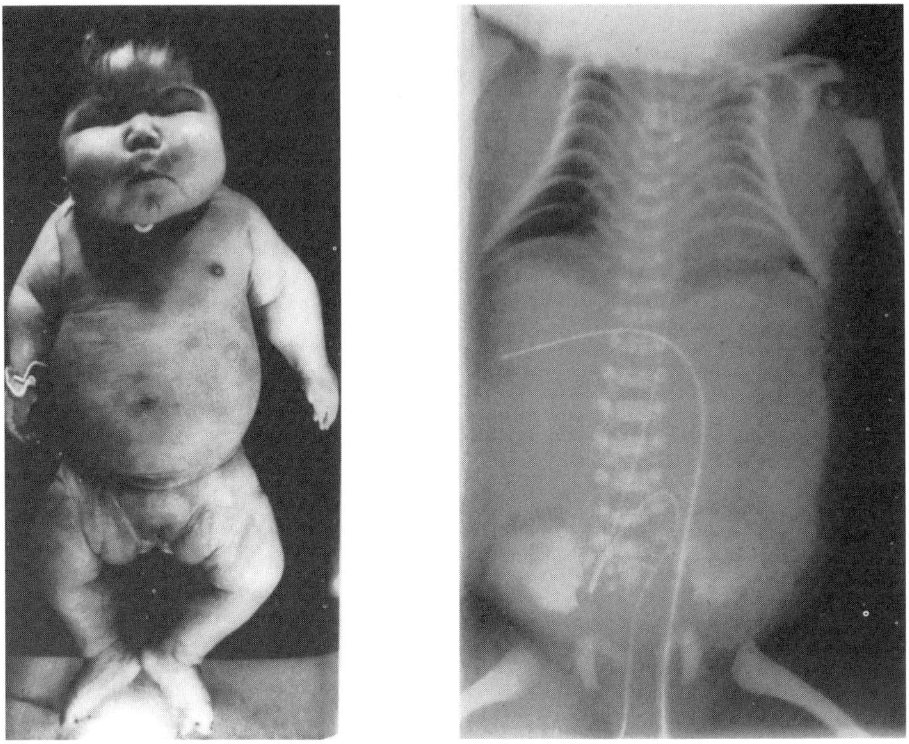

Figure 3 *Whole body and radiological appearance in a 6-month-old baby with paternal UPD14 (from Cotter et al., 1997. Reproduced with permission).*

thin, with long fibulae and small proximal tibial epiphyses (Walter et al., 1996) or hypoplastic ribs, and elongated clavicles on X-rays.

Of the four cases, two apparently resulted from gamete complementation as both parents coincidentally possessed a translocation involving chromosome 14 (Wang et al., 1991; Cotter et al., 1997); the two other cases had a *de novo* translocation t(14;14) inherited from the father along with the absence of the maternal chromosome 14 (Papenhausen et al., 1995; Walter et al., 1996).

The phenotypic expression of paternal and maternal UPD14 implies that both maternally and paternally imprinted genes exist on chromosome 14. The lack of the maternally derived alleles appear to cause the more severe phenotype, as seen in the emerging paternal UPD14 syndrome. The human region 14q13-qter is homologous to mouse chromosome 12, which is imprinted (Cotter et al., 1997; Cattanach and Jones, 1994). Interestingly, a duplication 14(q23.3–q31) in a father and daughter, the former normal, the latter dysmorphic and retarded, also raises the possibility of an imprinted domain in this chromosomal region, which would cause an overdose of expressed paternal gene(s) (Robin et al., 1997). It is therefore possible that the pathology in UPD14 arises from a combination of over- and underexpressed genes.

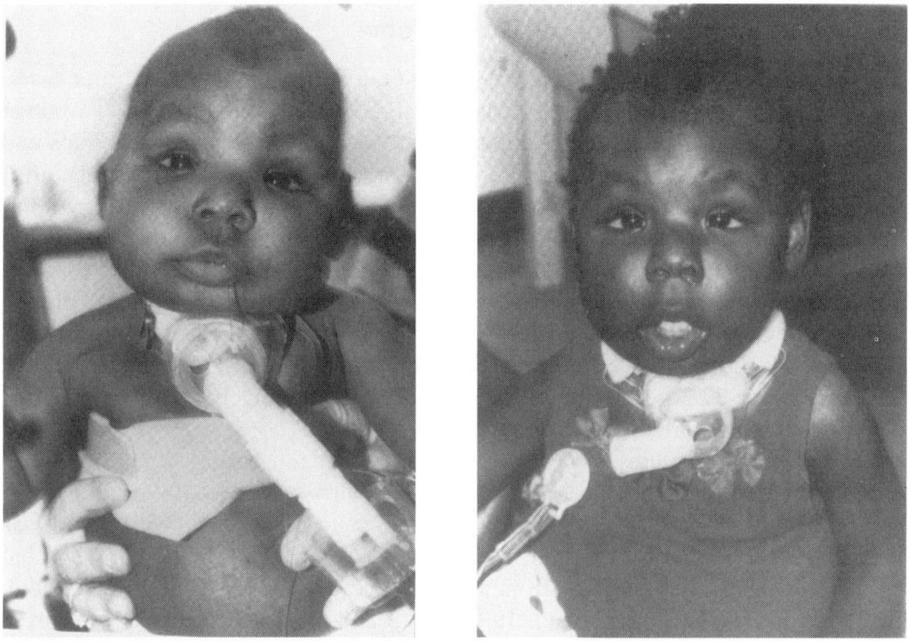

Figure 4 *Facial appearance of a child with paternal UPD14 at 8 and 20 months of age (Papenhausen et al., 1995. Reproduced with permission).*

TABLE 5 Main Facial Features in Paternal UPD14 Syndrome

Hirsute forehead	2/4
Blepharophimosis	3/4
Protruding philtrum	3/4
Puckered lips	2/4
Retrognathia	2/4
Small ears	3/4
Webbed neck	2/4

Source: From Cotter, P. D., et al., *Am J Med Genet* 70:74–79, 1997.

POTENTIALLY NEW UPD SYNDROMES

A subset of cases with maternal UPD2 and maternal UPD16 have been reported that present a pathology which could be considered into two distinct syndromes. Certainly, more cases are needed for the clear delineation of these two potentially new syndromes. Most of these cases are derived from trisomy rescue, and therefore some features may not be due to uniparental disomy but to the residual trisomy in various tissues.

1. The Potential Maternal UPD2 Syndrome

The clinical characteristics of maternal UPD2 (Table 6), as defined from four cases, are listed in (Harrison et al., 1995; Hansen et al., 1997; Shaffer et al., 1997; Webb et al., 1996). They include: (i) a gestational length shortened by oligohydramnios and frequent Cesarian section; (ii) severe IUGR, not related to gestational age; (iii) severe neonatal and postnatal pulmonary problems, requiring oxygen therapy up to the age of 1 year in one case; (iv) hypospadias in two of the three male cases and occasional malformations such as preauricular pits (Shaffer et al., 1997), patent ductus arteriosus, and pyloric stenosis (Webb et al., 1996); (v) persistent growth retardation with short stature, small head, and poor weight gain documented up to 8 years of age (Table 6); (vi) normal developmental patterns; (vii) an advanced maternal age of 32.2 years. Figure 5 demonstrates the relatively favorable outcome of the first case reported (Harrison et al., 1995).

Some of these features may be due to mosaicism for trisomy 2 that might be present in three of the four cases. In one case (Harrison et al., 1995), a 17-week amniocentesis showed trisomy 2 in one-third of mitoses, whereas cord blood was diploid. At birth, amnion, chorion, and villi revealed 12–32% trisomy 2 cells, but cord blood metaphases and cells from a buccal smear studied by FISH were diploid. In the second case (Hansen et al., 1997), CVS and term villi displayed trisomy 2 mosaicism, whereas second-trimester amniotic fluid, term amnion, and results from the study of four fetal tissues were only diploid. In the third case (Webb et al., 1996), CVS, amniotic fluid, and several term placenta samples showed trisomy/euploid mosaicism, but fetal blood at 29 weeks and cord blood and skin fibroblasts at delivery or shortly afterward were diploid. The fourth case was cytogenetically studied at 8 years only (Shaffer et al., 1997) and mosaicism had not been detected.

TABLE 6 Main Pre- and Postnatal Features in Five Cases of Maternal UPD2

	No. of Cases
Gestational data:	
Oligohydramnios: (severe) IUGR	4
Short gestation	4
Cesarean section	3 (at least)
Perinatal death	1
Clinical data:	
Pulmonary problems	3
Hypospadias	2 (of 3 males)
Preauricular pits	2
Patent ductus	1
Pyloric stenosis	1
Neonatal hypothyroidism	1
Persistent growth retardation	2 (at least)

In a sixth case (Heide et al., 2000), the only clinical information given is that the female child was phenotypically normal.

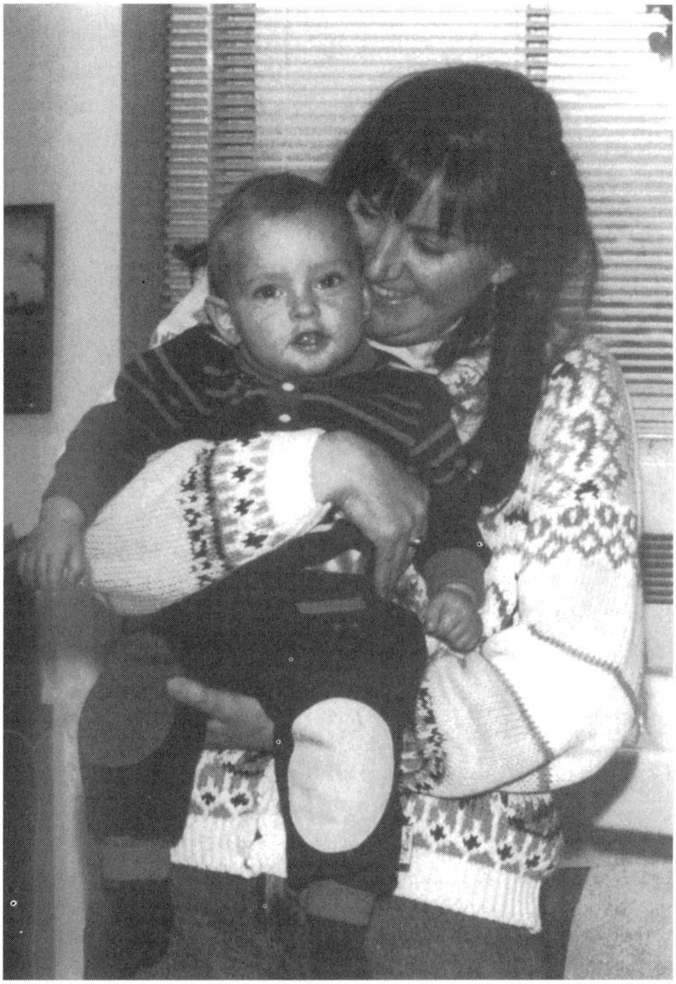

Figure 5 *Fourteen-month-old boy with maternal UPD2 (from Harrison et al., 1995. Reproduced with permission).*

Thus, in the above three cases studied both pre- and postnatally, at least two tissues from the probands (including lymphocytes and fibroblasts) had only diploid cells and maternal UPD2. This cannot disprove, however, the existence of trisomy 2 cells in other tissues.

In two cases of maternal UPD2, there was isodisomy 2 through isochromosome formation for each arm of chromosome 2, namely, presence in the karyotype of 46 chromosomes with discrete i(2p) + i(2q). This karyotype strongly suggests occurrence of a monosomy 2 conceptus converted, for one of its early viable blastomeres, into isodisomy for each arm of chromosome 2, through transverse splitting of the centromere of the singly inherited maternal chromosome. One of these patients, diagnosed at 36 years of age, was the only patient of maternal UPD2 with a normal phenotype, although she measured only 155 cm (Bernasconi et al., 1996). The other was one of the four cases considered fairly typical and used for the description of this presumptive syndrome. In these two cases, mosaicism for trisomy 2 is unlikely.

The phenotypes of patients with proven somatic trisomy 2/euploid mosaicism without UPD2 show striking heterogeneity that does not match the picture of maternal UPD2 cases. One severe case in which the diploid cells showed biparental inheritance for chromosome 2 had IUGR and multisystemic malformations in the brain, heart, chest, and feet (Sago et al., 1997). Another, of average weight and height, had cleft lip and palate and mild developmental delay (Pappas et al., 1995). A third displayed a "Pfeiffer-like" syndrome, scoliosis, and radio-ulnar hypoplasia (Cramer et al., 1993); a fourth had IUGR and mild anomalies (Casey et al., 1990). Additional cases of trisomy 2 mosaicism detected after prenatal diagnosis include one with transverse hemimelia and another with genital ambiguity (Bui et al., 1984).

It is possible that chromosome 2 contains imprinted genes. Human chromosome 2 is syntenic to segments of mouse chromosomes 1, 2, 6, and 11 that contain domains which had been shown to be imprinted (Harrison et al., 1995). Certainly, more cases of maternal UPD2 are needed to further confirm and define the syndrome. A case of a normal 36-year-old female with maternal UPD2 and isochromosomes for 2p and 2q is the only published evidence against imprinted genes on chromosome 2 (Bernasconi et al., 1996). The arguments for and against the involvement of imprinting in maternal UPD2 phenotype are presented in Table 7.

2. The Potential Maternal UPD16 Syndrome

Cases with maternal UPD16 are relatively common, because trisomy 16 is the most frequent trisomy observed in aborted conceptuses (Hassold, 1980) and "trisomy rescue" is a common mechanism of uniparental disomy. It is frequently diagnosed at chorionic villus sampling (CVS), thus demonstrating the potential for trisomy rescue. In the cases of maternal UPD16, several factors, aside from a putative imprinting of chromosome 16 genes, might contribute to the phenotype. These include placental insufficiency due to trisomy 16, obstetrical complications, severe

TABLE 7 Positive and Negative Considerations Regarding an Imprinting Effect in the Putative Maternal UPD2 Syndrome

Against an Imprinting Imbalance:
- Placental aneuploidy per se, causing nutritional, hormonal, and circulatory problems
- Possible, elusive residual trisomy 2 in solid tissues
- One case reported normal at 36 years

In Favor of an Imprinting Imbalance:
- Lasting pulmonary insufficiency
- Lasting growth deficiency
- Mouse homologous segments are imprinted
- Confirmed trisomy 2/euploid mosaics differ and vary

prematurity, and residual somatic trisomy 16. Comparison of the outcome between cases diagnosed at CVS with trisomy 16/euploid mosaicism and harboring uniparental versus biparental chromosome 16 should determine the existence of a maternal UPD16 syndrome.

The following features are reported in cases of maternal UPD16 (Tables 8 and 9) (Bennett et al., 1992; Dworniczak et al., 1992, Kalousek et al., 1993; Vaughan et al., 1994; Whiteford et al., 1995; O'Riordan et al., 1996; Schneider et al., 1996; Exeler et al., 1996; Woo et al., 1997; Wang et al., 1998): (i) the prenatal course is often abnormal and more so in uniparental than biparental chromosome 16 cases. There is major placental pathology (such as hematomas, fibrinous deposition, microinfarcts, two-vessel cords), prenatal and neonatal death, pervasive prematurity, and severe IUGR. Maternal serum screen anomalies are common and maternal health during pregnancy is often poor. (ii) Natural births are the minority, preempted by intrauterine deaths or C-section occurrences. (iii) The birthweight is barely over 1100 gm for a delivery occurring on average at 31 weeks. (iv) The malformations present in 11 cases (Table 10) are heart defects such as septal defects and AV canal (five cases), imperforated anus (three cases), hypospadias in males (two cases), and club foot (two cases); other malformations only present in one case each are left renal agenesis, hydronephrosis, inguinal hernia, scoliosis, and dislocation of the elbow. (v) When documented, postnatal growth remained at or below the 10th centile and postnatal death is not rare. As judged after 1 year of age, developmental landmarks in four cases were otherwise fair to good. (vi) It has been noted that a subtle but characteristic facies may exist, consisting of upslanted, almond-shaped eyes, thin nose with upturned nares, thin upper lip and downturned mouth corners (Figure 6). Some of these phenotypes could well be due to trisomy 16 mosaicism; however, the search for an aneuploid cell line has generally remained negative.

TABLE 8 Main Gestational and Clinical Characteristics in Patients with Maternal UPD16

Prenatal Course: • IUGR • Fetal demise • Neonatal death • Abnormal serum screen	*Placental Pathology:* • Hematomas • Fibrinous deposition • Microinfarctus • Two-vessel cord
Malformations: • *Heart defects:* AV canal septal defects • *UG defects:* renal agenesis, hypospadias	*GI defects:* • Imperforate anus • Inguinal hernia
Other features: • Facial dysmorphism • Little catch-up growth	*Skeletal Defects:* • Scoliosis • Talipes equinovarus • Clinodactyly
Etiopathogenic Differential Diagnosis: Placental aneuploidy; residual trisomy 16; isodisomy; genomic imprinting	

Source: from Schneider, A. S., et al. (1996), Wang, J. C., et al. (1997), and personal observations.

TABLE 9 Clinical Characteristics of the Potential Maternal UPD16 Syndrome

Clinical Features	Number
Maternal age	34.3 years
Abnormal placenta	5 of 16
Maternal serum biochemical abnormalities	4 of 16
Trisomy 16 at CVS	All
Trisomy 16 in term placenta	10 of 16
Premature birth, 35 weeks	6 of 10 liveborn
Cesarian section, 33 weeks	4 of 10 liveborn
Intrauterine death	4 of 16
Low birthweight	All
Malformations (see Table 10)	11 of 16
Growth in early childhood	$<$3rd percentile in 3 of 5
Good mental development	4 of 5

The clinical phenotype of cases where mosaicism for trisomy 16 has been documented in somatic cells (usually in fibroblasts) is much more severe (Gilbertson et al., 1990; Devi et al., 1993; Lindor et al., 1993; Pletcher et al., 1994; Greally et al., 1996). In six liveborn cases, there was IUGR (6/6), cranial asymmetry (2/6), eye anomalies (2/6), malformed ears (3/6), preauricular pits (2/6), scoliosis (2/6), thoracic asymmetry (2/6), heart defect (5/6), hypoplastic nipples (3/6), inguinal

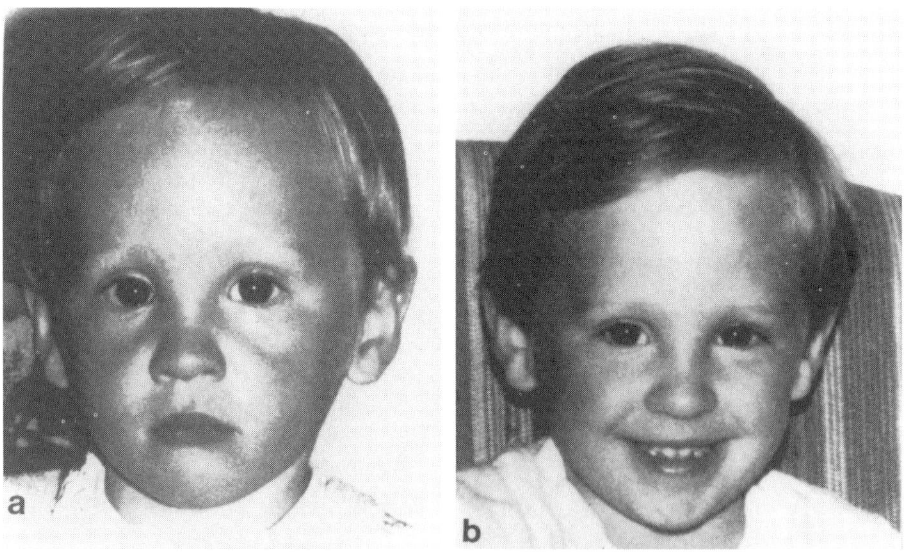

Figure 6 Facial appearance of a child with maternal UPD16 at 20 (a) and 36 (b) months of age (Schneider et al., 1996. Reproduced with permission).

TABLE 10 Major Malformations Observed in 16 Cases of Maternal UPD16

One Malformation at Least	11 Cases[a]
Heart defect	
Septal defect	4
AV canal	1
Imperforate anus	3
Glandular hypospadias	2
Club foot (unilateral)	2
Left renal agenesis	1
Left hydronephrosis (mild)	1
Inguinal hernia (unilateral)	1
Scoliosis (severe)	1
Elbow dislocation: other minor malformations sometimes reported	1

[a] Six male and five female patients; two or more malformations in three cases.

hernias (3/6), talipes calcaneo-valgus (2/6), hypospadias (1/6), single palmar crease (5/6), nail hypoplasia (3/6), and camptodactyly (3/6). A phenotype differing from that seen in maternal UPD16 has been proposed (Pletcher et al., 1994; Lindor et al., 1993).

The imprinting status of chromosome 16 in man is unknown, since mouse domains syntenic to the human counterpart have not yet been identified. For all practical purposes maternal UPD16 cases are more severe than biparental cases similarly derived from a trisomy 16, whereas cases with cytogenetically diagnosed trisomy 16/euploid mosaicism in somatic cells are generally worse than either of the above.

The arguments for and against a maternal UPD16 syndrome are shown in Table 11. Table 12 summarizes the association of UPD and imprinting in both the classical and newly recognized syndromes.

TABLE 11 Positive and Negative Considerations Regarding an Imprinting Effect in the Putative Maternal UPD16 Syndrome

Against an Imprinting Disorder:

- Severe placental vascular, hormonal, nutritional problems
- Elusive, scant, trisomy 16 lines, buried in somatic solid tissues

In Favor of an Imprinting Disorder:

- Scant, but stereotyped malformations
- Suggestive facies
- No catch-up growth
- Distinctive picture for cases of trisomy 16/euploid mosacism
- Mouse homologous segments may be imprinted

TABLE 12 Listing of the Main Syndromes Involving Uniparental Disomy Associated with an Imprinting Effect

UPD Type	Syndrome
Certain:	
Paternal 6	Neonatal diabetes (transient)
Maternal 7	Russell-Silver
Paternal 11	Beckwith-Wiedemann
Maternal 14	Growth failure—early puberty
Paternal 14	Dwarfism, rib cage hypoplasia
Maternal 15	Prader-Willi
Paternal 15	Angelmann
Probable:	
Maternal 2	Growth failure, bronchopulmonary dysplasia
Maternal 16	Growth failure and abnormalities

REFERENCES

Al-Fifi, S., Teebi, A. S. and Shevell, M. Autosomal dominant Russell-Silver syndrome [letter]. *Am J Med Genet* 61:96–97, 1996.

Antonarakis, S. E., Blouin, J. L., Maher, J., Avramopoulos, D., Thomas, G. and Talbot, C. C. J. Maternal uniparental disomy for human chromosome 14, due to loss of a chromosome 14 from somatic cells with t(13;14) trisomy 14. *Am J Hum Genet* 52:1145–1152, 1993.

Arthur, E. I., Zlotogora, J., Lerer, I., Dagan, J., Marks, K. and Abeliovich, D. Transient neonatal diabetes mellitus in a child with invdup(6)(q22q23) of paternal origin. *Eur J Hum Genet* 5:417–419, 1997.

Ayala-Madrigal, M. L., Shaffer, L. G. and Ramirez-Duenas, M. L. Silver-Russell syndrome and exclusion of uniparental disomy. *Clin Genet* 50:494–497, 1996.

Barton, D. E., McQuaid, S., Stallings, R., Griffin, E. and Geraghty, M. Further evidence for an emerging maternal uniparental disomy chromosome 14 syndrome: Analysis of a phenotypically abnormal de novo Robertsonian translocation t(13;14) carrier. *Am J Hum Genet* 59:A125,1996. (Abstract)

Bennett, P., Vaughan, J., Henderson, D., Loughna, S. and Moore, G. Association between confined placental trisomy, fetal uniparental disomy, and early intrauterine growth retardation [letter]. *Lancet* 340:1284–1285, 1992.

Bernasconi, F., Karaguzel, A., Celep, F., et al. Normal phenotype with maternal isodisomy in a female with two isochromosomes: i(2p) and i(2q) [see comments]. *Am J Hum Genet* 59:1114–1118, 1996.

Bottani, A., Robinson, W. P., DeLozier-Blanchet, C. D., et al. Angelman syndrome due to paternal uniparental disomy of chromosome 15: a milder phenotype? [see comments]. *Am J Med Genet* 51:35–40, 1994.

Bui, T. H., Iselius, L. and Lindsten, J. European collaborative study on prenatal diagnosis: mosaicism, pseudomosaicism and single abnormal cells in amniotic fluid cell cultures. *Prenat Diagn* 4 Spec No: 145–162, 1984.

Buiting, K., Dittrich, B., Gross, S., et al. Sporadic imprinting defects in Prader-Willi syndrome and Angelman syndrome: implications for imprint-switch models, genetic counseling, and prenatal diagnosis. *Am J Hum Genet* 63:170–180, 1998.

Burger, J., Kunze, J., Sperling, K. and Reis, A. Phenotypic differences in Angelman syndrome patients: imprinting mutations show less frequently microcephaly and hypopigmentation than deletions. *Am J Med Genet* 66:221–226, 1996.

Casey, J., Ketterer, D. M., Heisler, K. L., Daugherty, P. M., Prince, P. M. and Giles H. R. Prenatal diagnosis of trisomy 2 mosaicism confirmed in foreskin fibroblasts. *Am J Hum Genet* 47:A270, 1990. (Abstract)

Cassidy, S. B. Prader-Willi syndrome. *J Med Genet* 34:917–923, 1997.

Cassidy, S. B., Forsythe, M., Heeger, S., et al. Comparison of phenotype between patients with Prader-Willi syndrome due to deletion 15q and uniparental disomy 15. *Am J Med Genet* 68:433–440, 1997.

Cattanach, B. M. and Jones, J. Genetic imprinting in the mouse: implications for gene regulation. *J Inherit Metab Dis* 17:403–420, 1994.

Christensen, M. F. and Nielsen, J. Deletion short arm 18 and Silver-Russell syndrome. *Acta Paediatr Scand* 67:101–103, 1978.

Christian, S. L., Rich, B. H., Loebl, C., et al. Significance of genetic testing for paternal uniparental disomy of chromosome 6 in neonatal diabetes mellitus [see comments]. *J Pediatr* 134:42–46, 1999.

Cotter, P. D., Kaffe, S., McCurdy, L. D., Jhaveri, M., Willner, J. P. and Hirschhorn, K. Paternal uniparental disomy for chromosome 14: a case report and review. *Am J Med Genet* 70:74–79, 1997.

Coviello, D. A., Panucci, E., Mantero, M. M., et al. Maternal uniparental disomy for chromosome 14. *Acta Genet Med Gemellol (Roma)* 45:169–172, 1996.

Cramer, A. E., Richkind, K. E., Schlam, M., Muenke, M. and Amirkham, N. Tissue specific trisomy 2 mosaicism in an infant with Pfeiffer syndrome-like features. *Am J Hum Genet* 53:A538, 1993. (Abstract)

Cuisset, L., Le Stunff, C., Dupont, J. M., et al. PEG1 expression in maternal uniparental disomy 7. *Ann Genet* 40:211–215, 1997.

Desilets, V. A., Yong, S. L., Kalousek, D. K., Pantzar, T. J., Kwong, L. C. and Langlois S. Maternal uniparental disomy for chromosome 14. *Am J Hum Genet* 61:A122, 1997. (Abstract)

Devi, A. S., Velinov, M., Kamath, M. V., et al. Variable clinical expression of mosaic trisomy 16 in the newborn infant [see comments]. *Am J Med Genet* 47:294–298, 1993.

Duncan, P. A., Hall, J. G., Shapiro, L. R. and Vibert, B. K. Three-generation dominant transmission of the Silver-Russell syndrome. *Am J Med Genet* 35:245–250, 1990.

Dutly, F., Baumer, A., Kayserili, H., et al. Seven cases of Wiedmann-Beckwith syndrome, including the first reported case of mosaic paternal isodisomy along the whole chromosome 11. *Am J Med Genet* 79:347–353, 1998.

Dworniczak, B., Koppers, B., Kurlemann, G., Holzgreve, W., Horst, J. and Miny, P. Uniparental disomy with normal phenotype [letter]. *Lancet* 340:1285, 1992.

Eggermann, K., Wollmann, H. A., Binder, G., Kaiser, P., Ranke, M. B. and Eggermann, T. Biparental expression of IGFBP1 and IGFBP3 renders their involvement in the etiology of Silver-Russell syndrome unlikely. *Ann Genet* 42:117–121, 1999.

Eggermann, T., Eggermann, K., Mergenthaler, S., et al. Paternally inherited deletion of CSH1 in a patient with Silver-Russell syndrome. *J Med Genet* 35:784–786, 1998.

Eggermann, T., Wollmann, H. A., Kuner, R., et al. Molecular studies in 37 Silver-Russell syndrome patients: frequency and etiology of uniparental disomy. *Hum Genet* 100:415–419, 1997.

Engstrom, W., Lindham, S. and Schofield, P. Wiedemann-Beckwith syndrome. *Eur J Pediatr* 147:450–457, 1988.

Exeler, R., Dworniczak, B., Tercanli, S., et al. Physical and psychomotor development of three children with uniparental disomy 16. *Am J Hum Genet* 59:A92, 1996. (Abstract)

Fang, P., Lev-Lehman, E., Tsai, T. F., et al. The spectrum of mutations in UBE3A causing Angelman syndrome. *Hum Mol Genet* 8:129–135, 1999.

Fokstuen, S., Ginsburg, C., Zachmann, M. and Schinzel, A. Maternal uniparental disomy 14 as a cause of intrauterine growth retardation and early onset of puberty. *J Pediatr* 134:689–695, 1999.

Fosel, S. Transient and permanent neonatal diabetes. *Eur J Pediatr* 154:944–948, 1995.

Freeman, S. B., May, K. M., Pettay, D., Fernhoff, P. M. and Hassold, T. J. Paternal uniparental disomy in a child with a balanced 15;15 translocation and Angelman syndrome. *Am J Med Genet* 45:625–630, 1993.

Fujimoto, A., Allanson, J., Crowe, C. A., Lipson, M. H. and Johnson, V. P. Natural history of mosaic trisomy 14 syndrome. *Am J Med Genet* 44:189–196, 1992.

Gardner, R. J., Mackay, D. J., Mungall, A. J., et al. An imprinted locus associated with transient neonatal diabetes mellitus. *Hum Mol Genet* 9:589–596, 2000.

Gardner, R. J., Mungall, A. J., Dunham, I., et al. Localisation of a gene for transient neonatal diabetes mellitus to an 18.72 cR3000 (approximately 5.4 Mb) interval on chromosome 6q. *J Med Genet* 36:192–196, 1999.

Gardner, R. J., Robinson, D. O., Lamont, L., Shield, J. P. and Temple, I. K. Paternal uniparental disomy of chromosome 6 and transient neonatal diabetes mellitus. *Clin Genet* 54:522–525, 1998.

Gilbertson, N. J., Taylor, J. W. and Kovar, I. Z. Mosaic trisomy 16 in a live newborn infant. *Arch Dis Child* 65:388–389, 1990.

Gillessen-Kaesbach, G., Albrecht, B., Passarge, E. and Horsthemke, B. Further patient with Angelman syndrome due to paternal disomy of chromosome 15 and a milder phenotype [letter; comment]. *Am J Med Genet* 56:328–329, 1995.

Gillessen-Kaesbach, G., Demuth, S., Thiele, H., Theile, U., Lich, C. and Horsthemke, B. A previously unrecognized phenotype characterised by obesity, muscular hypotonia, and ability to speak in patients with Angelman syndrome caused by an imprinting defect [see comments]. *Eur J Hum Genet* 7:638–644, 1999.

Gillessen-Kaesbach, G., Robinson, W., Lohmann, D., Kaya-Westerloh, S., Passarge, E. and Horsthemke, B. Genotype-phenotype correlation in a series of 167 deletion and nondeletion patients with Prader-Willi syndrome. *Hum Genet* 96:638–643, 1995.

Greally, J. M., Neiswanger, K., Cummins, J. H., et al. A molecular anatomical analysis of mosaic trisomy 16. *Hum Genet* 98:86–90, 1996.

Grundy, P., Telzerow, P., Paterson, M. C., et al. Chromosome 11 uniparental isodisomy predisposing to embryonal neoplasms [letter]. *Lancet* 338:1079–1080, 1991.

Guerrini, R., De Lorey, T. M., Bonanni, P., et al. Cortical myoclonus in Angelman syndrome. *Ann Neurol* 40:39–48, 1996.

Hansen, W. F., Bernard, L. E., Langlois, S., et al. Maternal uniparental disomy of chromosome 2 and confined placental mosaicism for trisomy 2 in a fetus with intrauterine growth restriction, hypospadias, and oligohydramnios. *Prenat Diagn* 17:443–450, 1997.

Harrison, K., Eisenger, K., Anyane-Yeboa, K. and Brown, S. Maternal uniparental disomy of chromosome 2 in a baby with trisomy 2 mosaicism in amniotic fluid culture. *Am J Med Genet* 58:147–151, 1995.

Harrison, K. J., Allingham-Hawkins, D. J., Hummel, J., et al. Risk of uniparental disomy in Robertsonian translocation carriers: identification of UPD14 in a small cohort. *Am J Hum Genet* 63:A11, 1998. (Abstract)

Hassold, T. J. A cytogenetic study of repeated spontaneous abortions. *Am J Hum Genet* 32:723–730, 1980.

Hatada, I., Ohashi, H., Fukushima, Y., et al. An imprinted gene p57KIP2 is mutated in Beckwith-Wiedemann syndrome [see comments]. *Nat Genet* 14:171–173, 1996.

Healey, S., Powell, F., Battersby, M., Chenevix-Trench, G. and McGill, J. Distinct phenotype in maternal uniparental disomy of chromosome 14. *Am J Med Genet* 51:147–149, 1994.

Henry, I., Bonaiti-Pellie, C., Chehensse, V., et al. Uniparental paternal disomy in a genetic cancer-predisposing syndrome [see comments]. *Nature* 351:665–667, 1991.

Henry, I., Puech, A., Riesewijk, A., et al. Somatic mosaicism for partial paternal isodisomy in Wiedemann-Beckwith syndrome: a post-fertilization event. *Eur J Hum Genet* 1:19–29, 1993.

Hermann, R. and Soltesz, G. Paternal uniparental isodisomy of chromosome 6 in transient neonatal diabetes mellitus [letter; comment]. *Eur J Pediatr* 156:740, 1997.

Holm, V. A., Cassidy, S. B., Butler, M. G., et al. Prader-Willi syndrome: consensus diagnostic criteria. *Pediatrics* 91:398–402, 1993.

Hou, J. W., Wang, P. J. and Wang, T. R. Angelman syndrome assessed by neurological and molecular cytogenetic investigations. *Pediatr Neurol* 16:17–22, 1997.

Hoyme, H. E., Seaver, L. H., Jones, K. L., Procopio, F., Crooks, W. and Feingold, M. Isolated hemihyperplasia (hemihypertrophy): report of a prospective multicenter study of the incidence of neoplasia and review. *Am J Med Genet* 79:274–278, 1998.

Hu, J. F., Vu, T. H. and Hoffman, A. R. Genomic deletion of an imprint maintenance element abolishes imprinting of both insulin-like growth factor II and H19. *J Biol Chem* 272:20715–20720, 1997.

Ji, Y., Walkowicz, M. J., Buiting, K., et al. The ancestral gene for transcribed, low-copy repeats in the Prader-Willi/Angelman region encodes a large protein implicated in protein trafficking which is deficient in mice with neuromuscular and spermiogenic abnormalities. *Hum Mol Genet* 8:533–542, 1999.

Jiang, Y., Lev-Lehman, E., Bressler, J., Tsai, T. F. and Beaudet, A. L. Genetics of Angelman syndrome. *Am J Hum Genet* 65:1–6, 1999.

Jiang, Y., Tsai, T. F., Bressler, J. and Beaudet, A. L. Imprinting in Angelman and Prader-Willi syndromes. *Curr Opin Genet Dev* 8:334–342, 1998.

Kalousek, D. K., Langlois, S., Barrett, I., et al. Uniparental disomy for chromosome 16 in humans. *Am J Hum Genet* 52:8–16, 1993.

Kishino, T., Lalande, M. and Wagstaff, J. UBE3A/E6-AP mutations cause Angelman syndrome [published erratum appears in Nat Genet 1997 Apr;15(4):411]. *Nat Genet* 15:70–73, 1997.

Kobayashi, S., Kohda, T., Miyoshi, N., et al. Human PEG1/MEST, an imprinted gene on chromosome 7. *Hum Mol Genet* 6:781–786, 1997.

Kotzot, D., Schmitt, S., Bernasconi, F., et al. Uniparental disomy 7 in Silver-Russell syndrome and primordial growth retardation. *Hum Mol Genet* 4:583–587, 1995.

Lalande, M. Parental imprinting and human disease. *Annu Rev Genet* 30:173–195, 1996.

LaSalle, J. M. and Lalande, M. Domain organization of allele-specific replication within the GABRB3 gene cluster requires a biparental 15q11–13 contribution. *Nat Genet* 9:386–394, 1995.

LaSalle, J. M. and Lalande, M. Homologous association of oppositely imprinted chromosomal domains. *Science* 272:725–728, 1996.

Lee, M. P., Hu, R. J., Johnson, L. A. and Feinberg, A. P. Human KVLQT1 gene shows tissue-specific imprinting and encompasses Beckwith-Wiedemann syndrome chromosomal rearrangements [see comments]. *Nat Genet* 15:181–185, 1997.

Linck, L., McMilin, K., Popovich, B. and Magenis, R. E. Maternal uniparental disomy for chromosome 14. *Am J Hum Genet* 59:A124, 1996. (Abstract)

Lindor, N. M., Jalal, S. M., Thibodeau, S. N., Bonde, D., Sauser, K. L. and Karnes, P. S. Mosaic trisomy 16 in a thriving infant: maternal heterodisomy for chromosome 16. *Clin Genet* 44:185–189, 1993.

Malzac, P., Webber, H., Moncla, A., et al. Mutation analysis of UBE3A in Angelman syndrome patients. *Am J Hum Genet* 62:1353–1360, 1998.

Martin, R. A., Sabol, D. W. and Rogan, P. K. Maternal uniparental disomy of chromosome 14 confined to an interstitial segment (14q23–14q24.2) [In Process Citation]. *J Med Genet* 36:633–636, 1999.

Matsuura, T., Sutcliffe, J. S., Fang, P., et al. De novo truncating mutations in E6-AP ubiquitin-protein ligase gene (UBE3A) in Angelman syndrome. *Nat Genet* 15:74–77, 1997.

Meijers-Heijboer, E. J., Sandkuijl, L. A., Brunner, H. G., et al. Linkage analysis with chromosome 15q11–13 markers shows genomic imprinting in familial Angelman syndrome. *J Med Genet* 29:853–857, 1992.

Midro, A. T., Debek, K., Sawicka, A., Marcinkiewicz, D. and Rogowska, M. Second observation of Silver-Russel syndrome in a carrier of a reciprocal translocation with one breakpoint at site 17q25 [letter; comment]. *Clin Genet* 44:53–55, 1993.

Minassian, B. A., DeLorey, T. M., Olsen, R. W., et al. Angelman syndrome: correlations between epilepsy phenotypes and genotypes. *Ann Neurol* 43:485–493, 1998.

Mitchell, J., Schinzel, A., Langlois, S., et al. Comparison of phenotype in uniparental disomy and deletion Prader-Willi syndrome: sex specific differences. *Am J Med Genet* 65:133–136, 1996.

Miyoshi, N., Kuroiwa, Y., Kohda, T., et al. Identification of the Meg1/Grb10 imprinted gene on mouse proximal chromosome 11, a candidate for the Silver-Russell syndrome gene. *Proc Natl Acad Sci USA* 95:1102–1107, 1998.

Miyoshi, O., Hayashi, S., Fujimoto, M., Tomita, H., Sohda, M. and Niikawa, N. Maternal uniparental disomy for chromosome 14 in a boy with intrauterine growth retardation. *J Hum Genet* 43:138–142, 1998.

Moncla, A., Malzac, P., Livet, M. O., et al. Angelman syndrome resulting from UBE3A mutations in 14 patients from eight families: clinical manifestations and genetic counselling. *J Med Genet* 36:554–560, 1999.

Moncla, A., Malzac, P., Voelckel, M. A., et al. Phenotype-genotype correlation in 20 deletion and 20 nondeletion Angelman syndrome patients. *Eur J Hum Genet* 7:131–139, 1999.

Morichon-Delvallez, N., Segues, B., Pinson, M. P., et al. Maternal uniparental disomy for chromosome 14 by secondary non disjunction of a initial trisomy. *Am J Hum Genet* 55:A379, 1994. (Abstract)

Nicholls, R. D., Saitoh, S. and Horsthemke, B. Imprinting in Prader-Willi and Angelman syndromes. *Trends Genet* 14:194–200, 1998.

O'Riordan, S., Greenough, A., Moore, G. E., Bennett, P. and Nicolaides, K. H. Case report: uniparental disomy 16 in association with congenital heart disease. *Prenat Diagn* 16:963–965, 1996.

Ohta, T., Buiting, K., Kokkonen, H., et al. Molecular mechanism of Angelman syndrome in two large families involves an imprinting mutation. *Am J Hum Genet* 64:385–396, 1999.

Ohta, T., Gray, T. A., Rogan, P. K., et al. Imprinting-mutation mechanisms in Prader-Willi syndrome. *Am J Hum Genet* 64:397–413, 1999.

Papenhausen, P. R., Mueller, O. T., Johnson, V. P., Sutcliffe, M., Diamond, T. M. and Kousseff, B. G. Uniparental isodisomy of chromosome 14 in two cases: an abnormal child and a normal adult [see comments]. *Am J Med Genet* 59:271–275, 1995.

Pappas, J. G., Havens, G., Bogosian, V., Bhatt, K., Babu, A. and Penchaszadeh, V. B. Trisomy 2 mosaicism. *Am J Hum Genet* 57:A286, 1995. (Abstract)

Pentao, L., Lewis, R. A., Ledbetter, D. H., Patel, P. I. and Lupski, J. R. Maternal uniparental isodisomy of chromosome 14: association with autosomal recessive rod monochromacy. *Am J Hum Genet* 50:690–699, 1992.

Pettenati, M. J., Haines, J. L., Higgins, R. R., Wappner, R. S., Palmer, C. G. and Weaver, D. D. Wiedemann-Beckwith syndrome: presentation of clinical and cytogenetic data on 22 new cases and review of the literature. *Hum Genet* 74:143–154, 1986.

Pletcher, B. A., Sanz, M. M., Schlessel, J. S., et al. Postnatal confirmation of prenatally diagnosed trisomy 16 mosaicism in two phenotypically abnormal liveborns [see comments]. *Prenat Diagn* 14:933–940, 1994.

Prasad, C. and Wagstaff, J. Genotype and phenotype in Angelman syndrome caused by paternal UPD 15 [letter]. *Am J Med Genet* 70:328–329, 1997.

Preece, M. A., Price, S. M., Davies, V., et al. Maternal uniparental disomy 7 in Silver-Russell syndrome. *J Med Genet* 34:6–9, 1997.

Rainier, S., Johnson, L. A., Dobry, C. J., Ping, A. J., Grundy, P. E. and Feinberg, A. P. Relaxation of imprinted genes in human cancer. *Nature* 362:747–749, 1993.

Ralph, A., Scott, F., Tiernan, C., et al. Maternal uniparental isodisomy for chromosome 14 detected prenatally. *Prenat Diagn* 19:681–684, 1999.

Reik, W., Brown, K. W., Schneid, H., Le Bouc, Y., Bickmore, W. and Maher, E. R. Imprinting mutations in the Beckwith-Wiedemann syndrome suggested by altered imprinting pattern in the IGF2-H19 domain. *Hum Mol Genet* 4:2379–2385, 1995.

Riesewijk, A. M., Blagitko, N., Schinzel, A. A., et al. Evidence against a major role of PEG1/MEST in Silver-Russell syndrome. *Eur J Hum Genet* 6:114–120, 1998.

Riesewijk, A. M., Hu, L., Schulz, U., et al. Monoallelic expression of human PEG1/MEST is paralleled by parent-specific methylation in fetuses. *Genomics* 42:236–244, 1997.

Robin, N. H., Harari-Shacham, A., Schwartz, S. and Wolff, D. J. Duplication 14(q24.3–q31) in a father and daughter: delineation of a possible imprinted region. *Am J Med Genet* 71:361–365, 1997.

Robinson, W. P., Bernasconi, F., Basaran, S., et al. A somatic origin of homologous Robertsonian translocations and isochromosomes. *Am J Hum Genet* 54:290–302, 1994.

Robinson, W. P., Kuchinka, B. D., Bernasconi, F., et al. Maternal meiosis I non-disjunction of chromosome 15: dependence of the maternal age effect on level of recombination. *Hum Mol Genet* 7:1011–1019, 1998.

Robinson, W. P. and Langlois, S. Phenotype of maternal UPD(14) [letter; comment] [see comments]. *Am J Med Genet* 66:89, 1996.

Rogan, P. K., Seip, J. R., Driscoll, D. J., et al. Distinct 15q genotypes in Russell-Silver and ring 15 syndromes. *Am J Med Genet* 62:10–15, 1996.

Sago, H., Chen, E., Conte, W. J., et al. True trisomy 2 mosaicism in amniocytes and newborn liver associated with multiple system abnormalities. *Am J Med Genet* 72:343–346, 1997.

Saitoh, S., Buiting, K., Cassidy, S. B., et al. Clinical spectrum and molecular diagnosis of Angelman and Prader-Willi syndrome patients with an imprinting mutation. *Am J Med Genet* 68:195–206, 1997.

Saitoh, S., Harada, N., Jinno, Y., et al. Molecular and clinical study of 61 Angelman syndrome patients. *Am J Med Genet* 52:158–163, 1994.

Saracco, S., Abramowsky, C., Taylor, S., Silverman, R. A. and Berman, B. W. Spontaneously regressing adrenocortical carcinoma in a newborn. A case report with DNA ploidy analysis. *Cancer* 62:507–511, 1988.

Schinzel, A. A., Robinson, W. P., Binkert, F. and Fanconi, A. An interstitial deletion of proximal 8q (q11–q13) in a girl with Silver-Russell syndrome-like features. *Clin Dysmorphol* 3:63–69, 1994.

Schneider, A. S., Bischoff, F. Z., McCaskill, C., Coady, M. L., Stopfer, J. E. and Shaffer, L. G. Comprehensive 4-year follow-up on a case of maternal heterodisomy for chromosome 16. *Am J Med Genet* 66:204–208, 1996.

Shaffer, L. G., McCaskill, C., Egli, C. A., Baker, J. C. and Johnston, K. M. Is there an abnormal phenotype associated with maternal isodisomy for chromosome 2 in the presence of two isochromosomes? [letter; comment]. *Am J Hum Genet* 61:461–462, 1997.

Shield, J. P., Gardner, R. J., Wadsworth, E. J., et al. Aetiopathology and genetic basis of neonatal diabetes. *Arch Dis Child Fetal Neonatal Ed* 76:F39–F42, 1997.

Sirchia, S. M., De Andreis, C., Pariani, S., et al. Chromosome 14 maternal uniparental disomy in the euploid cell line of a fetus with mosaic 46,XX/47,XX,+14 karyotype. *Hum Genet* 94:355–358, 1994.

Slatter, R. E., Elliott, M., Welham, K., et al. Mosaic uniparental disomy in Beckwith-Wiedemann syndrome. *J Med Genet* 31:749–753, 1994.

Smith, A., Marks, R., Haan, E., Dixon, J. and Trent, R. J. Clinical features in four patients with Angelman syndrome resulting from paternal uniparental disomy. *J Med Genet* 34:426–429, 1997.

Smith, A., Marks, R., Haan, E., Dixon, J. and Trent, R. J. Clinical features in four patients with Angelman syndrome resulting from paternal uniparental disomy. *J Med Genet* 34:426–429, 1997.

Smith, A., Wiles, C., Haan, E., et al. Clinical features in 27 patients with Angelman syndrome resulting from DNA deletion. *J Med Genet* 33:107–112, 1996.

Splitt, M. P. and Goodship, J. A. Another case of maternal uniparental disomy chromosome 14 syndrome [letter]. *Am J Med Genet* 72:239–240, 1997.

Tamura, T., Tohma, T., Ohta, T., et al. Ring chromosome 15 involving deletion of the insulin-like growth factor 1 receptor gene in a patient with features of Silver-Russell syndrome. *Clin Dysmorphol* 2:106–113, 1993.

Taussig, L. M., Braunstein, G. D., White, B. J. and Christiansen, R. L. Silver-Russell dwarfism and cystic fibrosis in a twin. Endocrine, chromosomal, dermatoglyphic, and craniofacial studies. *Am J Dis Child* 125:495–503, 1973.

Temple, I. K., Cockwell, A., Hassold, T., Pettay, D. and Jacobs, P. Maternal uniparental disomy for chromosome 14. *J Med Genet* 28:511–514, 1991.

Temple, I. K., Gardner, R. J., Robinson, D. O., et al. Further evidence for an imprinted gene for neonatal diabetes localised to chromosome 6q22–q23. *Hum Mol Genet* 5:1117–1121, 1996.

Temple, I. K., James, R. S., Crolla, J. A., et al. An imprinted gene(s) for diabetes? [letter]. *Nat Genet* 9:110–112, 1995.

Tomkins, D. J., Roux, A. F., Waye, J., Freeman, V. C., Cox, D. W. and Whelan, D. T. Maternal uniparental isodisomy of human chromosome 14 associated with a paternal t(13q14q) and precocious puberty. *Eur J Hum Genet* 4:153–159, 1996.

Tonk, V., Schultz, R. A., Christian, S. L., Kubota, T., Ledbetter, D. H. and Wilson, G. N. Robertsonian (15q;15q) translocation in a child with Angelman syndrome: evidence of uniparental disomy. *Am J Med Genet* 66:426–428, 1996.

Turleau, C., de Grouchy, J., Chavin-Colin, F., Martelli, H., Voyer, M. and Charlas, R. Trisomy 11p15 and Beckwith-Wiedemann syndrome. A report of two cases. *Hum Genet* 67:219–221, 1984.

Vaughan, J., Ali, Z., Bower, S., Bennett, P., Chard, T. and Moore, G. Human maternal uniparental disomy for chromosome 16 and fetal development. *Prenat Diagn* 14:751–756, 1994.

Vgontzas, A. N., Kales, A., Seip, J., et al. Relationship of sleep abnormalities to patient genotypes in Prader-Willi syndrome. *Am J Med Genet* 67:478–482, 1996.

Wahlstrom, J., Holmgren, G., Albertsson-Wikland, K. and Tornhage, C. J. Silver-like syndrome and a small deletion on chromosome 13. *Acta Paediatr* 82:993–996, 1993.

Wakeling, E. L., Abu-Amero, S. N., Stanier, P., Preece, M. A. and Moore, G. E. Human EGFR, a candidate gene for the Silver-Russell syndrome, is biallelically expressed in a wide range of fetal tissues [published errata appear in Eur J Hum Genet 1998 Jul–Aug;6(4):415 and 1998 Sep–Oct;6(5):532]. *Eur J Hum Genet* 6:158–164, 1998.

Walgenbach, D. D., Yang, S. P., McCaskill, C., Shaffer, L. G. and Towner, D. R. Maternal uniparental disomy of chromosome 14 in an infant with growth restriction, minor dysmorphology, and confined placental mosaicism. *Am J Hum Genet* 61:A380, 1997. (Abstract)

Walter, C. A., Shaffer, L. G., Kaye, C. I., et al. Short-limb dwarfism and hypertrophic cardiomyopathy in a patient with paternal isodisomy 14: 45,XY,idic(14)(p11). *Am J Med Genet* 65:259–265, 1996.

Wang, J. C., Mamunes, P., Kou, S. Y., Schmidt, J., Mao, R. and Hsu, W. T. Centromeric DNA break in a 10;16 reciprocal translocation associated with trisomy 16 confined placental mosaicism and maternal uniparental disomy for chromosome 16. *Am J Med Genet* 80:418–422, 1998.

Wang, J. C., Passage, M. B., Yen, P. H., Shapiro, L. J. and Mohandas, T. K. Uniparental heterodisomy for chromosome 14 in a phenotypically abnormal familial balanced 13/14 Robertsonian translocation carrier. *Am J Hum Genet* 48:1069–1074, 1991.

Webb, A., Beard, J., Wright, C., Robson, S., Wolstenholme, J. and Goodship, J. A case of paternal uniparental disomy for chromosome 11. *Prenat Diagn* 15:773–777, 1995.

Webb, A. L., Sturgiss, S., Warwicker, P., Robson, S. C., Goodship, J. A. and Wolstenholme, J. Maternal uniparental disomy for chromosome 2 in association with confined placental mosaicism for trisomy 2 and severe intrauterine growth retardation. *Prenat Diagn* 16:958–962, 1996.

Whiteford, M. L., Coutts, J., al-Roomi, L., et al. Uniparental isodisomy for chromosome 16 in a growth-retarded infant with congenital heart disease. *Prenat Diagn* 15:579–584, 1995.

Woo, V., Bridge, P. J. and Bamforth, J. S. Maternal uniparental heterodisomy for chromosome 16: case report. *Am J Med Genet* 70:387–390, 1997.

Zori, R. T., Hendrickson, J., Woolven, S., Whidden, E. M., Gray, B. and Williams, C. A. Angelman syndrome: clinical profile. *J Child Neurol* 7:270–280, 1992.

Chapter 6

The Prader-(Labhardt)-Willi Syndrome

Prader-Willi syndrome (PWS) and Angelman syndrome (AS) are two neurobehavioral disorders resulting mainly from paternal deletion or maternal UPD of the 15q11-q13 segment in PWS (Ledbetter et al., 1981; Nicholls et al., 1989) and maternal deletion or paternal UPD in AS (Magenis et al., 1987; Kaplan et al., 1987; Malcolm et al., 1991). Thus, the loss of paternally expressed loci on proximal 15q causes PWS, whereas AS results from the loss of maternally expressed genes on the same chromosomal region.

The fact that gene expression at some loci is selectively paternal or maternal, and cannot be achieved by a substitution with a contribution from the other parent—as evidenced by UPD—uncovers the existence and importance of the secondary (epigenetic) changes of the parental DNA, which irreversibly modify the gene activity in somatic cells.

Such parent-of-origin specific molecular alterations are known as genomic imprinting and imply the superimposition of so-called epigenetic changes to the DNA (see Chapter 3). These epigenetic changes result from specific maternal or paternal molecular "epiglyphes" (as E.E., one of the authors, has coined), such as superimposed methylation at some sites. These genomic epiglyphes have been partially characterized in somatic tissues (i.e., blood cells) (van den Ouweland et al., 1995) and must be part of a complex, signal network aimed at modifying the DNA of some germinal and somatic nuclear domains. In a sense, the pathology of a syndrome such as PWS is the result of an exclusion of paternal expression such that certain genes are silent or inadequately expressed because they are absent (by deletion or UPD) or because there is inadequacy or misinstruction for expression. Why there is a need for parental-specific gene expression of certain genes and genomic regions still remains a mystery.

CLINICAL DESCRIPTION OF THE PWS

The condition was first described some 40 years ago (Prader et al., 1956). The clinical phenotype of PWS is characterized by neonatal hypotonia and developmental delay, followed by hyperphagia and major obesity, short stature, secondary hypogonadism, mild dysmorphism, small hands and feet, and mild to moderate mental retardation with learning disability. Consensus diagnostic criteria have been established and proven to be satisfactory for the clinicians (Holm et al., 1993). The frequency of the syndrome is on the order of 1 in 10,000 to 1 in 15,000 (Cassidy, 1984).

There is wide variability in severity and a discrepancy between the clinical presentation in infancy and that in childhood or adulthood (Mascari et al., 1992). Initially, in the neonatal period hypotonia, impaired sucking reflex and failure to thrive attract parental and clinical attention; feeding difficulties are thus a major problem. This sharply contrasts with the second phase of the syndrome, usually between ages 2 and 4, marked by impressive hyperphagia, leading to severe obesity combined with short stature and small hands and feet. The facial appearance is said to be characteristic, with almond-shaped eyes, narrow forehead, and bitemporal narrowing.

Aside from a ravenous appetite, several behavior problems are noted, such as frequent temper tantrums, obsessive-compulsive behavior, stubbornness. There is a variable degree of mental retardation, usually not severe; the average IQ is around 60, with a range of 20–80. Obesity is the main clinical and psychosocial problem. As a result, a tendency to develop type II diabetes mellitus is seen in adolescence, and cardiovascular problems and hypertension may occur (Webb et al., 1994). There is hypogenitalism and hypogonadism that is more obvious in males. The hypogonadism is of hypothalamic origin. There is also growth hormone deficiency and short stature; the average adult height is around 155 cm in males and 148 cm in females. Hypopigmentation is seen in a proportion of (mostly deletion) cases (Webb et al., 1994).

THE MOLECULAR CATEGORIES OF PWS

The following different categories of patients with PWS exist according to the findings of molecular and cytogenetic analyses of chromosome 15q11-q13 (Table 2 of Chapter 5 and Figure 1 in this chapter): (1) patients with deletions of about 4 Mb of the paternal chromosome 15; (2) patients with maternal UPD15; (3) patients with imprinting mutations with either small deletions or no identified deletions; (4) rare cases with chromosomal abnormalities that involve 15q11-q13.

MATERNAL UPD15 IN PWS

Frequency

Maternal UPD15 is responsible for 20–30% of cases of PWS; some clinical series, however, fall short of this proportion (Webb et al., 1994; Kokkonen et al., 1995). In

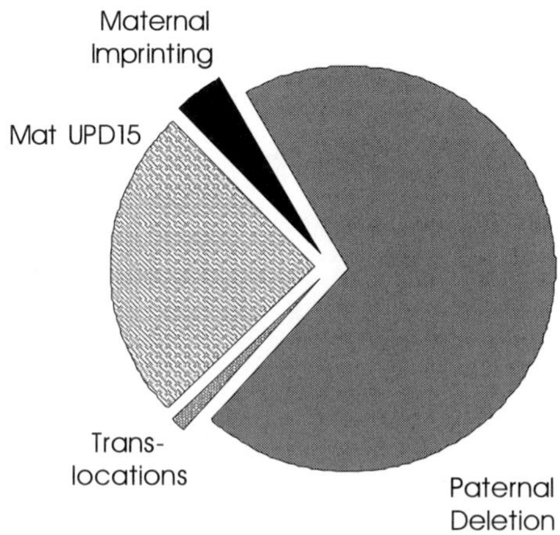

Figure 1 *Classes of PWS according to the molecular etiology.*

the first series of patients ascertained in Zurich, Switzerland, where this clinical syndrome was initially recognized (Prader et al., 1956), 21 of 29 clinically diagnosed cases had a deletion and seven others (or 25%) showed UPD15. In an Australian series, eight of 30 (27%) displayed maternal UPD15; of these, four were heterodisomic and one isodisomic (Woodage et al., 1994). In an early Japanese series (Hamabe et al., 1991), the incidence of UPD15 was much lower, which is not surprising, since several potential (nondeletion) cases were uninformative for the parental origin of chromosomes 15, whereas others were, perhaps erroneously, considered to be deleted by the cytogenetic means available at that time (Saitoh et al., 1994).

A study designed to document the maternal UPD15 frequency only considered cases with no cytogenetic deletion at the 550 band level (Mascari et al., 1992). Among 30 PWS patients without cytogenetically recognized deletion on 15q11-q13, there were eight with a deletion detectable by molecular techniques; 18 of the 22 remaining patients were subsequently found to have UPD15 on the basis of DNA polymorphism analysis (Mascari et al., 1992). Of the four exceptions with an intact biparental genome, three on further evaluation "...did not have the classic features of Prader-Willi syndrome." Considering that some 60% of PWS may have a cytogenetic deletion, the authors estimated that one-fifth of all PWS patients may have maternal UPD for chromosome 15. A British study found a lower proportion of mat UPD15, namely, 7 in 52 (adults only), i.e., 13% in this series (Webb et al., 1994). A similar proportion, four UPD15 cases among 27 (14%) clinically typical PWS, was observed in a Finnish study (Kokkonen et al., 1995).

In summary, the prevalence of maternal UPD15 in PWS varies from 14–30%, according to the series considered. These differences may well reflect ascertainment biases, the small size of most series (29 to 50 subjects), variable sophistication of the molecular techniques employed, and geographical differences in breeding patterns (the older the parental population, the greater the number of UPD15 versus deletion cases). It is now well established that the frequency of maternal UPD15 in PWS is approximately 25%.

The Origin of Maternal UPD15 in PWS (Heterodisomy versus Isodisomy)

Analysis of data from several series of PWS cases indicated that about 80% of the cases of maternal UPD15 were not only heterodisomic, but also had occurred at meiosis 1 (Me1) division, as shown by heterozygosity in the centromeric or pericentromeric regions (Robinson et al., 1991; Mascari et al., 1992; Webb et al., 1994; Kokkonen et al., 1995; Woodage et al., 1994). Of 18 informative cases in the three initial series, only four appeared isodisomic, and one would expect this proportion to decline when more polymorphic markers are used on the entire chromosome 15.

The above analysis has revealed a high frequency of Me1 nonsegregation of chromosome 15 in oogenesis. The study of polymorphic markers closer to the centromere of chromosome 15 further provided evidence for heterodisomy (Mutirangura et al., 1993). Heterozygosity for a polymorphic locus close to the centromere supports the conclusion of Me1 origin nondisjunction of the maternal chromosomes 15. The most decisive information on the meiotic stage of the nondisjunction error in maternal meiosis and the resulting UPD15 came from a collaborative study involving participants from the United States, Canada, Germany, and Switzerland (Robinson et al., 1996); this study added 66 unrelated cases to the others already published.

Thus, of a total of 124 cases of maternal UPD15, 88 were concluded to be of Me1 origin, 14 of Me2 origin, and 18 of somatic origin, while the origin of four cases remained inconclusive. If we consider only the informative cases, the maternal Me1 errors accounted for 73% of cases, whereas the maternal Me2 and somatic errors accounted for 12 and 15% of the cases, respectively.

On the premise of a PWS frequency of 1/15,000 and a 25% rate of maternal UPD15 as the etiology, an estimated population frequency of 1/60,000 was calculated for this specific maternal disomy. As expected, the maternal age increased (34.9 versus 28 years in the general population) in the cases of Me1 errors, leading to maternal UPD15 (Robinson et al., 1998) (see also Chapter 4).

TYPICAL PATERNAL 15q11-q13 DELETION

The deletion in 15q11-q13 found in about 70% of patients with PWS was initially studied with Southern blot analysis after restriction endonuclease digestion of DNA and hybridization with several probes that map on chromosome 15. Analysis of the intensity of hybridization revealed the copy number of these DNA fragments. Many probes were used over the years such as in order from the 15cen to 15qter: D15S18,

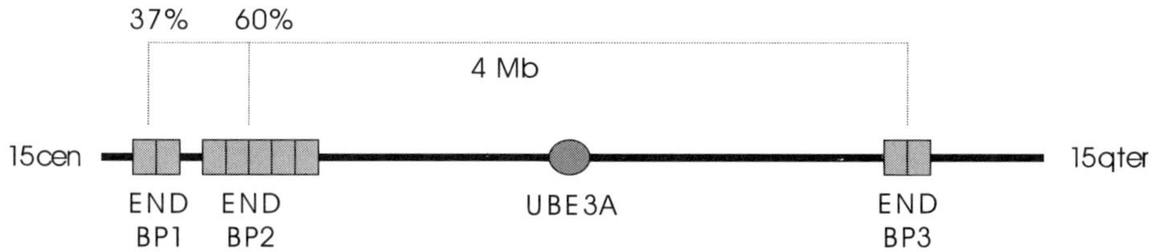

Figure 2 *Schematic representation of the 15q11-q13 region that is prone to deletions in PWS and AS. The repeats that predispose to unequal homologous recombination are shown as gray boxes. The UBE3A gene mutated in the AS (see Chapter 7) is shown as a filled circle. Approximately 60% of the deletions occur between BP2 and BP3 repeats and 37% between B1 and B2 repeats. The estimated distance between the proximal and distal repeats is 4 Mb (figure modified from Ji et al., 2000).*

D15S9, D15S13, D15S11, D15S63, SNRPN, D15S10, GABR3, GABR5, and D15S24 (Donlon et al., 1986; Tantravahi et al., 1989; Nicholls et al., 1989). Most of the PWS patients showed a deletion that extended for approximately four megabases that encompasses most of these markers (Figure 2). It was hypothesized early on that this might reflect the existence of specific hot spots of meiotic recombination and unequal crossing-over, leading to deletions; these sites could also represent regions prone to accidental breakage. The same region of deletion was found responsible for both PWS and AS; the deletion is paternal in origin in PWS and maternal in origin in AS (Kuwano et al., 1992; Carrozzo et al., 1997; Smith et al., 1992; Kokkonen et al., 1995; Robinson et al., 1991; Hamabe et al., 1991; Woodage et al., 1994; Buiting et al., 1993). A compilation of the results of these studies reveals that 80 of 94 patients with documented deletions appear to have the same set of DNA markers deleted and therefore the same size of deletion. The remaining cases had a larger or smaller deletion defect. The deleted area was sized to be 4 Mb by the construction of a YAC (yeast artificial chromosome) config (Mutirangura et al., 1993). Approximately 70–75% of both PWS and AS cases had this deletion (Glenn et al., 1997). There is one common breakpoint region in the distal end of the deletion; in contrast, there are two common centromeric breakpoints (proximal and distal to D15S18) that define two classes of deletions, class I and II (Christian et al., 1995; Knoll et al., 1990; Kuwano et al., 1992).

It was therefore tempting to determine why or how a rather consistent sequence length of chromosome material could accidentally be deleted from the whole chromosome as would, to use a comparison, a chocolate square easily be split off from the whole plate! As a result of this search, “hot spots” prone or predisposed to repeated breakage accidents were characterized by studying the breakpoint regions in YAC clones (Kuwano et al., 1992), the identification of restriction fragments positive for the same genomic DNA probes (Amos-Landgraf et al., 1999), or by FISH analyses in which more than one hybridizing signal was detected after hybridization with the same probe (Carrozzo et al., 1997).

The mechanisms of maternal and paternal deletions do not seem to be different; both interchromosomal and intrachromosomal rearrangements have been observed in AS and PWS (Carrozzo et al., 1997; Robinson et al., 1998). Remarkably, there are at least 10 copies of a complex DNA sequence termed the END repeat (or END duplicon) that maps to the three common deletion breakpoint regions on chromosome 15 (Figure 2) (Buiting et al., 1992, 1998; Amos-Landgraf et al., 1999; Ji et al., 1999). The END repeats are mostly composed of duplications of a large gene, HERC2, recessive mutations of which cause a neurological and developmental syndrome in mice (Ji et al., 1999). A considerable number of multiple rearrangements within the END repeat have also occurred during evolution (Amos-Landgraf et al., 1999; Ji et al., 1999, 2000). The simplest model to account for the AS and PWS deletions is via homologous misalignment and meiotic recombination between different END repeat copies in proximal and distal 15q11-q13 breakpoints (Amos-Landgraf et al., 1999). These duplicated areas that predispose to unequal "homologous" recombination have also been described independently as breakpoint BP1 to BP3 (Christian et al., 1999). BP1 and BP2 correspond to the first and second proximal END repeats, respectively, whereas BP3 corresponds to the distal END repeat.

In the same 15q region, the breakpoints of rearrangements other than deletions are also similar to those of the deletions. A study using YAC clones within 15q11-q13 as probes for FISH analysis localized the breakpoint between D15S144 and D15S118 in the nine cases with the largest-sized invdup(15), which is also called a class III abnormality (Wandstrat et al., 1998). In these patients, a 100–200 kb region was found to contain the breakpoints near the distal boundary of the PWS/AS chromosomal region.

The molecular mechanism of three patients with *de novo* dup(15) was studied and was particularly supportive of the mechanisms of deletions in the PWS/AS region. The patients, two males and one female (ages 3–21 years), had autistic behavior, hypotonia, and variable degrees of mental retardation. The extent, orientation, and parental origin of the duplications were assessed by fluorescent *in situ* hybridization, microsatellite analyses, and methylation status at D15S63. Two patients had large direct duplications of 15q11-q13 [dir dup(15)(q11-q13)] that extended through the entire AS/PWS chromosomal region. Their proximal and distal breaks were comparable to those found in the common AS/PWS deletions (Repetto et al., 1998). This suggests that duplications and deletions may be the reciprocal product of an unequal recombination event. These two duplications were maternally derived; in one patient, the duplication originated from two different maternal chromosomes, while in the other patient, it arose from the same maternal chromosome. The third patient had a much smaller duplication and parental origin could not be determined.

IMPRINTING MUTATIONS AND SMALLEST REGION OF OVERLAP (PWS-SRO) IN PAT 15 DELETIONS

A minority of PWS patients (<5%) probably have mutations that affect the mechanisms involved in the resetting of the parental imprint during gametogenesis. These patients do not have either large deletions in the paternally derived PWS/AS region, or maternal UPD15. Such patients, however, show a PWS pattern of methylation of 15q11-q13 loci after hybridization with specific probes (maternal only type imprint). Some of the patients have been found to have small deletions that span the so-called imprinting center (IC). Actually, the region important for imprinting has been defined as the smallest region of overlap (PWS-SRO) among the deletions in the rare PWS patients. The PWS-SRO has been defined as a region of only 4 kb by the analysis of seven microdeletions, the size of which ranges from 6–200 kb (Buiting et al., 1995). This region includes exon 1 and the CpG island promoter of the SNRPN gene (Nicholls et al., 1998). One of these deletions was silently transmitted from the proband's grandmother to a son who fathered a PWS child, or from the proband's grandfather through a daughter who gave birth to an AS offspring (Buiting et al., 1995). The IC molecular lesion appears to result in a failure to reset the imprint of chromosome 15 in the germ cells in accordance with the parental sex-of-origin. The AS-SRO, which is the smallest region of overlap in the imprinting center mutations associated with AS, is about 2 kb and is located 35 kb upstream of the SNRPN promoter (Nicholls et al., 1998) (Figure 3). The cases that contributed to the PWS-SRO were reported in (Robinson et al., 1991; Buiting et al., 1993; Glenn et al., 1993, 1996; Nicholls et al., 1998; Reis et al., 1994; Sutcliffe et al., 1994).

When a maternally derived imprinting mutation is transmitted from a male, the maternal "epiglyphe" or epigenotype cannot be reestablished in the germ line to become the normal paternal epigenotype. Therefore, the mutation blocks the ability to switch the maternal to paternal imprint in the male germ line and that results in the transmission of a maternal imprint to half the gametes. These offspring inherit the

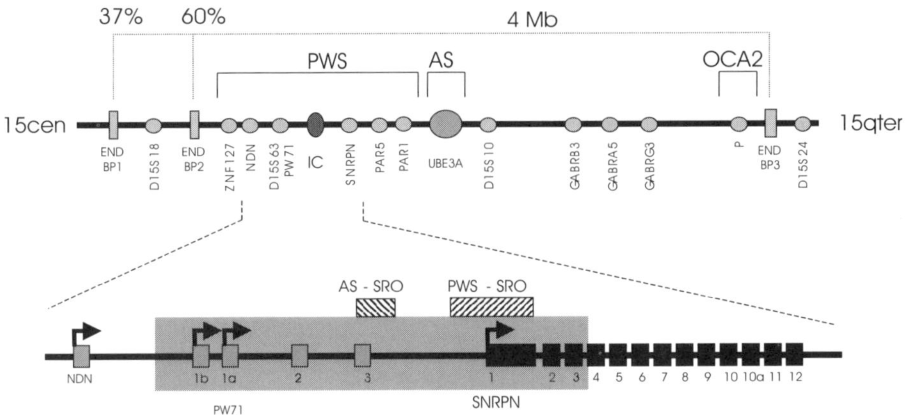

Figure 3 *Schematic representation and partial gene content of the 15q11-q13 chromosomal region involved in PWS and AS. Top: Different loci are shown as small circles. The area of the imprinting center (IC) is shown as a vertical oval on the left and the UBE3A gene as a large circle in the middle. Bottom: Enlargement of the area of the imprinting center (IC; background). PWS-SRO and AS-SRO are the minimum regions of overlap from imprinting mutations with micro-deletions in PWS and AS, respectively. The exons of the SNRPN gene are numbered and arrows indicate the transcription initiation sites.*

abnormal maternal-type epigenotype from the *father* and another maternal epigenotype from the *mother*. They are therefore homozygous for the maternal epigenotype and have the phenotype of PWS (Nicholls et al., 1998).

A particularly instructive case was one in which several relatives with AS belonged to a family where a small deletion encompassed D15S10, D15S13, and GABR3. When passed through the paternal line, the deletion did not cause PWS, whereas it caused AS when maternally transmitted. Thus, since the AS cluster was distal to that of PWS, the proximal AS breakpoint in this family indicated that there were two different areas that, if deleted, cause one and not the other syndrome (Saitoh et al., 1992). These areas are schematically shown in Figure 3.

THE IMPRINTING BOX

The studies of DNA sequences that are important in the imprinting mechanisms could be primarily done in animal models and, in particular, in mice. Transgenic mice have been generated that contained a mini-transgene composed of the sequences of AS-SRO and PWS-SRO. The exact DNA sequences were 200 bp of the promoter/exon 1 of mouse SNRPN and 1 kb of the human AS-SRO, which is normally located approximately 35 kb upstream of the SNPRN promoter (Shemer et al., 2000). This transgene amazingly confers imprinting in the mouse as judged by differential methylation, parent-of-origin specific transcription, and asynchronous replication. This 1.2-kb minitransgene therefore carries all the *cis* elements essential for establishing and maintaining differential methylation of the Snrpn promoter. A critical HhaI site was methylated in oocytes and unmethylated in sperm; in addition, the 1.2-kb minitransgene protects the maternal allele from demethylation in preimplantation embryos. Finally, the paternal allele is protected from *de novo* methylation at the postimplantation stage. The investigators concluded that this minitransgene sequence constitutes an imprinting box. A model for the regional control of PWS/AS imprinting has been developed and is shown in Figure 4 (Shemer et al., 2000).

CHROMOSOME ABNORMALITIES OR BALANCED TRANSLOCATIONS IN PWS

A series of chromosome abnormalities including translocations have been reported in rare patients with PWS. Less that 2% of PWS patients harbor various chromosome abnormalities other than the common interstitial deletions. Table 1 lists the chromosomal abnormalities found among clinically typical PWS cases.

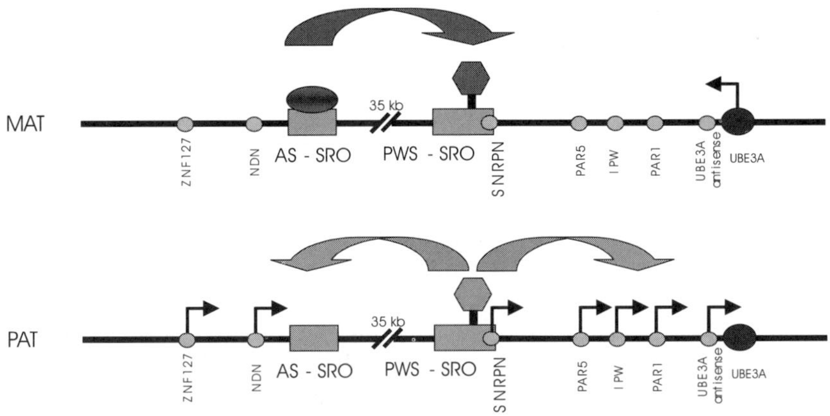

Figure 4 *Regional control of the PWS/AS imprinting box (modified from Shemer et al., 2000). Interaction of gamete-specific imprinting factors (large oval, upper left) with the AS-SRO on the maternal allele results in methylation of the SNRPN promoter in the PWS-SRO (top dark hexagon), which is consequently silenced. The SNRPN promoter of the paternal allele remains unmethylated (bottom gray hexagon) and active. The active state of the paternal allele is spread (mechanism unknown) and the genes nearby are in a similar active state. This includes the UBE3A antisense gene (Rougeulle et al., 1998), the transcript of which prevents the expression of the UBE3A from the paternal allele. The UBE3A gene is only therefore expressed from the maternal allele.*

TABLE 1 Chromosome Rearrangements Other than Deletions 15q11- q13 Observed in PWS

Chromosomal Rearrangements	References
46,XX,t(15;20)(q13;q13)	(Robinson et al., 1991)
46,XY,t(9;15)(p24.3;q13)	(Robinson et al., 1991)
45,X,del(15)(pter → q11::q12 → qter)/46,X,ibid+mar	(Hamabe et al., 1991)
45,XX,t(15;15)(qter → p11.1::q12 → qter)	(Hamabe et al., 1991)
45,XY,t(15;15)(qter → p11.1::q12 → qter)	(Hamabe et al., 1991)
47,del(15)(pter → q11::q12 → qter+invdup(pter → q11::q11 → pter)	(Hamabe et al., 1991)
46,t(8;18)(p24.1;q23)	(Mascari et al., 1992)
46/47+mar	
46inv(15)(p11;q13)	(Webb, 1994)
46t(Y;15)(q12;q11.2)	(Webb et al., 1994)
45,XY,-9,- 15,+t(9;15)(q34;q13)	(Woodage et al., 1994)
45,XX,t(14q15q)der mat.	(Woodage et al., 1994)
45,XY,t(13q15q)der mat.	(Woodage et al., 1994)
46,XY/47,XY,+del(15)(pter → q13)	(Woodage et al., 1994)

GENES AND LOCI WITHIN THE PWS CRITICAL REGION

Unlike the AS in which mutations in a single gene UBE3A were identified, there is no evidence for a single mutation that causes the PWS phenotype. It is therefore assumed that PWS results from the coordinated or serial alteration of the expression of several loci localized within the deleted region of 15q11-q13 (Wevrick et al., 1994). The other suggestive evidence against a single-gene involvement in PWS is (unlike AS) the absence of cases that show with no deletion, no UPD15, and no methylation defect.

Several genes have been cloned and characterized in the PWS/AS deleted region of chromosome 15 and are hemizygously expressed from the paternal chromosome only. These genes are SNRPN, PAR-1, PAR-5, IPW, ZNF127, MAGEL2, NCN, and PWCR1 (Figure 3). ZNF127 and PW71 at D15S63, are more proximal and, along with SNRPN, demonstrate methylated "epiglyphes," i.e., imprints specific to the parent-of-origin, whose analysis is useful in the molecular diagnosis of PWS and AS. At PW71 and ZNF127, methylation is specifically more dense on the maternally derived haplotype. The mechanisms by which these methylation-modified loci influence the expression of other parental genes is unknown, particularly since the active state of expression could be spread over many kilobases to other (*cis*) loci on the same chromosome. Some characteristics of these loci are briefly reviewed below.

ZNF127/FNZ127

These two human genes map proximally within the 15q11-q13 region, at approximately 1 megabase (Mb) centromeric to SNRPN (Jong et al., 1999a) (Figure 3). Both belong to an intron-less sequence that encodes them as overlapping antisense transcripts, an arrangement also conserved in the mouse (Jong et al., 1999b; Nicholls et al., 1993; Glenn et al., 1997).

The ZNF127 gene encodes a putative 507-amino acid protein with a RING (C3HC4) zinc finger motive and multiple C3H zinc finger motives. These motives predict ribonucleoprotein function for the ZNF127 polypeptide (Jong et al., 1999a). Ring zinc finger motives are involved in protein–protein interaction. ZNF127 (or Macorin 3; MKRN3; OMIM 603856) is maternally imprinted and, as for SNRPN, paternal expression is seen in control and AS cells while PWS cells lack such expression (Nicholls et al., 1993).

The DNA methylation imprint that distinguishes the inactive maternal from the functional paternal ZNF127 allele was first recognized by Driscoll et al. (1992) and has been proposed as a pilot diagnostic test. This differential methylation imprint is probably more pronounced in a human brain, as it is in the developing mouse brain (Barr et al., 1995), and the gene is also highly expressed in the testes of both species.

The paternal-only expression of ZNF127 has been shown in the brain, heart, and kidney, while it is biallelic in the liver and spleen (Jong et al., 1999a, 1999b). The above tissue-specific expression is characteristic of other mouse imprinted genes such as the IGF2 (DeChiara et al., 1991) and the Mas oncogene (Villar and Pedersen, 1994).

The antisense gene FNZ127 even more widely imprinted, and expressed from the paternal allele during development, is however not expressed in the brain and in the other tissues of high ZNF127 expression. The regulation for differential expression and imprinting of these two opposite transcripts is unknown (Glenn et al., 1997).

SNRPN: Small Nucleoribonucleoprotein Polypeptide N

The SNRPN (OMIM 182279) gene is located at 15q11-q13; this gene spans at least 25 kb of genomic DNA and is composed of more than 10 exons (Figure 3) (Dittrich et al., 1996; Buiting et al., 1997). Exon1 is part of a CpG island, which is preferentially methylated on the maternal nonexpressed allele and unmethylated on the expressed paternal allele By contrast, CpG sites at intron-7 are preferentially methylated on the expressed paternal allele, in both somatic tissues and male germ cells (Glenn et al., 1993).

SNRPN is a member of a gene family that encodes proteins involved in pre-m-RNA splicing. It is an abundantly expressed gene, with a 1.3-kb transcript, and high levels of expression in the brain, heart, striated muscle, kidney, pancreas, lung, and liver (Sutcliffe et al., 1994; Luhrmann et al., 1990; Ozcelik et al., 1992).

In the mouse, Snrpn maps in a region of chromosome 7 (Leff et al., 1992) that is syntenic to the PWS human critical region at 15q11-q13 (Nicholls et al., 1993; Chaillet et al., 1991).

Like other genes in the PWS area of humans, the mouse Snrpn is expressed from the paternal allele, as shown by RNAse protection assays (Leff et al., 1992), and mice having maternal UPD7 for the relevant area do not express the Snrpn gene and die shortly after birth (Cattanach et al., 1992). Additional mouse loci encode related splicing factors to Snrpn such as SmB; the latter, however, is no substitute for Snrpn (Glenn et al., 1997). Studies of the Snrpn knock-out mouse showed, however, that the neuronal-specific alternative splicing of the RNAs encoding several different classes of protein proceeds normally (Huntriss et al., 1994).

Human SNRPN is also paternally active and maternally imprinted as shown by RT- PCR assays on RNA of lymphoblasts and fibroblasts of normal individuals (who showed biparental SNRPN alleles), of AS patients (who display presence of paternal SNRPN alleles only), or of PWS patients (with presence of maternal SNRPN alleles, only). Lack of expression of this gene in the PWS patients proved that one paternal allele was at least needed for proper activity (Glenn et al., 1996; Sutcliffe et al., 1994; Nakao et al., 1994). More than one alternatively spliced exon has been identified (Schmauss et al., 1992; Glenn et al., 1996). It has been hypothesized that imprint switching on chromosome 15 may involve alternative transcripts of the SNRPN gene (Dittrich et al., 1996).

PAR5 and PAR1

The two genes PAR5 (OMIM 600162) and PAR1 (OMIM 600161) (for Prader-Willi/Angelman region) are also expressed from the paternal allele only, as judged from cultured human fibroblasts and lymphoblasts (Sutcliffe et al., 1994). With

SNRPN and IPW genes, they define an imprinted transcriptional domain of some 200 kb. PAR5 detects a 12-kb transcript, predominant in skeletal muscle, and PAR1 also detects a large transcript mostly in the adult brain; both genes, however, are ubiquitously expressed.

IPW

The IPW (for imprinted Prader-Willi gene; OMIM 601491) is transcribed to an RNA, which is spliced and polyadenylated and potentially encodes a putative polypeptide of only 45 amino acids (Wevrick et al., 1994). Alternatively, this transcript could function as an RNA, similar to H19 and XIST (Erdmann et al., 1999). A mouse sequence, in part homologous to IPW, has been found on mouse chromosome 7, which contains no long open reading frame, is alternatively spliced, and includes multiple copies of a 147-bp repeat, interrupted by the insertion of an intracisternal A particle sequence, predominantly expressed in the brain (Wevrick and Francke, 1997). IPW is located in between PAR5 and PAR1, some 250 kb distal to the SNRPN gene (Glenn et al., 1997).

NDN, Necdin

The NDN gene (OMIM 602117) maps between the ZNF127 and AW-SRO and represents one of the paternally expressed alleles of the PWS deletion region (MacDonald and Wevrick, 1997; Jay et al., 1997). In the mouse, the Ndn is located in the corresponding homologous area of chromosome 7, near Snrpn, and seems to govern the permanent arrest of cell growth of postmitotic neurons of the developing nervous system (MacDonald and Wevrick, 1997). The use of an intragenic polymorphism determined that in normal human fibroblasts, NDN expression occurs from the paternal allele. The NDN displays several characteristics of an imprinted locus, including allelic DNA methylation and asynchronous DNA replication. A complete lack of NDN expression in PWS brain and fibroblasts further indicates that the gene is expressed exclusively from the paternal allele.

MAGEL2; MAGE-like 2 Gene

Another gene named MAGEL2 (also known as NDNL1; OMIM 605283) with 51% amino acid sequence similarity to NDN and located 41 kb distal to it in the PWS deletion region has been characterized. It is expressed predominantly in brain and in several fetal tissues. MAGEL2 is imprinted with monoallelic expression in the control normal brain and paternal-only expression in the central nervous system, as demonstrated by its lack of expression in the brain from a PWS-affected individual. The orthologous mouse gene (Magel2) is located within 150 kb of Ndn, is imprinted with paternal-only expression, and is expressed predominantly in late developmental stages and the adult brain. Both the mouse and human genes are intronless and contain a CpG island in the 5′ flanking region (Lee et al., 2000; Boccaccio et al., 1999).

PWCR1

A gene named PWCR1 (Prader-Willi critical region 1) was also recently described; it is an imprinted gene within the PWS/AS region and its mouse orthologue, Pwcr1, maps to the syntenic region on mouse chromosome 7 (de Los et al., 2000). These genes are expressed only from the paternal allele. They are also intronless and do not appear to encode a protein product. High human/mouse sequence similarity (87% identity) is limited to a 99-bp region called "HMCR" (for "human-mouse conserved region"). This sequence has features of a C/D box small nucleolar RNA (snoRNA). Located in nucleoli, snoRNAs serve as methylation guidance RNAs in the modification of ribosomal RNA and other small nuclear RNAs. In addition to the nonpolyadenylated small RNAs, larger polyadenylated PWCR1 transcripts are found in most human tissues. Genomic sequence analysis revealed the presence of multiple copies of PWCR1 and Pwcr1 that are organized within local tandem-repeat clusters (de Los et al., 2000).

COMMENTS ON THE METHYLATION STATUS OF CHROMOSOME 15 LOCI IN PWS

Parental origin-specific methylation has been documented for loci such as ZNF127 (Driscoll et al., 1992), PW71 (Gillessen-Kaesbach et al., 1995; van den Ouweland et al., 1995; Dittrich et al., 1993), and SNRPN (Glenn et al., 1993, 1996). In AS (with the exception of the UBE3A mutation cases), the methylation pattern instead of being biparental becomes exclusively paternal; in contrast, there is a maternal-only type of methylation status in PWS (whether caused by deletion, UPD15, or imprinting center mutations). Therefore, the analysis of methylation is able to detect practically all cases of PWS.

Locus ZNF127

The methyl-sensitive enzymes HpaII and HhaI identify distinct differences in DNA methylation of the parental alleles at the D15S9/ZNF127 locus (Driscoll et al., 1992).

Locus PW71

The PW71 or D15S63 locus was first identified as a microdisection clone, which maps 130 kb centromeric to the SNRPN gene within the PWS region (Buiting et al., 1993). Subsequently, it was found that an HpaII site was methylated on the maternally derived and unmethylated on the paternally derived chromosome 15 (Dittrich et al., 1993). Southern blot hybridization of Hind III and HpaII digested DNA from the blood of normal subjects revealed a 6.6-kb fragment derived from the maternal chromosome and a 4.7-kb fragment derived from the paternal chromosome. DNA from blood cells of patients with PWS lacks the 4.7-kb fragment, while patients with AS lack the 6.6-kb fragment. This difference quickly became a useful and efficient diagnostic test for PWS. In the first diagnostic study, 29 of 58 patients considered to have the clinical diagnosis of PWS were tested for the lack of a paternal methylation status (Gillessen-Kaesbach et al., 1995). Twenty-eight of 29, indeed, lacked the paternal PW71 band. The 29th case had a normal methylation pattern and, on further clinical evaluation, was diagnosed to have "Ohdo-like blepharophimosis syndrome." Other studies the same year confirmed these results (van den Ouweland et al., 1995).

Locus SNRPN

This locus is also differentially methylated according to the parental origin of the allele. For example, when Hind III/HhaI digests were hybridized with an SNRPN exon 2–8 cDNA probe in PWS, AS, and normal blood cell DNA, consistent 6.0 and 5.0 kb bands were seen in all three types of samples, whereas bands of 3.5 and 2.5 kb were also present in normal and PWS patients but not in AS patients, indicating an HhaI site methylated on the maternal chromosome (Glenn et al., 1993, 1996; Sutcliffe et al., 1994). The results further indicated that SNRPN was expressed exclusively from the paternal chromosome and thus inactive in PWS patients. Further studies identified additional parent-of-origin dependent DNA methylation imprints in exon 1 that contains a CpG island. The probe of exon 1 has now practical validity as a diagnostic for the methylation test (Kubota et al., 1996). The comparison of methylation tests using either PW71 or exon 1 of SNRPN probes in 495 cases did not reveal any differences between the two. The authors of this large comparative study stressed that DNA methylation analyses were the "...most efficient single test for ruling out PWS diagnosis" and advised one "...to follow up abnormal methylation results by FISH and microsatellite-analyses to determine the precise etiology...for genetic counseling and recurrence risk assessment" (Kubota et al., 1996).

PHENOTYPIC DIFFERENCES BETWEEN 15q DELETIONS AND MATERNAL UPD15 IN PWS

A large series of 167 cases was comprised of 116 deletions and 51 nondeletion cases (Gillessen-Kaesbach et al., 1995). The classification relied on the PWS methylation test, followed by quantitative Southern blot analysis for SNRPN and search for UPD15. Essentially no substantial phenotypic differences were found between the two groups. Of note were the following features (Table 2). The birthweight and length of PWS patients were significantly lower when compared with healthy newborn. An increase in maternal age in nondeletion cases was confirmed as expected, since both trisomy rescue and gamete complementation follows maternal 15 meiotic nondisjunction. Table 2 shows the phenotypic differences of PWS patients with UPD15 versus 15q deletion.

TABLE 2 Phenotypic Differences of PWS Patients with Maternal UPD15 as Compared to Those with 15q Deletion

Facies less typical
Hypopigmentation less frequent
Birthweight and length closer to normal
Higher pain threshold
Skin picking less frequent
Aptitude with puzzles not striking
Possible excess of males
More benign phenotype in females
Increased maternal age

The occurrence of hypopigmentation was different within each molecular class; it was present in about 50% of deletion cases (48 of 93) and only 25% (9 of 39) nondeletion cases (Gillessen-Kaesbach et al., 1995).

A later study that used more stringent laboratory criteria of diagnosis of paternal deletions or maternal UPD15 included 54 PWS patients (37 with deletion and 17 with UPD15) (Cassidy et al., 1997). Again, there were no major phenotypic differences in PWS patients with deletion or maternal UPD15 except for two: (i) hypopigmentation was present in 50% of cases of deletion compared to 15% with UPD15; (ii) the other feature (which may reflect the unique clinical experience of the senior author of this study, Dr. S. Cassidy) relates to "characteristic" facial appearance that was less frequent in UPD15 (10 of 17 cases) than in deletion cases (33 of 36) (Figure 5). Furthermore, skin picking, increased threshold for pain, and skill with jigsaw puzzle were more commonly represented in the deletion cases and joint abnormalities also seemed more frequent but not quite statistically

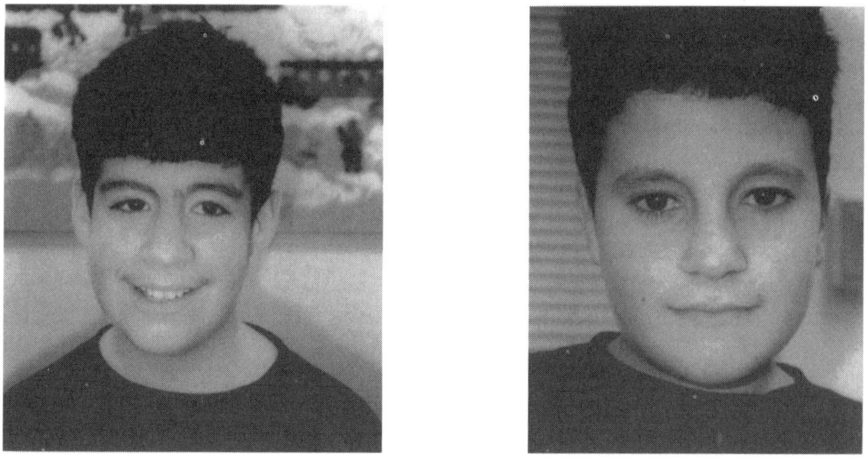

Figure 5 *Facial characteristics of two boys with PWS due to maternal UPD15 (from Cassidy et al., 1997. Reproduced with permission).*

significant in deletions. Maternal age was advanced among the patients with UPD15, 34.4 compared to 26.8 years in deletions.

The bias in sex-ratio found in another study (Mitchell et al., 1996), i.e., 68% males with maternal UPD15, was not borne out in other studies.

The differences in phenotypes of patients with different genomic alterations might be the result of different factors such as: (i) preservation of the biallelic expression of nonimprinted genes in UPD (in regions of heterodisomy) versus hemizygous expression in deletion; (ii) double dose of the active allele of an imprinted locus in UPD, as compared to the hemizygous effect in deletion; (iii) preclusion of necessary interactions between homologous imprinted domains in the case of deletions (LaSalle and Lalande, 1995, 1996).

In a third study of phenotypic differences among PWS patients with deletion (43 cases) or UPD15 (79 cases) (Mitchell et al., 1996), significant differences were observed between female PWS with deletion versus UPD15. Female UPD15 patients were found to be less severely affected than female deletion patients in terms of length of gavage feeding and the later onset of hyperphagia.

In general, the need of gavage feeding of either deletion or UPD15 was definitely shorter for female than male babies (Cassidy et al., 1997). The age at diagnosis was also delayed in UPD15 (9 years versus 4 in females; 9 versus 2 in males) (Gunay-Aygun et al., 1997).

Study of sleep anomalies, including sleep-onset REM periods and excessive daytime sleepiness, were not consistently associated with a particular genotype in PWS, but were possibly more common in patients with paternal deletions as documented from only a small cohort of patients (Vgontzas et al., 1996).

LABORATORY TESTS IN PWS

Recommendations for these tests have been published by the American Society of Human Genetics/American College of Medical Genetics Test and Technology Transfer Committee (1996) (Anonymous 1996) and can also be found in the excellent Web article of Cassidy and Schwartz in GeneClinics (http://www.geneclinics.org/profiles/praderwilli).

The laboratory tests are the following:

(i) *Parent-specific methylation status*. In over 99% of cases of PWS, there is abnormal, maternal-only pattern methylation status at the SNRPN or PW71 loci on 15q11-q13. There is only methylated DNA instead of an equal mixture of methylated and unmethylated CpG SNRPN island. This abnormal methylation could be determined using Southern blot hybridization with appropriate probes after digestion of DNA with methylation-sensitive enzymes such as HpaII or HhaI (Glenn et al., 1996).

Alternatively, the methylation status could be determined by PCR using methylation-sensitive oligonucleotide primers (Kubota et al., 1997; Zeschnigk et al., 1997). The methylation test is also useful in the detection of AS cases. The PWS patients with abnormal, maternal-only methylation pattern could have deletions in paternal 15q11-q13, maternal UPD15, or an imprinting mutation. Further analysis to determine the exact cause includes:

(ii) *DNA polymorphisms to detect UPD15.* Informative microsatellite polymorphisms of chromosome 15 should be used in the DNAs of the patients and their parents to establish the origin and inheritance patterns of the polymorphic alleles.

(iii) *A careful high-resolution karyotype.* This should be performed to detect visible deletions or to rule out structural chromosome rearrangements. High-resolution chromosome analyses at the 650-band level are necessary. Parental karyotypes need also to be performed in cases of chromosomal rearrangement breaking within 15q11-q13.

(iv) *DNA analysis by FISH.* Probes within the commonly deleted region of 3–4 Mb could be used. In addition, probes including exon 1 of SNRPN are necessary to detect smaller deletions in the imprinting center. Alternatively, the deletion analysis could be performed using the inheritance of DNA polymorphic alleles in 15q11-q13. The introduction of FISH analysis contributed to the accuracy of the diagnosis of both PWS and AS. In a study of 19 cases, for example (14 PWS and 5 AS), four of the diagnosed deletions by high-resolution chromosomal analysis could not be confirmed by FISH. Furthermore, FISH identified a deletion not detected by HRCB. False-positive deletions by cytogenetic analysis might be attributed to a "satellite" or band polymorphism (Delach et al., 1994). Additional similar results were reported in (Bettio et al., 1995; Toth-Fejel et al., 1995). FISH analysis also detected rare cases of mosaicism for the deletion in 15q11-q13 (Mowery-Rushton et al., 1996).

GENETIC COUNSELING IN PWS

The genetic counseling of parents of a child with PWS depends on the cause of PWS in this particular family. This could be diagnosed after the battery of laboratory tests listed above. In about 99% of patients, the molecular mechanisms of PWS could be identified and therefore fetal testing could be offered, and risks to family members could be predicted. The categories of known molecular mechanisms include:

(i) De novo *deletion of paternal 15q11-q13.* This is the most common etiology of PWS and accounts for approximately 70% of cases. The recurrence risk (RR) is less than 1%. Fetal diagnosis could be offered by DNA polymorphisms in the deleted region, or by FISH analysis using probes from the deleted region. Prenatal fetal diagnosis is usually not recommended but could be offered for reassurance.

(ii) *Maternal UPD15.* This is the second most common cause of PWS and accounts for approximately 25% of cases. The RR for future pregnancies in these cases is negligible (less than 1%). Karyotypes of patients and parents should be performed to exclude a paternally inherited Robertsonian translocation. Theoretically for reassurance, UPD15 analysis using appropriate chromosome 15 polymorphisms could be used in fetal diagnosis.

(iii) *Imprinting mutations*. There are either microdeletions of the imprinting center (IC) or an abnormal, maternal-only DNA methylation pattern (Buiting et al., 1995; Ohta et al., 1999; Reis et al., 1994; Sutcliffe et al., 1994; Saitoh et al., 1996) and account for less than 5% of PWS cases. Recurrence of PWS in the siblings of such cases has been observed (Buiting et al., 1994; Nicholls, 1994) and the theoretical RR could be as high as 50% (if an IC deletion is present in the asymptomatic father, e.g.). First-degree relatives of these fathers may also have a risk for children or grandchildren with PWS. Parents who have one PWS child caused by an imprinting center mutation could be offered fetal diagnosis. The testing could either be aimed at the detection of a deletion or the methylation status of the critical chromosomal region.

(iv) *Structural chromosomal rearrangements*. These are rare and account for less than 1% of PWS. A careful and detailed cytogenetic analysis in the parents is required to establish the nature and inheritance of the chromosomal rearrangement. In the *de novo* cases, the RR is negligible, but in the inherited cases, the RR is considerable and up to 25%. A cytogenetic analysis in fetal cells could be offered in subsequent pregnancies of parents with such a PWS child.

Unlike the case in AS, there are no mutations in specific genes associated with PWS. Furthermore, there are very few (if any) PWS patients without a detectable 15q11-q13 deletion or UPD15 or methylation abnormality.

REFERENCES

Anonymous. Diagnostic testing for Prader-Willi and Angleman syndromes: Report of the ASHG/ACMG Test and Technology Transfer Committee. *Am J Hum Genet* 58:1085–1088, 1996.

Amos-Landgraf, J. M., Ji, Y., Gottlieb, W., et al. Chromosome breakage in the Prader-Willi and Angelman syndromes involves recombination between large, transcribed repeats at proximal and distal breakpoints. *Am J Hum Genet* 65:370–386, 1999.

Barr, J. A., Jones, J., Glenister, P. H. and Cattanach, B. M. Ubiquitous expression and imprinting of Snrpn in the mouse. *Mamm Genome* 6:405–407, 1995.

Bettio, D., Rizzi, N., Giardino, D., et al. FISH analysis in Prader-Willi and Angelman syndrome patients. *Am J Med Genet* 56:224–228, 1995.

Boccaccio, I., Glatt-Deeley, H., Watrin, F., Roeckel, N., Lalande, M. and Muscatelli, F. The human MAGEL2 gene and its mouse homologue are paternally expressed and mapped to the Prader-Willi region. *Hum Mol Genet* 8:2497–2505, 1999.

Buiting, K., Dittrich, B., Endele, S. and Horsthemke, B. Identification of novel exons 3′ to the human SNRPN gene. *Genomics* 40:132–137, 1997.

Buiting, K., Dittrich, B., Gross, S., et al. Molecular definition of the Prader-Willi syndrome chromosome region and orientation of the SNRPN gene. *Hum Mol Genet* 2:1991–1994, 1993.

Buiting, K., Dittrich, B., Robinson, W. P., et al. Detection of aberrant DNA methylation in unique Prader-Willi syndrome patients and its diagnostic implications [published erratum appears in Hum Mol Genet 1994 Nov;3(11):2092]. *Hum Mol Genet* 3:893–895, 1994.

Buiting, K., Greger, V., Brownstein, B. H., et al. A putative gene family in 15q11-q13 and 16p11.2: possible implications for Prader-Willi and Angelman syndromes. *Proc Natl Acad Sci USA* 89:5457–5461, 1992.

Buiting, K., Gross, S., Ji, Y., Senger, G., Nicholls, R. D. and Horsthemke, B. Expressed copies of the MN7 (D15F37) gene family map close to the common deletion breakpoints in the Prader-Willi/Angelman syndromes. *Cytogenet Cell Genet* 81:247–253, 1998.

Buiting, K., Saitoh, S., Gross, S., et al. Inherited microdeletions in the Angelman and Prader-Willi syndromes define an imprinting centre on human chromosome 15 [published erratum appears in Nat Genet 1995 Jun;10(2):249]. *Nat Genet* 9:395–400, 1995.

Carrozzo, R., Rossi, E., Christian, S. L., et al. Inter- and intrachromosomal rearrangements are both involved in the origin of 15q11-q13 deletions in Prader-Willi syndrome [letter]. *Am J Hum Genet* 61:228–231, 1997.

Cassidy, S. B. Prader-Willi syndrome. *Curr Probl Pediatr* 14:1–55, 1984.

Cassidy, S. B., Forsythe, M., Heeger, S., et al. Comparison of phenotype between patients with Prader-Willi syndrome due to deletion 15q and uniparental disomy 15. *Am J Med Genet* 68:433–440, 1997.

Cattanach, B. M., Barr, J. A., Evans, E. P., et al. A candidate mouse model for Prader-Willi syndrome which shows an absence of Snrpn expression. *Nat Genet* 2:270–274, 1992.

Chaillet, J. R., Knoll, J. H., Horsthemke, B. and Lalande, M. The syntenic relationship between the critical deletion region for the Prader- Willi/Angelman syndromes and proximal mouse chromosome 7. *Genomics* 11:773–776, 1991.

Christian, S. L., Fantes, J. A., Mewborn, S. K., Huang, B. and Ledbetter, D. H. Large genomic duplicons map to sites of instability in the Prader- Willi/Angelman syndrome chromosome region (15q11-q13). *Hum Mol Genet* 8:1025–1037, 1999.

Christian, S. L., Robinson, W. P., Huang, B., et al. Molecular characterization of two proximal deletion breakpoint regions in both Prader- Willi and Angelman syndrome patients. *Am J Hum Genet* 57:40–48, 1995.

de Los, S., Schweizer, J., Rees, C. A. and Francke, U. Small evolutionarily conserved RNA, resembling C/D box small nucleolar RNA, is transcribed from PWCR1, a novel imprinted gene in the Prader-Willi deletion region, which is highly expressed in brain [In Process Citation]. *Am J Hum Genet* 67:1067–1082, 2000.

DeChiara, T. M., Robertson, E. J. and Efstratiadis, A. Parental imprinting of the mouse insulin-like growth factor II gene. *Cell* 64:849–859, 1991.

Delach, J. A., Rosengren, S. S., Kaplan, L., Greenstein, R. M., Cassidy, S. B. and Benn, P. A. Comparison of high resolution chromosome banding and fluorescence in situ hybridization (FISH) for the laboratory evaluation of Prader-Willi syndrome and Angelman syndrome [see comments]. *Am J Med Genet* 52:85–91, 1994.

Dittrich, B., Buiting, K., Gross, S. and Horsthemke, B. Characterization of a methylation imprint in the Prader-Willi syndrome chromosome region. *Hum Mol Genet* 2:1995–1999, 1993.

Dittrich, B., Buiting, K., Korn, B., et al. Imprint switching on human chromosome 15 may involve alternative transcripts of the SNRPN gene [see comments]. *Nat Genet* 14:163–170, 1996.

Donlon, T. A., Lalande, M., Wyman, A., Bruns, G. and Latt, S. A. Isolation of molecular probes associated with the chromosome 15 instability in the Prader-Willi syndrome [published erratum appears in Proc Natl Acad Sci USA 1986 Sep;83(18):6964]. *Proc Natl Acad Sci USA* 83:4408–4412, 1986.

Driscoll, D. J., Waters, M. F., Williams, C. A., et al. A DNA methylation imprint, determined by the sex of the parent, distinguishes the Angelman and Prader-Willi syndromes. *Genomics* 13:917–924, 1992.

Erdmann, V. A., Szymanski, M., Hochberg, A., de Groot, N. and Barciszewski, J. Collection of mRNA-like non-coding RNAs. *Nucleic Acids Res* 27:192–195, 1999.

Gillessen-Kaesbach, G., Gross, S., Kaya-Westerloh, S., Passarge, E. and Horsthemke, B. DNA methylation based testing of 450 patients suspected of having Prader-Willi syndrome. *J Med Genet* 32:88–92, 1995.

Gillessen-Kaesbach, G., Robinson, W., Lohmann, D., Kaya-Westerloh, S., Passarge, E. and Horsthemke, B. Genotype-phenotype correlation in a series of 167 deletion and non-deletion patients with Prader-Willi syndrome. *Hum Genet* 96:638–643, 1995.

Glenn, C. C., Driscoll, D. J., Yang, T. P. and Nicholls, R. D. Genomic imprinting: potential function and mechanisms revealed by the Prader-Willi and Angelman syndromes. *Mol Hum Reprod* 3:321–332, 1997.

Glenn, C. C., Porter, K. A., Jong, M. T., Nicholls, R. D. and Driscoll, D. J. Functional imprinting and epigenetic modification of the human SNRPN gene. *Hum Mol Genet* 2:2001–2005, 1993.

Glenn, C. C., Saitoh, S., Jong, M. T., et al. Gene structure, DNA methylation, and imprinted expression of the human SNRPN gene. *Am J Hum Genet* 58:335–346, 1996.

Gunay-Aygun, M., Heeger, S., Schwartz, S. and Cassidy, S. B. Delayed diagnosis in patients with Prader-Willi syndrome due to maternal uniparental disomy 15. *Am J Med Genet* 71:106–110, 1997.

Hamabe, J., Fukushima, Y., Harada, N., et al. Molecular study of the Prader-Willi syndrome: deletion, RFLP, and phenotype analyses of 50 patients. *Am J Med Genet* 41:54–63, 1991.

Holm, V. A., Cassidy, S. B., Butler, M. G., et al. Prader-Willi syndrome: consensus diagnostic criteria. *Pediatrics* 91:398–402, 1993.

Huntriss, J. D., Barr, J. A., Horn, D. A., Williams, D. G. and Latchman, D. S. Mice lacking Snrpn expression show normal regulation of neuronal alternative splicing events. *Mol Biol Rep* 20:19–25, 1994.

Jay, P., Rougeulle, C., Massacrier, A., et al. The human necdin gene, NDN, is maternally imprinted and located in the Prader-Willi syndrome chromosomal region. *Nat Genet* 17:357–361, 1997.

Ji, Y., Eichler, E. E., Schwartz, S. and Nicholls, R. D. Structure of chromosomal duplicons and their role in mediating human genomic disorders. *Genome Res* 10:597–610, 2000.

Ji, Y., Rebert, N. A., Joslin, J. M., Higgins, M. J., Schultz, R. A. and Nicholls, R. D. Structure of the highly conserved HERC2 gene and of multiple partially duplicated paralogs in human. *Genome Res* 10:319–329, 2000.

Ji, Y., Walkowicz, M. J., Buiting, K., et al. The ancestral gene for transcribed, low-copy repeats in the Prader-Willi/Angelman region encodes a large protein implicated in protein trafficking, which is deficient in mice with neuromuscular and spermiogenic abnormalities. *Hum Mol Genet* 8:533–542, 1999.

Jong, M. T., Carey, A. H., Caldwell, K. A., et al. Imprinting of a RING zinc-finger encoding gene in the mouse chromosome region homologous to the Prader-Willi syndrome genetic region. *Hum Mol Genet* 8:795–803, 1999b.

Jong, M. T., Gray, T. A., Ji, Y., et al. A novel imprinted gene, encoding a RING zinc-finger protein, and overlapping antisense transcript in the Prader-Willi syndrome critical region. *Hum Mol Genet* 8:783–793, 1999a.

Kaplan, L. C., Wharton, R., Elias, E., Mandell, F., Donlon, T. and Latt, S. A. Clinical heterogeneity associated with deletions in the long arm of chromosome 15: report of 3 new cases and their possible genetic significance. *Am J Med Genet* 28:45–53, 1987.

Knoll, J. H., Nicholls, R. D., Magenis, R. E., et al. Angelman syndrome: three molecular classes identified with chromosome 15q11-q13-specific DNA markers. *Am J Hum Genet* 47:149–155, 1990.

Kokkonen, H., Kahkonen, M. and Leisti, J. A molecular and cytogenetic study in Finnish Prader-Willi patients. *Hum Genet* 95:568–571, 1995.

Kubota, T., Das, S., Christian, S. L., Baylin, S. B., Herman, J. G. and Ledbetter, D. H. Methylation-specific PCR simplifies imprinting analysis [letter]. *Nat Genet* 16:16–17, 1997.

Kubota, T., Sutcliffe, J. S., Aradhya, S., et al. Validation studies of SNRPN methylation as a diagnostic test for Prader-Willi syndrome. *Am J Med Genet* 66:77–80, 1996.

Kuwano, A., Mutirangura, A., Dittrich, B., et al. Molecular dissection of the Prader-Willi/Angelman syndrome region (15q11-q13) by YAC cloning and FISH analysis [published erratum appears in Hum Mol Genet 1992 Dec;1(9):784]. *Hum Mol Genet* 1:417–425, 1992.

LaSalle, J. M. and Lalande, M. Domain organization of allele-specific replication within the GABRB3 gene cluster requires a biparental 15q11-q13 contribution. *Nat Genet* 9:386–394, 1995.

LaSalle, J. M. and Lalande, M. Homologous association of oppositely imprinted chromosomal domains. *Science* 272:725–728, 1996.

Ledbetter, D. H., Riccardi, V. M., Airhart, S. D., Strobel, R. J., Keenan, B. S. and Crawford, J. D. Deletions of chromosome 15 as a cause of the Prader-Willi syndrome. *N Engl J Med* 304:325–329, 1981.

Lee, S., Kozlov, S., Hernandez, L., et al. Expression and imprinting of MAGEL2 suggest a role in Prader-Willi syndrome and the homologous murine imprinting phenotype [In Process Citation]. *Hum Mol Genet* 9:1813–1819, 2000.

Leff, S. E., Brannan, C. I., Reed, M. L., et al. Maternal imprinting of the mouse Snrpn gene and conserved linkage homology with the human Prader-Willi syndrome region. *Nat Genet* 2:259–264, 1992.

Luhrmann, R., Kastner, B. and Bach, M. Structure of spliceosomal snRNPs and their role in pre-mRNA splicing. *Biochim Biophys Acta* 1087:265–292, 1990.

MacDonald, H. R. and Wevrick, R. The necdin gene is deleted in Prader-Willi syndrome and is imprinted in human and mouse. *Hum Mol Genet* 6:1873–1878, 1997.

Magenis, R. E., Brown, M. G., Lacy, D. A., Budden, S. and LaFranchi, S. Is Angelman syndrome an alternate result of del(15)(q11-q13)? *Am J Med Genet* 28:829–838, 1987.

Malcolm, S., Clayton-Smith, J., Nichols, M., et al. Uniparental paternal disomy in Angelman's syndrome. *Lancet* 337:694–697, 1991.

Mascari, M. J., Gottlieb, W., Rogan, P. K., et al. The frequency of uniparental disomy in Prader-Willi syndrome. Implications for molecular diagnosis. *N Engl J Med* 326:1599–1607, 1992.

Mitchell, J., Schinzel, A., Langlois, S., et al. Comparison of phenotype in uniparental disomy and deletion Prader-Willi syndrome: sex specific differences. *Am J Med Genet* 65:133–136, 1996.

Mowery-Rushton, P. A., Hanchett, J. M., Zipf, W. B., Rogan, P. K. and Surti, U. Identification of mosaicism in Prader-Willi syndrome using fluorescent in situ hybridization. *Am J Med Genet* 66:403–412, 1996.

Mutirangura, A., Greenberg, F., Butler, M. G., et al. Multiplex PCR of three dinucleotide repeats in the Prader-Willi/Angelman critical region (15q11-q13): molecular diagnosis and mechanism of uniparental disomy. *Hum Mol Genet* 2:143–151, 1993.

Mutirangura, A., Jayakumar, A., Sutcliffe, J. S., et al. A complete YAC contig of the Prader-Willi/Angelman chromosome region (15q11-q13) and refined localization of the SNRPN gene. *Genomics* 18:546–552, 1993.

Nakao, M., Sutcliffe, J. S., Durtschi, B., Mutirangura, A., Ledbetter, D. H. and Beaudet, A. L. Imprinting analysis of three genes in the Prader- Willi/Angelman region: SNRPN, E6-associated protein, and PAR-2 (D15S225E). *Hum Mol Genet* 3:309–315, 1994.

Nicholls, R. D. New insights reveal complex mechanisms involved in genomic imprinting [editorial; comment]. *Am J Hum Genet* 54:733–740, 1994.

Nicholls, R. D., Gottlieb, W., Russell, L. B., Davda, M., Horsthemke, B. and Rinchik, E. M. Evaluation of potential models for imprinted and nonimprinted components of human chromosome 15q11-q13 syndromes by fine- structure homology mapping in the mouse. *Proc Natl Acad Sci USA* 90:2050–2054, 1993.

Nicholls, R. D., Knoll, J. H., Butler, M. G., Karam, S. and Lalande, M. Genetic imprinting suggested by maternal heterodisomy in nondeletion Prader-Willi syndrome. *Nature* 342:281–285, 1989.

Nicholls, R. D., Knoll, J. H., Glatt, K., et al. Restriction fragment length polymorphisms within proximal 15q and their use in molecular cytogenetics and the Prader-Willi syndrome. *Am J Med Genet* 33:66–77, 1989.

Nicholls, R. D., Saitoh, S. and Horsthemke, B. Imprinting in Prader-Willi and Angelman syndromes. *Trends Genet* 14:194–200, 1998.

Ohta, T., Buiting, K., Kokkonen, H., et al. Molecular mechanism of Angelman syndrome in two large families involves an imprinting mutation. *Am J Hum Genet* 64:385–396, 1999.

Ohta, T., Gray, T. A., Rogan, P. K., et al. Imprinting-mutation mechanisms in Prader-Willi syndrome. *Am J Hum Genet* 64:397–413, 1999.

Ozcelik, T., Leff, S., Robinson, W., et al. Small nuclear ribonucleoprotein polypeptide N (SNRPN), an expressed gene in the Prader-Willi syndrome critical region. *Nat Genet* 2:265–269, 1992.

Prader, A., Labhart, A. and Willi, H. Ein syndrom von adipositas, kleinwuchs, kryptorchismus, und oligophrenie nach myotoniertigen zustang in neugeborenatler. *Schweiz Med Wochenschr* 86:1260–1261, 1956.

Reis, A., Dittrich, B., Greger, V., et al. Imprinting mutations suggested by abnormal DNA methylation patterns in familial Angelman and Prader-Willi syndromes [see comments]. *Am J Hum Genet* 54:741–747, 1994.

Repetto, G. M., White, L. M., Bader, P. J., Johnson, D. and Knoll, J. H. Interstitial duplications of chromosome region 15q11-q13: clinical and molecular characterization. *Am J Med Genet* 79:82–89, 1998.

Robinson, W. P., Bottani, A., Xie, Y. G., et al. Molecular, cytogenetic, and clinical investigations of Prader-Willi syndrome patients. *Am J Hum Genet* 49:1219–1234, 1991.

Robinson, W. P., Dutly, F., Nicholls, R. D., et al. The mechanisms involved in formation of deletions and duplications of 15q11-q13. *J Med Genet* 35:130–136, 1998.

Robinson, W. P., Kuchinka, B. D., Bernasconi, F., et al. Maternal meiosis I non-disjunction of chromosome 15: dependence of the maternal age effect on level of recombination. *Hum Mol Genet* 7:1011–1019, 1998.

Robinson, W. P., Langlois, S., Schuffenhauer, S., et al. Cytogenetic and age-dependent risk factors associated with uniparental disomy 15. *Prenat Diagn* 16:837–844, 1996.

Rougeulle, C., Cardoso, C., Fontes, M., Colleaux, L. and Lalande, M. An imprinted antisense RNA overlaps UBE3A and a second maternally expressed transcript [letter]. *Nat Genet* 19:15–16, 1998.

Saitoh, S., Buiting, K., Rogan, P. K., et al. Minimal definition of the imprinting center and fixation of chromosome 15q11-q13 epigenotype by imprinting mutations. *Proc Natl Acad Sci USA* 93:7811–7815, 1996.

Saitoh, S., Kubota, T., Ohta, T., et al. Familial Angelman syndrome caused by imprinted submicroscopic deletion encompassing GABAA receptor beta 3-subunit gene [letter]. *Lancet* 339:366–367, 1992.

Saitoh, S., Mutirangura, A., Kuwano, A., Ledbetter, D. H. and Niikawa, N. Isochromosome 15q of maternal origin in two Prader-Willi syndrome patients previously diagnosed erroneously as cytogenetic deletions. *Am J Med Genet* 50:64–67, 1994.

Schmauss, C., Brines, M. L. and Lerner, M. R. The gene encoding the small nuclear ribonucleoprotein-associated protein N is expressed at high levels in neurons. *J Biol Chem* 267:8521–8529, 1992.

Shemer, R., Hershko, A. Y., Perk, J., et al. The imprinting box of the Prader-Willi/Angelman syndrome domain. *Nat Genet* 26:440–443, 2000.

Smith, J. C., Webb, T., Pembrey, M. E., Nichols, M. and Malcolm, S. Maternal origin of deletion 15q11-q13 in 25/25 cases of Angelman syndrome. *Hum Genet* 88:376–378, 1992.

Sutcliffe, J. S., Nakao, M., Christian, S., et al. Deletions of a differentially methylated CpG island at the SNRPN gene define a putative imprinting control region [see comments]. *Nat Genet* 8:52–58, 1994.

Tantravahi, U., Nicholls, R. D., Stroh, H., et al. Quantitative calibration and use of DNA probes for investigating chromosome abnormalities in the Prader-Willi syndrome. *Am J Med Genet* 33:78–87, 1989.

Toth-Fejel, S., Magenis, R. E., Leff, S., et al. Prenatal diagnosis of chromosome 15 abnormalities in the Prader-Willi/Angelman syndrome region by traditional and molecular cytogenetics. *Am J Med Genet* 55:444–452, 1995.

van den Ouweland, A. M., van der Est, M. N., Wesby-van, S. E., et al. DNA diagnosis of Prader-Willi and Angelman syndromes with the probe PW71 (D15S63). *Hum Genet* 95:562–567, 1995.

Vgontzas, A. N., Kales, A., Seip, J., et al. Relationship of sleep abnormalities to patient genotypes in Prader-Willi syndrome. *Am J Med Genet* 67:478–482, 1996.

Villar, A. J. and Pedersen, R. A. Parental imprinting of the Mas protooncogene in mouse. *Nat Genet* 8:373–379, 1994.

Wandstrat, A. E., Leana-Cox, J., Jenkins, L. and Schwartz, S. Molecular cytogenetic evidence for a common breakpoint in the largest inverted duplications of chromosome 15. *Am J Hum Genet* 62:925–936, 1998.

Webb, T. Invdup(15) supernumerary marker chromosomes. *J Med Genet* 31:585–594, 1994.

Webb, T., Hardy, C. A., Dahlitz, M., Watkiss, E. and Clarke, D. Family studies in Prader-Willi syndrome. *Genet Couns* 5:329–336, 1994.

Wevrick, R. and Francke, U. An imprinted mouse transcript homologous to the human imprinted in Prader-Willi syndrome (IPW) gene. *Hum Mol Genet* 6:325–332, 1997.

Wevrick, R., Kerns, J. A. and Francke, U. Identification of a novel paternally expressed gene in the Prader-Willi syndrome region. *Hum Mol Genet* 3:1877–1882, 1994.

Woodage, T., Deng, Z. M., Prasad, M., et al. A variety of genetic mechanisms are associated with the Prader-Willi syndrome. *Am J Med Genet* 54:219–226, 1994.

Zeschnigk, M., Lich, C., Buiting, K., Doerfler, W. and Horsthemke, B. A single-tube PCR test for the diagnosis of Angelman and Prader-Willi syndrome based on allelic methylation differences at the SNRPN locus. *Eur J Hum Genet* 5:94–98, 1997.

Chapter 7

The Angelman Syndrome (AS)

The study and understanding of the Angelman syndrome that, to use a comparison with chemistry, is genomically "isomeric" to the Prader-Willi syndrome have always been somewhat lagging, relative to PWS. This may not be too surprising, considering that PWS started its clinical course in 1956 from Switzerland and AS in 1965 from England.

Thus, in succession, it was in 1987 that the first 15q11-q13 deletions in AS were observed (versus 1981 in PWS); in 1989 that the deletion appeared to preferentially involve the maternal chromosome (versus 1983 for preferential paternal 15 involvement in PWS); in 1991 that paternal UPD 15 became recognized as a cause of AS (versus 1989, for maternal UPD 15, as a cause of PWS).

CLINICAL PROFILE

In 1965 Harry Angelman reported the cases of "three children with flat heads, jerky movements, protruding tongues and bouts of laughter giving them a superficial resemblance to puppets, an unscientific name, but one which may provide for easy identification" (Angelman, 1965).

The presentation of the clinical picture (Zori et al., 1992; Clayton-Smith and Pembrey, 1992; Magenis et al., 1990) in the first year of life is somewhat similar to that of PWS and hardly considered typical of the diagnosis. Pregnancy and delivery are usually uneventful, with a birthweight slightly less than that of other siblings.

Clinical signs mostly consist of failure to thrive, feeding problems, seizures, and developmental difficulties. The feeding problem consists of frequent spitting or food refusal, which, only infrequently, results from inappropriate motor control of chewing and swallowing. Seizures are often present by 6 months, although, in some cases, they may develop later but mostly within the first 2 years. They could be of several types: myoclonic, motor generalized, or with absences and drop attacks. The epilepsy is initially severe and difficult to control in half the cases, but tends to become milder and more manageable in later childhood. Abnormalities of the EEG consist of diffuse sharp and slow waves of the 2–4 Hz types. No consistent CNS anomaly is seen on neuroimaging; cortical atrophy, leukomalacia, and demyelinization may be documented. Mental retardation is profound. By 1 year, in most cases, a notable cheerfulness is accompanied by outbursts of laughter. The first abnormalities noted in these children are delayed gross motor milestones, muscular hypotonia, and speech delay. There is also hypermotoric activity, usually without aggressive behavior. AS children are also overexcitable with short attention spans. Their limited need for sleep is quite disruptive to their familial environment.

Affected children are tongue-thrusting, mouthing, hand-flapping and have a rigid, jerky ataxia with stiff, hypertonic legs and hypotonic trunk, causing a "puppet-like" demeanor as they walk. They become ambulatory, often between 2 and 5 years of age. Verbal communication is generally limited to a very few words and the children display better nonverbal than verbal communications, using gestures to communicate. Microbrachycephaly (usually below the 25th centile) is noted and the face presents deep-set eyes, thin upper lip, and midface hypoplasia, along with macrognathia, macrosomia, and tongue protrusion, a clinical picture quite different from that of PWS.

Occular and cutaneous hypopigmentation is often noted and eye and ear problems are common, consisting of strabismus, optic atrophy, and keratoconus. Hearing difficulties are related to middle-ear diseases.

The genetic/molecular causes of the syndrome, which are heterogenous, are discussed below (Figure 1 of this chapter and Table 3 of Chapter 5).

The differential diagnosis includes PWS, in the first year of life, and Rett syndrome, because of developmental stagnation, seizures, a jerky walk, and a

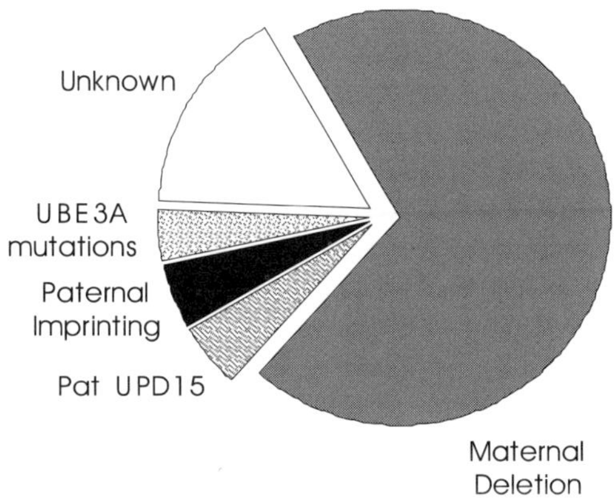

Figure 1 *Classes of AS according to their molecular etiology.*

behavior that can also be pleasant and demonstrates sleep disturbances (Zori et al., 1992; Clayton-Smith and Pembrey, 1992; Williams et al., 1995). Other differential diagnoses include the ATR-X syndrome, an X-linked thalassemia with mental retardation.

The frequency of the syndrome is estimated to be on the order of 1/30,000 and is poorly documented. Table 3 of Chapter 5 summarizes the current molecular subclassification of AS. These are the different classes (Jiang et al., 1999): (i) maternal deletion 15q11-q13 (65–75% of cases); (ii) complex rearrangements involving 15q11-q13 (<1%); (iii) paternal UPD15 (3–5%); (iv) imprinting mutation with or without IC deletion (3–5%); (v) UBE3A mutations (4–6%); and (vi) cases with no detectable molecular abnormality. The relative frequency of these classes of AS can also be seen in Figure 1.

PATERNAL UNIPARENTAL DISOMY 15 IN AS

Type and Frequency of Paternal UPD15 in AS

UPD is significantly less frequent than in PWS and, in general, occurs by a different nondisjunctional mechanism. The paternal UPD15, foreseen in theory by virtue of the cytogenetic features of AS as compared to PWS, was not easily discovered. An initial search (Knoll et al., 1991) failed probably because of a bias in selecting the study cases, 7 out of 10 that were familial (Engel, 1991). Shortly afterward, the same group of authors (Malcolm et al., 1991), however, presented the first two cases to result from paternal UPD15. This added strengthening evidence for genomic imprinting in humans since paternal-in-origin disomy 15 produced AS, but maternal-in-origin resulted in PWS. One of the above AS cases showed paternal heterodisomy and the other isodisomy; these conclusions were also substantiated by cytogenetic evaluation of chromosome 15 parental heteromorphisms.

Within a year, another case was reported (Nicholls et al., 1992), the first to be studied with several probes within the 15q11-q13 critical region, which, altogether with a telomeric probe, showed uniform homozygosity for paternal alleles, suggesting paternal isodisomy of the entire chromosome 15. This was further confirmed by cytogenetic observations of heteromorphisms, whereby both probands' chromosome copies appeared derived from the same single paternal chromosome (Nicholls et al., 1992). Other cases were occasionally reported afterwards; they can be subdivided into those with two free chromosome 15 and those with 15;15 translocation. The frequency of paternal UPD15 in the cases of AS is between 1–5%.

Paternal UPD15 with Free Chromosomes 15

Both cases reported by Bottani et al. (1994), from a group of 14 sporadic cases of AS, had paternal isodisomy 15; a milder phenotype was suggested (Figure 2). The case of Gillessen-Kaesbach (1995) was also milder and the inherited paternal pair was considered isodisomic. Several other reports contained such cases of paternal

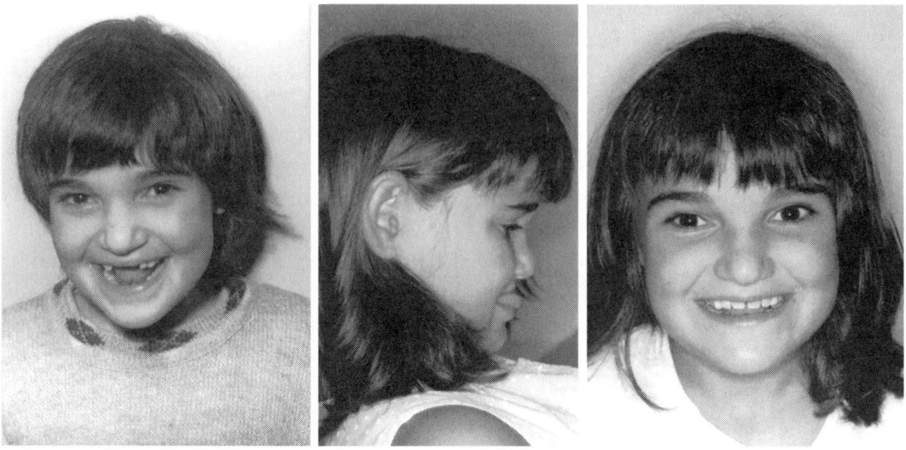

Figure 2 Angelman syndrome patient with paternal UPD15, shown at ages 8 and 10 years. Note characteristic but relatively mild facial phenotype (same patient as in Bottani et al., 1994).

UPD15 with free chromosomes 15 (Exeler et al., 1996; Fridman et al., 2000; Gyftodimou et al., 1999).

Paternal UPD15 with t(15;15) [i(15q)]

Although cases with this chromosomal complexion had been documented in the past, the case of Freeman et al. (1993) was the first one shown to derive from a potential i(15q) and to be isodisomic. The karyotype was 45,XY,t(15q;15q). Results at four different loci demonstrated that the patient had inherited chromosomal material 15 from his father only, and that he was homozygous (or hemizygous) at the three loci for which his father was heterozygous. Densitometry showed that he had indeed two identical copies of the polymorphic alleles derived from the father. It was concluded that this represented a paternal i(15q), rather than a rob t(15q;15q).

Absence of the maternal chromosome could have resulted from ovular nullisomy 15 or postzygotic loss, followed by selection of the cell line only containing the paternal i(15q). The clinical features were typical for AS.

In another case (Ramsden et al., 1996), the AS proband showed a 15;15 translocation with no visible deletion of the critical proximal area by FISH. A polymorphic microsatellite for the GABRA5 gene ruled out a maternal contribution at this locus. Parent-of-origin methylation studies at PW71 were consistent with a lack of maternal inheritance at 15q11-q13. There were no studies to determine whether a Robertsonian translocation or an isodisomic formation explained the centric fusion of chromosomes 15.

In still another case, 45,XY,t(15q;15q) (Tonk et al., 1996), FISH analysis indicated diploidy in the critical region of chromosome 15. The use of highly polymorphic microsatellite markers demonstrated paternal UPD15 for chromosome 15. In addition, all markers showed homozygosity, indicating isodisomy 15. The child's phenotype left no doubt about the diagnosis, but the facial appearance was uncharacteristic and there were no seizures, suggesting again a milder phenotype.

Mechanism of Paternal UPD15 in AS

It was apparent from the above cases that the paternally inherited pair 15 [involving either free chromosomes 15 or t(15;15)] was more often isodisomic. In a more recent collective study to examine the mechanisms of nondisjunction involved in paternal UPD15 in AS, a total of 17 cases were analyzed. Of the 17 cases (several already featured in the cases reviewed in this chapter), a somatic event accounted for paternal UPD15 in 11 and meiosis 1 origin in four, whereas the origin in two cases was unknown (Robinson et al., 1996). Two cases of a meiosis 2 nondisjunctional event in the father resulting in paternal UPD15 and AS were also recently reported (Gyftodimou et al., 1999; Fridman and Koiffmann, 2000). Even though the numbers are small, they are probably representative since paternal UPD15 is not a common cause of AS. In serial studies, there were three paternal UPD15 cases among 93 AS probands (Chan et al., 1993); none was observed in another study of 61 AS cases (Saitoh et al., 1994), suggesting a frequency around 2% or less. From the available sample, nearly 3/4 of the cases arise from somatic duplication of paternal 15, either by somatic chromatid nondisjunction or centromeric misdivision. All happens as if a maternal 15 nullisomy or loss in the early conceptus is compensated for by duplication of the singly inherited paternal member. If so, a maternal age effect might be detected in these instances. We computed average parental age for seven cases of paternal UPD15 from the literature: three with chromosome 15 centric fusion (Freeman et al., 1993; Ramsden et al., 1996; Tonk et al., 1996) and four with a free pair 15 (Nicholls et al., 1992; Bottani et al., 1994; Gillessen-Kaesbach et al., 1995). Average maternal age (seven mothers) was 31 and average paternal age (six fathers) was close to 34. An increased paternal age had already been reported in AS (Robinson et al., 1993). In a more recent study, the origin of the extrachromosome 15 was determined in 21 AS patients with paternal UPD15 (Robinson et al., 2000). Only four of 21 paternal UPD15 cases could be clearly attributed to a meiotic error. The mean maternal and paternal ages of 33.4 and 39.4 years, respectively, for paternal UPD15 cases were increased, as compared with normal controls. This may be simply the consequence of an age association with maternal nondisjunction leading to nullisomy for chromosome 15 in the oocyte, although the higher paternal age in paternal UPD15 as compared with maternal UPD15 cases is suggestive that paternal age may also play a role in the origin of paternal UPD15. These data are in agreement with those reported in (Fridman and Koiffmann, 2000).

MATERNAL DELETION OF 15q11-q13 IN AS

Cytogenetic Studies

The initial report of deletions in some cases of the AS syndrome came from two different cytogenetic laboratories. Kaplan et al. (1987) noted the clinical heterogeneity of three cases associated with cytogenetic deletions in the long arm of chromosome 15, one of whom had AS. The same year, Magenis et al. (1987) described two girls with features of AS and an interstitial deletion affecting the same 15q12 region. Similar cytogenetic abnormalities were then confirmed by investigators in the UK (Pembrey et al., 1989), establishing the area 15q11-q13 as critical for AS. They examined another 10 AS cases, confirming four *de novo* interstitial deletions; they also found a case of inv(15)(p11q13) transmitted by a phenotypically unaffected mother! From these cytogenetic data, it appeared that 40% of AS patients exhibited a 15q proximal deletion. Another paper the same year (Williams et al., 1989) reported results of prometaphase studies at the 850-band level, showing a similar interstitial deletion at 15q12 in five of 12 typical AS patients, or 40% of the cases. Molecular studies utilizing as markers five chromosome 15q11-q13 specific polymorphisms (3-21, 1R43R, 189-1, 34, 1R10-1) further established the existence of a molecular deletion of this area (Knoll et al., 1989), which did not appear to be different or distinct from that of PWS cases, confirming an earlier molecular study by Donlon (1988). The important difference, however, was that in the AS cases the DNA marker analysis showed the deletion to be of maternal origin. Such an origin had already been cytogenetically detected in some previous cases, mainly by Magenis et al., and it was considered that the maternal origin of the deletion in AS might be the decisive element for such a highly different phenotype in AS, as compared to PWS. The authors therefore considered parental imprinting as a possible factor in the clinical expression of either AS or PWS, depending on the side of parental inheritance (Knoll et al., 1989). The maternal origin of the 15q11-q13 deletion in AS was further verified in four cases with informative cytogenetic heteromorphisms and the potential role of genomic imprinting was reemphasized (Williams et al., 1990).

Magenis et al. (1990) attempted a cytogenic comparison of the size of the deletion in both syndromes, in seven cases of PWS and 10 of AS. The same proximal band was deleted (15q11.2) in both syndromes. In general, the deletion in patients with Angelman syndrome was larger, although variable, and included bands q12 and part of q13. All the studied deletions in patients with AS were of maternal origin. This contrasted with the predominant paternal origin of the deletion in patients with PWS. Two possible reasons for these observations were postulated: (1) The deleted regions were different at the cytologic and/or molecular level; (2) the deleted regions were essentially the same, but differential expression (imprinting) of the genes in the homologous chromosome 15 had occurred. Based on genomic homologies between human chromosome 15 and mouse chromosome 7, and information from mouse uniparental for chromosome 7, it was stated that " . . . AS and PWS may be the result of lack of the maternal component (AS) versus lack of the paternal component (PWS), in the same critical segment of this apparently unstable region of chromosome 15" (Magenis et al., 1990).

Molecular Studies

Molecular studies (see Figures 2–4, Chapter 6) soon dominated the search for the elucidation of the intricacies of the various genomic alterations resulting in AS.

Knoll et al. (1990) investigated the size of the molecular deletion in AS and compared it to that in PWS. They observed that the deletions were similar and often spanned loci from D15S9 to IR10, within 15q11-q13 (they termed these class I patients); for class II deletion patients (both AS and PWS), the locus IR39 was included. In a third class of cases, no molecular deletion could be uncovered in AS, which occurs also occasionally in PWS.

In a different study (Hamabe et al., 1991), 14 AS patients, nine sporadic and five familial (among them three siblings), were analyzed. There was a variable deletion of the five markers tested at 15q11-q13: There was loss of only one locus (D15S11) in one sporadic case, loss of another single locus (D15S10) in the three siblings, or loss of three loci (D15S11-D15S10-D15S12) in another sporadic patient, and loss of four loci in the other three sporadic cases (D15S9-D15S11-D15S10-D15S12). The findings were consistent with the genomic imprinting hypothesis, that lack of maternally derived DNA leads to AS. Remarkably, the maternal grandfather and mother of the AS siblings, also had the deletion of locus D15S10, but they did not display any phenotypic characteristics of AS.

The proof of the parent-of-origin effect of the deletion 15q21-q13 was further strengthened in a relatively large series of 25 deletion cases of AS, all maternally derived, 18 considered fully informative, either by cytogenetic (seven cases) or DNA polymorphisms (11 cases) (Smith et al., 1992).

A final confirmation was obtained from the description of an unusually large pedigree. Indeed, familial cases of AS appear more often than in PWS. These may result from a detectable submicroscopic deletion or from an undetectable lesion (point mutation?) of a crucial AS locus in other pedigrees (Meijers-Heijboer et al., 1992). The family described in this report was an unusually large pedigree with segregation of AS through maternal inheritance and apparent asymptomatic transmission through several male ancestors. Deletion and paternal disomy at 15q11-q13 were excluded. However, the genetic defect was still located in this region, as a maximum lod score of 5.40 for linkage to GABRB3 and D15S10 loci was obtained, which has been mapped within or adjacent to the AS critical region at 15q11-q13.

Larger studies were subsequently performed to determine the frequency of the different defects associated with AS. The initial cytogenetic analyses could detect deletions in the crucial proximal segment of chromosome 15 in somewhat less than 50% of typical AS patients (Pembrey et al., 1989; Williams et al., 1990). The use of improved chromosomal resolution yielded additional cases with chromosome 15 deletions (Zackowski et al., 1993). Of the 24 AS probands out of 28 for whom conclusive cytogenetic analyses were possible, four were characterized as normal and 20 as deleted. In the same study, 19 of the 22 probands analyzed molecularly had a deletion. The correlation between both approaches was remarkably accurate; molecular studies revealed a deletion in two more cases cytogenetically suspected but inconclusive. In this series of AS probands, therefore, the rate of documented deletions approached 85%. Two classes of (large) deletions were detected by molecular means: one involving the five more proximal probes, the other (two cases only) including a 6th, more distal probe. In 13 informative cases, the deleted 15 was of maternal origin (Zackowski et al., 1993).

The Deletion Prone 15q11-q13 Region

Early on, investigators were considering what makes this proximal region of chromosome 15 so liable to rearrangements and so variably elusive to cytogenetic deciphering. As discussed by various authors, considerable normal variation was observed in 15q11.2; there is a significant excess of translocation breakpoints within this region as well as cloning instability of DNA segments from the 15q11-q13 area (Knoll et al., 1993; Pembrey et al., 1989; Magenis et al., 1990). In addition, statements such as the following were included in the literature in the early 1990s:

> Higher rates of cytological meiotic chiasma, and sister chromatid exchanges have been observed. The nature of the chromatin in chromosome 15 differs from the other acrocentrics in its cytochemical properties. Its constitutive heterochromatin, composed of all of the satellite DNAs, is rich in 5-methylcytosine, a finding consistent with 5-methylcytidine antibody specificity, distamycin/DAPI stainability and failure to condense after treatment with 5-azacytidine. Specific DNA sequences leading to instability of this region have been suggested.... (Knoll et al., 1993)

Cytogenetically, it was difficult to reliably detect a small deletion due to the heteromorphic nature of 15q11.2, the minute size of the region, the variability of condensation between homologues in this area (one of which replicates later), and frequent bending in the centromeric vicinity (Knoll et al., 1993).

Thus, it was not easy to determine the presence or the size of 15q deletions and locations of breakpoints. There were also some discrepancies in the high-resolution banding in this region. Chromosomal band 15q12 (at the 550-band resolution) was split in two gray bands with a light band in between (q12.1;q12.2;q12.3) at the 850-band resolution; this was different from ISCN (Magenis et al., 1990). These considerations resulted in the extensive use of FISH to document deficits and other rearrangements in this area. It is now understood that hot spots with low copy repeats appear responsible for mismatched recombinational events causing such a regional liability to deletions and duplications.

Approximately 70% of AS cases (a same proportion as in PWS cases) are due to an $\sim$4-Mb deletion of 15q11-q13 of maternal origin (as discussed elsewhere, there is paternal origin of the deletion in PWS) (Nicholls et al., 1998; Jiang et al., 1999). The vast majority (>95%) of patients with deletions have clustered breakpoint regions. There is a single distal breakpoint and two proximal breakpoint regions (Christian et al., 1995; Amos-Landgraf et al., 1999). The mechanisms of maternal and paternal deletions do not seem to be different; both interchromosomal and intrachromosomal rearrangements have been observed in AS and PWS (Carrozzo et al., 1997; Robinson et al., 1998). Remarkably, there are at least 10 copies of a complex DNA sequence termed the END repeat (or END duplicon) that maps to the three common deletion breakpoint regions on chromosome 15 (Figure 2 of Chapter

6) (Buiting et al., 1992, 1998; Amos-Landgraf et al., 1999; Ji et al., 1999). The END repeats are mostly composed of duplications of a large gene, HERC2, recessive mutations of which cause a neurological and developmental syndrome in mice (Ji et al., 1999). A considerable number of multiple rearrangements within the END repeat have also occurred during evolution (Amos-Landgraf et al., 1999; Ji et al., 1999, 2000). The most simple model to account for the AS and PWS deletions is via homologous misalignment and meiotic recombination between different END repeat copies in proximal and distal 15q11-q13 breakpoints (Amos-Landgraf et al., 1999). These duplicated areas that predispose to unequal "homologous" recombination have also been described independently as breakpoint BP1 to BP3 (Christian et al., 1999). BP1 and BP2 correspond to the first and second proximal END repeats, respectively, while BP3 corresponds to the distal END repeat.

THE VARIOUS CAUSES OF AS

The first large study to determine the deletions in AS contained 93 patients (Chan et al., 1993). The deletions were assessed using RFLP and (CA)n repeat polymorphisms. A total of 60 patients were found to have a chromosomal deletion. At least 40 of the 60 AS patients had a *de novo* deletion, since their mothers were heterozygous for one or several of their hemizygous loci. Among these 93 AS, 13 were familial cases from six kinships. In 12 of these, there was a critical dominant mutation, limited in expression by a *cis* maternal origin. All of these cases were cytogenetically normal, and none of the tested molecular loci were deleted, while at least one of them showed heterozygosity within the AS critical region. DNA polymorphism analysis indicated that related affected children shared the same maternal chromosome. No molecular deletion could be documented in 15 patients, five of whom, monoallelic in the ASCR locus, were considered to have an undetected deletion (or somatic segmental recombination involving the AS critical region). The other 10 cases, however, were heterozygous for at least one of the loci examined, suggesting the existence of a point mutation in a critical gene or very small deletion as in the above familial cases. Finally, only three of the 93 AS patients resulted from paternal UPD15, all proven by DNA polymorphic analyses. One of these UPD15 cases showed heterodisomy, while the two others had isodisomy for the entire chromosome 15.

The second important large study included patients of Japanese origin (Saitoh et al., 1994). A total of 61 AS patients were studied. There were 53 sporadic cases (87%) and eight familial (13%), including three siblings and two half-siblings of different fathers and first cousins. The investigation included high-resolution cytogenetic and DNA polymorphisms analysis and quantitative Southern blots. Among the sporadic cases, there were 37 with deletion and 16 without; among the familial cases, there were three with deletion and five without. Therefore, deletions account for 65% of cases. The deletion in the three siblings and relatives of the familial cases was much smaller than those in sporadic cases. The deletion without exception affected the maternally derived chromosome in all 28 informative cases. No case of paternal UPD15 was identified in this study. It was also obvious from this study that the diagnostic evaluation of AS should include DNA analysis. Of 23 cytogenetically "normal" cases, 10 had a deletion diagnosed by molecular probes (a total of 43% false negatives).

These two first large series of AS were also of particular interest, as they emphasized the relevance of methylation studies in the diagnosis of AS. In this disorder, a sizable subgroup of patients in high-risk families have a biparental chromosome 15, no detectable deletion, and do not display an altered methylation pattern (van den Ouweland et al., 1995). Methylation analyses using probe PW71 were compared with the results of DNA polymorphisms in 85 families with suspected PWS (see Chapter 4) and 73 families with suspected AS probands. In the AS group, 20 cases in the 73 families had a maternal deletion and were confirmed as only possessing a paternal pattern of methylation with the PW71 probe; in one case, paternal UPD15 was diagnosed and only the unmethylated paternal allele was detected. In 44 cases, both DNA polymorphisms and methylation analysis identified the presence of a maternal allele. In a multiplex family where the AS phenotypes showed linkage to markers of chromosome 15q11-q13 (Meijers-Heijboer et al., 1992), there was neither deletion, nor paternal UPD15 nor abnormal methylation; the conclusion was that a mutation in a putative AS locus could also cause the disorder (a prediction that was later vindicated by the discovery of mutations in the UBE3A gene) (Matsuura et al., 1997; Kishino et al., 1997).

Of particular interest was also the case of a typical AS patient, with normal biparental inheritance of 15q11-q13 markers but with an abnormal methylation pattern using the PW71 probe; only the paternal methylation pattern was observed. This observation supported the conclusion that a class of AS patients may be due to an aberrant imprinting process, and this could be determined by the study of the methylation pattern. In addition, there are AS patients with normal methylation pattern. In seven cases of suspected AS, DNA marker analyses were inconclusive: none had detectable deletion, three had only paternal methylation at PW71, probably by UPD15, and four maintained biparental methylation. Finally, of the 48 normally methylated cases, nine were affected with AS according to recommended criteria. Once again, cytogenetic "routine" analysis showed inconsistencies as compared to polymorphic and methylation studies. Of 17 deletions so documented (10 in PWS, 7 in AS), five could not be confirmed by DNA analysis. Conversely, of 29 molecularly detected deletions, only 12 had been identified using chromosome analysis. In two earlier cases with a ring 15 of paternal origin (Glenn et al., 1993), and in an additional case with ring 15 of the series here summarized, there was a normal PW71 methylation but an abnormal one with probe D15S9 or ZNF127.

It is of interest to note that the test for the methylation status shows some differences among different cells (van den Ouweland et al., 1995). Blood leucocytes are the tissue of choice for the detection of the methylation differences. DNA of first-trimester chorionic villi and of fibroblast cultures are not really suitable for analysis with PW71 because of different methylation patterns. On amniotic fluid cells, the methylated and unmethylated fragments show the same pattern as in blood cells (Dittrich et al., 1993).

The results of the large series indicated that AS could be explained by four types (Figure 1) of genomic lesions: (i) 15q11-q13 deletion in 75% of cases; (ii) paternal UPD15 in 2%; (iii) imprinting center (IC) mutation in 1–5%; and (iv) putative mutation in a single gene within the AS region in approximately 20% of cases. The first three causes could be detected by methylation studies (using, e.g., ZnF127, PW71, SNRPN exon alpha as probes). The putative mutations in a single gene and the IC mutations were most likely to recur within families.

MUTATIONS IN THE UBE3A GENE

The gene E6AP ubiquitin protein ligase (UBE3A), although mapping to the critical AS region, was considered an unlikely candidate for the syndrome, in view of its expression from both parental alleles in lymphocytes and fibroblasts of AS and PWS patients (Matsuura et al., 1997). However, mutation analysis of this gene in AS patients without large 15q deletions, paternal UPD15, or imprinting mutations has led to the identification of loss of function mutations in several cases (Kishino et al., 1997; Matsuura et al., 1997).

The first UBE3A mutations identified in AS patients were a *de novo* duplication of five nucleotides, a maternally inherited splice site mutation, a *de novo* two nucleotide deletion, and a *de novo* amino acid substitution (Kishino et al., 1997; Matsuura et al., 1997). The first three mutations result in frameshifts and premature translation termination. Subsequently, many more UBE3A mutations have been identified in patients with AS (Table 1) (Kishino et al., 1997; Matsuura et al., 1997; Fang et al., 1999; Malzac et al., 1998; Baumer et al., 1999; Tsai et al., 1998; Russo et al., 2000).

The UBE3A gene had been known to encode the E6-associated protein, a protein interacting with the human papilloma virus, to promote the nonlysosomal degradation of the p53 protein (Huibregtse et al., 1991; Scheffner et al., 1990). The gene encodes an E3 ubiquitin protein ligase that is probably involved in the ubiquitination of a diverse range of proteins (Scheffner et al., 1993; Huibregtse et al., 1993). The UBE3A gene encodes a member of a family of functionally related proteins defined by a conserved C-terminal 350- amino acid "hect" domain. Hect E3 proteins appear to be important in substrate recognition and in ubiquitin transfer. RT-PCR experiments provided evidence for several isoforms differing at their *N* termini. Each of the mRNAs was expressed in all cell lines tested. Additional 5′ untranslated exons were also identified (Kishino and Wagstaff, 1998). In fact, at least 16 exons were identified, including six exons that encode the 5′ UTR. The gene spans approximately 120 kb, with transcription oriented from telomere to centromere. Two processed UBE3A pseudogenes to chromosomes 2 and 21 have also been identified (Kishino and Wagstaff, 1998).

TABLE 1 Mutations in UBE3A in AS

	Nucleotide Substitutions	
TGT → TAT	Cys21Tyr	(Matsuura et al., 1997)
GCT → ACT	Ala178Tur	(Matsuura et al., 1997)
TGG → TAG	Trp305Ter	(Fang et al., 1999)
TCG → CCG	Ser34Pro	(Malzac et al., 1998)
CGA → TGA	Arg417Ter	(Matsuura et al., 1997)
CGA → TGA	Arg482Ter	(Malzac et al., 1998)
CGT → TGT	Arg506Cys	(Baumer et al., 1999)
TAT → TAG	Tyr533Ter	(Fang et al., 1999)
TGG → TGA	Trp768Ter	(Tsai et al., 1998)
ATA → AAA	Ile804Lys	(Fang et al., 1999)
GAA → TAA	Glu167Ter	(Russo et al., 2000)
	Splicing Mutations	
IVS9 − 8A > G		(Kishino et al., 1997)
	Small Deletions	
1 nt del codon 89		(Fang et al., 1999)
5 nt del codon 107		(Fang et al., 1999)
2 nt del codon 131		(Fang et al., 1999)
14 nt del codon 291		(Malzac et al., 1998)
1 nt del codon 311		(Malzac et al., 1998)
1 nt del codon 321		(Fang et al., 1999)
7 nt del codon 324		(Fang et al., 1999)
4 nt del codon 369		(Fang et al., 1999)
2 nt del codon 447		(Matsuura et al., 1997)
10 nt del codon 483		(Malzac et al., 1998)
1 nt del codon 596		(Baumer et al., 1999)
3 nt del codon 781		(Fang et al., 1999)
4 nt del codon 835		(Fang et al., 1999)
15 nt del codon 851		(Fang et al., 1999)
	Small Insertions	
1 nt ins codon 20		(Malzac et al., 1998)
2 nt ins codon 59		(Baumer et al., 1999)
1 nt ins codon 105		(Russo et al., 2000)
1 nt ins codon 237		(Baumer et al., 1999)
1 nt ins codon 648		(Malzac et al., 1998)
1 nt ins codon 654		(Russo et al., 2000)
4 nt ins codon 662		(Fang et al., 1999)
3 nt ins codon 803		(Malzac et al., 1998)
1 nt ins codon 816		(Malzac et al., 1998)
5 nt ins codon 836		(Kishino et al., 1997)
16 nt ins codon 845		(Baumer et al., 1999)
	Complex Rearrangement	
26 nt del and 1 nt ins (nt 2230)		(Malzac et al., 1998)

The public databases contain the following UBE3A mutations in AS (http://archive.uwcm.ac.uk/uwcm/mg/search/228487.html).

The early studies on the expression of UBE3A concluded that there was no evidence of imprinting (of any alternatively spliced form) in the tissues studied (Kishino et al., 1997; Matsuura et al., 1997; Nakao et al., 1994). Subsequent studies, however, clearly demonstrated that in mice and humans there is imprinted expression of UBE3A and that there is maternal-only expression in specific regions of the brain, particularly in the hippocampus, cerebellum, Purkinje cells, and cells of the olfactory bulb in mice (Albrecht et al., 1997; Vu and Hoffman, 1997; Rougeulle et al., 1997). Animals exhibiting partial paternal UPD for the syntenic AS region had striking reduction of Ube3a expression in the above cells, with variably decreased levels in other brain parts, as compared to controls. If extrapolated to the same human tissues, it was thought that such differences might explain retardation, seizures, and ataxia in AS patients (Albrecht et al., 1997). Indeed, when UBE3A expression in various tissues including brain samples of normal, PWS, and AS subjects was compared, the results confirmed major reductions of all UBE3A alternatively spliced mRNA in AS brain samples (Rougeulle et al., 1997).

Transgenic mice with the maternal or paternal UBE3A genes knocked out have been generated (Jiang et al., 1998). These mice were compared with their wild- type (m+/p+) littermates. Mice with paternal deficiency (m+/p−) were essentially similar to wild-type mice. The phenotype of mice with a maternal deficiency (m−/p+) resembles that of human AS with motor dysfunction, inducible seizures, and a context-dependent learning deficit. The absence of detectable expression of UBE3A in hippocampal neurons and Purkinje cells in m−/p+ mice, indicating imprinting with silencing of the paternal allele, correlated well with the neurologic and cognitive impairments. Long-term potentiation in the hippocampus was severely impaired. The cytoplasmic abundance of p53 was found to be greatly increased in Purkinje cells and in a subset of hippocampal neurons in m−/p+ mice, as well as in a deceased AS patient. The authors suggested that failure of UBE3A to ubiquitinate target proteins and promote their degradation could be a key aspect of the pathogenesis of AS (Jiang et al., 1998).

Not all “3 Nos” biparental AS cases (no deletion, no paternal UPD15, no IC mutations) have proven to have a mutation in the exons of UBE3A (Figure 1). Even after the identification of all the exons of UBE3A, the determination of the several alternatively spliced isoforms, and extensive mutation analyses, only about 5% of AS patients have mutations in the UBE3A gene (or about 20% of the AS patients with no deletion, no UPD, no IC mutations). It is therefore possible that mutations in other genes within the critical region could cause AS, or that these patients have a phenotype similar to AS but due to genes or genomic regions outside of chromosome 15q11- q13. For example, AS and Rett syndrome may have similar phenotypes and it is possible that some cases diagnosed as AS may have mutations in the recently identified Rett syndrome gene MECP2 (Amir et al., 1999).

IMPRINTING MUTATIONS IN AS

A considerable minority (2–5%) of patients with AS have mutations that affect the mechanisms involved in the resetting of the parental imprint during gametogenesis.

There is no large deletion 15q, or paternal UPD15, or UBE3A mutation in these cases. These are called imprinting mutations and their diagnosis could be done using methylation assays of selected regions of 15q11-q13 after hybridization with specific probes (SNRPN, PW71). Transmission of a paternally derived imprinting mutation through a female causes a failure to switch the imprint and results in the inheritance of an abnormal (paternal-type) pattern of imprinting from the mother. This, along with the normal paternal imprint, results in "homozygosity" for parental-type imprint and Angelman syndrome (Nicholls et al., 1998). The opposite scenario leads to PWS.

The study of the small deletions on chromosome 15 associated with the imprinting mutations in a handful of informative AS patients resulted in the definition of the shortest region of microdeletion overlap (AS-SRO) (Saitoh et al., 1996) (Figure 3, Chapter 6). The AS-SRO spans only 2 kb and maps in intron 3 of the SNRPN gene (Dittrich et al., 1996). The SNRPN exons surrounding intron 3 are 5′ exons of alternatively spliced SNRPN isoforms that are expressed about 100 times less than the main isoform transcript that initiates 3′ to these exons (Dittrich et al., 1996). The AS-SRO is also called the imprinting center (IC). The SNRPN isoform transcript that spans the IC is imprinted and only paternally expressed in the heart and brain (Dittrich et al., 1996). It is of interest that the PWS-SRO associated with the deletions of imprinting mutations in PWS does not overlap with that of AS-SRO, but it is located several kilobases downstream and includes the first exon of the most abundant isoform of SNRPN gene (Reis et al., 1994; Sutcliffe et al., 1994; Saitoh et al., 1996). The recent study of an English patient (AS-LO) further narrowed down the AS-SRO to 880 bp located 35 kb upstream to the SNRPN promoter (Buiting et al., 1999).

THE IMPRINTING BOX

For a discussion of the concept of the imprinting box that includes both AS-SRO and PWS-SRO, see Figures 3 and 4, Chapter 6.

PHENOTYPE-GENOTYPE CORRELATIONS IN AS

The initial observation that paternal UPD15 cases of AS display a milder phenotype (Bottani et al., 1994) has been confirmed by other reports (Gillessen-Kaesbach et al., 1995; Smith et al., 1997; Fridman et al., 2000). There is still no unanimous consensus regarding a less severe facial dysmorphism in the patUPD15 cases (Figure 2) (Bottani et al., 1994; Smith et al., 1997).

To summarize, Table 2 lists the differences noted between 19 AS cases of paternal UPD15 and 21 cases of deletions of maternal 15q (Smith et al., 1996, 1997; Saitoh et al., 1994; Fridman et al., 2000). It is clear that there is milder severity of the AS due to paternal UPD15.

TABLE 2 Phenotypic Characteristics of Angelman Syndrome Patients with Paternal UPD15 as Compared to Deletion Cases

Features	Del(15q)mat	patUPD15
Age at diagnosis	4 3/12	7 3/12
Microcephaly	58% (11/19)	16% (3/19)
Height < 25th centile	32% (6/19)	17% (3/18)
Weight > 75th centile	20% (4/20)	71% (12/17)
Absence of speech	89% (16/18)	48% (9/19)
Seizures	82% (18/21)	42% (8/19)
Independent gait (year)	4 6/12	2 9/12

A recent study of 14 AS patients with UBE3A mutations attempted to identify potential phenotypic characteristics of this class of AS (Moncla et al., 1999). Consistent manifestations were psychomotor delay, a happy disposition, a hyperexcitable personality, EEG abnormalities, and mental retardation with severe speech impairment. The other main manifestations of AS, ataxia, epilepsy, and microcephaly, were either milder or absent in various combinations among the patients. In addition, myoclonus of cortical origin was frequently observed with severe myoclonic seizures. The majority of the patients were overweight. This study concluded that ataxia, myoclonus, EEG abnormalities, speech impairment, characteristic behavioral phenotype, and abnormal head circumference are attributable to a deficiency in the maternally inherited UBE3A allele (Moncla et al., 1999). Obviously, more extensive studies are needed to substantiate potential clinical differences in this class of patients.

In the category of AS with IC mutation, nine cases were compared to nine age-matched AS patients caused by maternal deletions of 15q (Burger et al., 1996). All had mental retardation, delayed motor development, and absent speech. However, hypopigmentation and microcephaly were only present in one of the IC mutation patients, as opposed to seven in nine of the deletion control group. The authors suggested that IC mutations cause incomplete loss of gene function or that deletions involve also genes not subject to imprinting (Burger et al., 1996).

Recently, seven patients were reported who lacked most of the features of AS, but presented with obesity, muscular hypotonia, and mild mental retardation. Based on this clinical presentation, these patients were initially suspected of having PWS, and DNA methylation analyses of SNRPN and D15S63 were performed. To the surprise of the investigators, however, the test resulted in an AS methylation pattern, i.e., paternal-only methylation (the maternal band was faint or absent) identical to that of AS (Gillessen-Kaesbach et al., 1999). Cytogenetic studies and microsatellite analysis demonstrated apparently normal chromosomes 15 of biparental inheritance. The authors concluded that these patients have an imprinting defect and a previously unrecognized form of AS. The mild phenotype may be explained by an incomplete imprinting defect or by cellular mosaicism.

LABORATORY TESTS IN AS

Recommendations for these tests have been published by the American Society of Human Genetics/American College of Medical Genetics Test and Technology Transfer Committee (1996) (Anonymous 1996) and can also be found in the excellent Web article of Williams et al. in Gene Clinics (http://www.geneclinics.org/profiles/angelman/).

The laboratory tests are as follows:

Parent-specific Methylation Status

In 80% of AS cases, there is abnormal paternal-only pattern methylation status at the SNRPN locus. There is only unmethylated DNA instead of an equal mixture of methylated and unmethylated CpG SNRPN island. This abnormal methylation could be determined using Southern blot hybridization and SNRPN probes (Glenn et al., 1996). The methylation test is also useful in the detection of PWS cases. The AS cases with paternal-only methylation could have deletions in maternal 15q11-q13, paternal UPD15, or an imprinting mutation.

Further analysis to determine the exact cause includes:

(i) *DNA polymorphisms to detect UPD15.* Informative microsatellite polymorphisms of chromosome 15 should be used in the DNAs of the patients and their parents to establish the origin and inheritance patterns of the polymorphic alleles.

(ii) *DNA analysis by FISH.* Probes D15S10 or SNRPN could be used to detect large deletions of 3–4 Mb, or other probes including UBE3A to detect smaller deletions in the imprinting center. Alternatively, the deletion analysis could be performed using the inheritance of DNA polymorphic alleles in 15q11-q13.

(iii) *A careful high-resolution karyotype.* This should be performed in the patients and their parents to rule out structural chromosomal rearrangements.

Mutation Analysis of UBE3A

In the 20% of cases without parental specific methylation, absence of 15q11-q13 deletion and/or UPD15 (the "3 Nos!"), a detailed analysis of the UBE3A gene should be performed to detect point mutations. The UBE3A cDNA is 2.7 kb long and encodes an 865 amino acid predicted protein. The genomic DNA spans ~120 kb. These are 15 coding exons and an additional 6 to 9 exons in the 5′UTR. Alternative splicing of the 5′UTR accounts for several different transcripts that could be translated in more than one different protein isoforms (Yamamoto et al., 1997; Kishino et al., 1997; Vu and Hoffman, 1997; Kishino and Wagstaff, 1998). The experience to date shows that the vast majority of UBE3A mutations in AS are severe null protein truncating mutations. The potential phenotypic consequence of milder mutations is unknown.

GENETIC COUNSELING IN AS

The genetic counseling of parents with a child with AS depends on the cause of AS in this particular family. The cause of AS could be diagnosed after a battery of laboratory tests listed above. In about 80% of patients, the molecular mechanism of AS could be identified and therefore prenatal testing could be offered. In 20% of patients, however, the molecular mechanism could not be elucidated and therefore empiric recurrence risks are usually used. In these cases, accurate prenatal diagnosis is not possible.

The categories of known molecular mechanism include (Figure 1) the following:

De novo Deletion of Maternal 15q11-q13

This is the most common etiology of AS and accounts for 70–75% of all cases. The recurrence risk (RR) is less than 1% (Connerton-Moyer et al., 1997) and prenatal diagnosis could be offered by DNA polymorphisms in the deleted region or by FISH analysis using probes from the deleted region. These probes are usually D15S10 and SNRPN.

Mutations in the UBE3A Gene

This is the second most common cause of AS (~5%, i.e., ~20% of the "3 Nos" cases, no deletion, no UPD15, no methylation abnormality). If the mutation is present in the mother, the RR is 50%, the same as for an autosomal dominant disease. Other members of the mother's family may be at risk of having affected children. The UBE3A mutations in AS are not inherited from the fathers. In the case of presumed *de novo* mutations, there is always the possibility of maternal mosaicism (Malzac et al., 1998; Matsuura et al., 1997; Kishino et al., 1997). Fetal diagnosis could be offered by UBE3A mutation analysis of fetal DNA.

Paternal UPD15

A small fraction, approximately 2–5% of AS patients, show paternal UPD15. The RR in future pregnancies in these cases is less than 1%. Karyotypes of patients and parents should be performed to exclude a paternally inherited Robertsonian translocation. Theoretically, UPD analysis using appropriate chromosome 15 polymorphisms could be used in fetal diagnosis.

Imprinting Mutations

These are either (a) microdeletions of the imprinting center (IC) that are between 6 and 200 kb in length, or (b) abnormal, paternal-only DNA methylation pattern (Buiting et al., 1995, 1998; Saitoh et al., 1996) and account for 2–5% of these cases. If the IC deletion is present in the asymptomatic mother, the RR is 50%. First-degree relatives of these mothers may also be at risk for children or grandchildren with AS.

If the IC deletion is not present in the mother, then the RR is less than 1%. Fetal diagnosis could be done using probes in the deleted region.

Structural Chromosomal Rearrangements

These are rare and account for less than 1% of AS cases (Chan et al., 1993). A careful and detailed cytogenetic analysis in the mother is required to establish the nature and the inheritance of the chromosomal rearrangement. In the *de novo* cases, the RR is negligible, but in the inherited cases, the RR is considerable and up to 50%. A cytogenetic analysis in fetal cells could be offered in subsequent pregnancies.

In the remaining category of 20% of AS patients, there is no detectable molecular or methylation abnormality. In these cases, no accurate RR could be given and no specific prenatal test could be offered. The empiric RR is not, however, negligible.

REFERENCES

Anonymous. Diagnostic testing for Prader-Willi and Angleman syndromes: Report of the ASHG/ACMG Test and Technology Transfer Committee. *Am J Hum Genet* 58:1085–1088, 1996.

Albrecht, U., Sutcliffe, J. S., Cattanach, B. M., et al. Imprinted expression of the murine Angelman syndrome gene, Ube3a, in hippocampal and Purkinje neurons. *Nat Genet* 17:75–78, 1997.

Amir, R. E., Van, d. V., I, Wan, M., Tran, C. Q., Francke, U. and Zoghbi, H. Y. Rett syndrome is caused by mutations in X-linked MECP2, encoding methyl- CpG-binding protein 2 [see comments]. *Nat Genet* 23:185–188, 1999.

Amos-Landgraf, J. M., Ji, Y., Gottlieb, W., et al. Chromosome breakage in the Prader-Willi and Angelman syndromes involves recombination between large, transcribed repeats at proximal and distal breakpoints. *Am J Hum Genet* 65:370–386, 1999.

Angelman, H. "Puppet" children: A report of three cases. *Dev Med Child Neurol* 7:681–683, 1965.

Baumer, A., Balmer, D. and Schinzel, A. Screening for UBE3A gene mutations in a group of Angelman syndrome patients selected according to non- stringent clinical criteria. *Hum Genet* 105:598–602, 1999.

Bottani, A., Robinson, W. P., DeLozier-Blanchet, C. D., et al. Angelman syndrome due to paternal uniparental disomy of chromosome 15: a milder phenotype? [see comments]. *Am J Med Genet* 51:35–40, 1994.

Buiting, K., Dittrich, B., Gross, S., et al. Sporadic imprinting defects in Prader-Willi syndrome and Angelman syndrome: implications for imprint-switch models, genetic counseling, and prenatal diagnosis. *Am J Hum Genet* 63:170–180, 1998.

Buiting, K., Greger, V., Brownstein, B. H., et al. A putative gene family in 15q11-q13 and 16p11.2: possible implications for Prader-Willi and Angelman syndromes. *Proc Natl Acad Sci USA* 89:5457–5461, 1992.

Buiting, K., Gross, S., Ji, Y., Senger, G., Nicholls, R. D. and Horsthemke, B. Expressed copies of the MN7 (D15F37) gene family map close to the common deletion breakpoints in the Prader-Willi/Angelman syndromes. *Cytogenet Cell Genet* 81:247–253, 1998.

Buiting, K., Lich, C., Cottrell, S., Barnicoat, A. and Horsthemke, B. A 5-kb imprinting center deletion in a family with Angelman syndrome reduces the shortest region of deletion overlap to 880 bp. *Hum Genet* 105:665–666, 1999.

Buiting, K., Saitoh, S., Gross, S., et al. Inherited microdeletions in the Angelman and Prader-Willi syndromes define an imprinting centre on human chromosome 15 [published erratum appears in Nat Genet 1995 Jun;10(2):249]. *Nat Genet* 9:395–400, 1995.

Burger, J., Kunze, J., Sperling, K. and Reis, A. Phenotypic differences in Angelman syndrome patients: imprinting mutations show less frequently microcephaly and hypopigmentation than deletions. *Am J Med Genet* 66:221–226, 1996.

Carrozzo, R., Rossi, E., Christian, S. L., et al. Inter- and intrachromosomal rearrangements are both involved in the origin of 15q11-q13 deletions in Prader-Willi syndrome [letter]. *Am J Hum Genet* 61:228–231, 1997.

Chan, C. T., Clayton-Smith, J., Cheng, X. J., et al. Molecular mechanisms in Angelman syndrome: a survey of 93 patients. *J Med Genet* 30:895–902, 1993.

Christian, S. L., Fantes, J. A., Mewborn, S. K., Huang, B. and Ledbetter, D. H. Large genomic duplicons map to sites of instability in the Prader-Willi/Angelman syndrome chromosome region (15q11-q13). *Hum Mol Genet* 8:1025–1037, 1999.

Christian, S. L., Robinson, W. P., Huang, B., et al. Molecular characterization of two proximal deletion breakpoint regions in both Prader-Willi and Angelman syndrome patients. *Am J Hum Genet* 57:40–48, 1995.

Clayton-Smith, J. and Pembrey, M. E. Angelman syndrome. *J Med Genet* 29:412–415, 1992.

Connerton-Moyer, K. J., Nicholls, R. D., Schwartz, S., et al. Unexpected familial recurrence in Angelman syndrome. *Am J Med Genet* 70:253–260, 1997.

Dittrich, B., Buiting, K., Gross, S. and Horsthemke, B. Characterization of a methylation imprint in the Prader-Willi syndrome chromosome region. *Hum Mol Genet* 2:1995–1999, 1993.

Dittrich, B., Buiting, K., Korn, B., et al. Imprint switching on human chromosome 15 may involve alternative transcripts of the SNRPN gene [see comments]. *Nat Genet* 14:163–170, 1996.

Donlon, T. A. Similar molecular deletions on chromosome 15q11.2 are encountered in both the Prader-Willi and Angelman syndromes [see comments]. *Hum Genet* 80:322–328, 1988.

Engel, E. Chromosome 15 uniparental disomy is not frequent in Angelman syndrome [letter]. *Am J Hum Genet* 49:459–460, 1991.

Exeler, R., Dworniczak, B., Tercanli, S., et al. Physical and psychomotor development of three children with uniparental disomy 16. *Am J Hum Genet* 59:A92, 1996. (Abstract)

Fang, P., Lev-Lehman, E., Tsai, T. F., et al. The spectrum of mutations in UBE3A causing Angelman syndrome. *Hum Mol Genet* 8:129–135, 1999.

Freeman, S. B., May, K. M., Pettay, D., Fernhoff, P. M. and Hassold, T. J. Paternal uniparental disomy in a child with a balanced 15;15 translocation and Angelman syndrome. *Am J Med Genet* 45:625–630, 1993.

Fridman, C. and Koiffmann, C. P. Origin of uniparental disomy 15 in patients with Prader-Willi or Angelman syndrome [In Process Citation]. *Am J Med Genet* 94:249–253, 2000.

Fridman, C., Varela, M. C., Kok, F., Diament, A. and Koiffmann, C. P. Paternal UPD15: further genetic and clinical studies in four Angelman syndrome patients. *Am J Med Genet* 92:322–327, 2000.

Fridman, C., Varela, M. C., Kok, F., Diament, A. and Koiffmann, C. P. Paternal UPD15: further genetic and clinical studies in four Angelman syndrome patients. *Am J Med Genet* 92:322–327, 2000.

Gillessen-Kaesbach, G., Albrecht, B., Passarge, E. and Horsthemke, B. Further patient with Angelman syndrome due to paternal disomy of chromosome 15 and a milder phenotype [letter; comment]. *Am J Med Genet* 56:328–329, 1995.

Gillessen-Kaesbach, G., Demuth, S., Thiele, H., Theile, U., Lich, C. and Horsthemke, B. A previously unrecognized phenotype characterized by obesity, muscular hypotonia, and ability to speak in patients with Angelman syndrome caused by an imprinting defect [see comments]. *Eur J Hum Genet* 7:638–644, 1999.

Glenn, C. C., Nicholls, R. D., Robinson, W. P., et al. Modification of 15q11-q13 DNA methylation imprints in unique Angelman and Prader-Willi patients. *Hum Mol Genet* 2:1377–1382, 1993.

Glenn, C. C., Saitoh, S., Jong, M. T., et al. Gene structure, DNA methylation, and imprinted expression of the human SNRPN gene. *Am J Hum Genet* 58:335–346, 1996.

Gyftodimou, J., Karadima, G., Pandelia, E., Vassilopoulos, D. and Petersen, M. B. Angelman syndrome with uniparental disomy due to paternal meiosis II nondisjunction [see comments]. *Clin Genet* 55:483–486, 1999.

Hamabe, J., Kuroki, Y., Imaizumi, K., et al. DNA deletion and its parental origin in Angelman syndrome patients. *Am J Med Genet* 41:64–68, 1991.

Huibregtse, J. M., Scheffner, M. and Howley, P. M. A cellular protein mediates association of p53 with the E6 oncoprotein of human papillomavirus types 16 or 18. *EMBO J* 10:4129–4135, 1991.

Huibregtse, J. M., Scheffner, M. and Howley, P. M. Cloning and expression of the cDNA for E6-AP, a protein that mediates the interaction of the human papillomavirus E6 oncoprotein with p53. *Mol Cell Biol* 13:775–784, 1993.

Ji, Y., Rebert, N. A., Joslin, J. M., Higgins, M. J., Schultz, R. A. and Nicholls, R. D. Structure of the highly conserved HERC2 gene and of multiple partially duplicated paralogs in human. *Genome Res* 10:319–329, 2000.

Ji, Y., Walkowicz, M. J., Buiting, K., et al. The ancestral gene for transcribed, low-copy repeats in the Prader-Willi/Angelman region encodes a large protein implicated in protein trafficking, which is deficient in mice with neuromuscular and spermiogenic abnormalities. *Hum Mol Genet* 8:533–542, 1999.

Jiang, Y., Lev-Lehman, E., Bressler, J., Tsai, T. F. and Beaudet, A. L. Genetics of Angelman syndrome. *Am J Hum Genet* 65:1–6, 1999.

Jiang, Y. H., Armstrong, D., Albrecht, U., et al. Mutation of the Angelman ubiquitin ligase in mice causes increased cytoplasmic p53 and deficits of contextual learning and long-term potentiation [see comments]. *Neuron* 21:799–811, 1998.

Kaplan, L. C., Wharton, R., Elias, E., Mandell, F., Donlon, T. and Latt, S. A. Clinical heterogeneity associated with deletions in the long arm of chromosome 15: report of 3 new cases and their possible genetic significance. *Am J Med Genet* 28:45–53, 1987.

Kishino, T., Lalande, M. and Wagstaff, J. UBE3A/E6-AP mutations cause Angelman syndrome [published erratum appears in Nat Genet 1997 Apr;15(4):411]. *Nat Genet* 15:70–73, 1997.

Kishino, T. and Wagstaff, J. Genomic organization of the UBE3A/E6-AP gene and related pseudogenes. *Genomics* 47:101–107, 1998.

Knoll, J. H., Glatt, K. A., Nicholls, R. D., Malcolm, S. and Lalande, M. Chromosome 15 uniparental disomy is not frequent in Angelman syndrome. *Am J Hum Genet* 48:16–21, 1991.

Knoll, J. H., Nicholls, R. D., Magenis, R. E., et al. Angelman syndrome: three molecular classes identified with chromosome 15q11q13-specific DNA markers. *Am J Hum Genet* 47:149–155, 1990.

Knoll, J. H., Nicholls, R. D., Magenis, R. E., Graham, J. M. J., Lalande, M. and Latt, S. A. Angelman and Prader-Willi syndromes share a common chromosome 15 deletion but differ in parental origin of the deletion. *Am J Med Genet* 32:285–290, 1989.

Knoll, J. H., Wagstaff, J. and Lalande, M. Cytogenetic and molecular studies in the Prader-Willi and Angelman syndromes: an overview. *Am J Med Genet* 46:2–6, 1993.

Magenis, R. E., Brown, M. G., Lacy, D. A., Budden, S. and LaFranchi, S. Is Angelman syndrome an alternate result of del(15)(q11q13)? *Am J Med Genet* 28:829–838, 1987.

Magenis, R. E., Toth-Fejel, S., Allen, L. J., et al. Comparison of the 15q deletions in Prader-Willi and Angelman syndromes: specific regions, extent of deletions, parental origin, and clinical consequences. *Am J Med Genet* 35:333–349, 1990.

Malcolm, S., Clayton-Smith, J., Nichols, M., et al. Uniparental paternal disomy in Angelman's syndrome. *Lancet* 337:694–697, 1991.

Malzac, P., Webber, H., Moncla, A., et al. Mutation analysis of UBE3A in Angelman syndrome patients. *Am J Hum Genet* 62:1353–1360, 1998.

Matsuura, T., Sutcliffe, J. S., Fang, P., et al. De novo truncating mutations in E6-AP ubiquitin-protein ligase gene (UBE3A) in Angelman syndrome. *Nat Genet* 15:74–77, 1997.

Meijers-Heijboer, E. J., Sandkuijl, L. A., Brunner, H. G., et al. Linkage analysis with chromosome 15q11-q13 markers shows genomic imprinting in familial Angelman syndrome. *J Med Genet* 29:853–857, 1992.

Moncla, A., Malzac, P., Livet, M. O., et al. Angelman syndrome resulting from UBE3A mutations in 14 patients from eight families; clinical manifestations and genetic counselling. *J Med Genet* 36:554–560, 1999.

Nakao, M., Sutcliffe, J. S., Durtschi, B., Mutirangura, A., Ledbetter, D. H. and Beaudet, A. L. Imprinting analysis of three genes in the Prader-Willi/Angelman region: SNRPN, E6-associated protein, and PAR-2 (D15S225E). *Hum Mol Genet* 3:309–315, 1994.

Nicholls, R. D., Pai, G. S., Gottlieb, W. and Cantu, E. S. Paternal uniparental disomy of chromosome 15 in a child with Angelman syndrome. *Ann Neurol* 32:512–518, 1992.

Nicholls, R. D., Saitoh, S. and Horsthemke, B. Imprinting in Prader-Willi and Angelman syndromes. *Trends Genet* 14:194–200, 1998.

Pembrey, M., Fennell, S. J., van den Berghe, J., et al. The association of Angelman's syndrome with deletions within 15q11-q13 [see comments]. *J Med Genet* 26:73–77, 1989.

Ramsden, S., Gaunt, L., Seres-Santamaria, A. and Clayton-Smith, J. A case of Angelman syndrome arising as a result of a de novo Robertsonian translocation. *Acta Genet Med Gemellol (Roma)* 45:255–261, 1996.

Reis, A., Dittrich, B., Greger, V., et al. Imprinting mutations suggested by abnormal DNA methylation patterns in familial Angelman and Prader-Willi syndromes [see comments]. *Am J Hum Genet* 54:741–747, 1994.

Robinson, W. P., Christian, S. L., Kuchinka, B. D., et al. Somatic segregation errors predominantly contribute to the gain or loss of a paternal chromosome leading to uniparental disomy for chromosome 15. *Clin Genet* 57:349–358, 2000.

Robinson, W. P., Dutly, F., Nicholls, R. D., et al. The mechanisms involved in formation of deletions and duplications of 15q11-q13. *J Med Genet* 35:130–136, 1998.

Robinson, W. P., Langlois, S., Schuffenhauer, S., et al. Cytogenetic and age-dependent risk factors associated with uniparental disomy 15. *Prenat Diagn* 16:837–844, 1996.

Robinson, W. P., Lorda-Sanchez, I., Malcolm, S., et al. Increased parental ages and uniparental disomy 15: a paternal age effect? *Eur J Hum Genet* 1:280–286, 1993.

Rougeulle, C., Glatt, H. and Lalande, M. The Angelman syndrome candidate gene, UBE3A/E6-AP, is imprinted in brain [letter]. *Nat Genet* 17:14–15, 1997.

Russo, S., Cogliati, F., Viri, M., et al. Novel mutations of ubiquitin protein ligase 3A gene in Italian patients with Angelman syndrome. *Hum Mutat (Online)* 15:387, 2000.

Saitoh, S., Buiting, K., Rogan, P. K., et al. Minimal definition of the imprinting center and fixation of chromosome 15q11-q13 epigenotype by imprinting mutations. *Proc Natl Acad Sci USA* 93:7811–7815, 1996.

Saitoh, S., Harada, N., Jinno, Y., et al. Molecular and clinical study of 61 Angelman syndrome patients. *Am J Med Genet* 52:158–163, 1994.

Scheffner, M., Huibregtse, J. M., Vierstra, R. D. and Howley, P. M. The HPV-16 E6 and E6-AP complex functions as a ubiquitin-protein ligase in the ubiquitination of p53. *Cell* 75:495–505, 1993.

Scheffner, M., Werness, B. A., Huibregtse, J. M., Levine, A. J. and Howley, P. M. The E6 oncoprotein encoded by human papillomavirus types 16 and 18 promotes the degradation of p53. *Cell* 63:1129–1136, 1990.

Smith, A., Marks, R., Haan, E., Dixon, J. and Trent, R. J. Clinical features in four patients with Angelman syndrome resulting from paternal uniparental disomy. *J Med Genet* 34:426–429, 1997.

Smith, A., Wiles, C., Haan, E., et al. Clinical features in 27 patients with Angelman syndrome resulting from DNA deletion. *J Med Genet* 33:107–112, 1996.

Smith, J. C., Webb, T., Pembrey, M. E., Nichols, M. and Malcolm, S. Maternal origin of deletion 15q11-q13 in 25/25 cases of Angelman syndrome. *Hum Genet* 88:376–378, 1992.

Sutcliffe, J. S., Nakao, M., Christian, S., et al. Deletions of a differentially methylated CpG island at the SNRPN gene define a putative imprinting control region [see comments]. *Nat Genet* 8:52–58, 1994.

Tonk, V., Schultz, R. A., Christian, S. L., Kubota, T., Ledbetter, D. H. and Wilson, G. N. Robertsonian (15q;15q) translocation in a child with Angelman syndrome: evidence of uniparental disomy. *Am J Med Genet* 66:426–428, 1996.

Tsai, T. F., Raas-Rothschild, A., Ben-Neriah, Z. and Beaudet, A. L. Prenatal diagnosis and carrier detection for a point mutation in UBE3A causing Angelman syndrome [letter]. *Am J Hum Genet* 63:1561–1563, 1998.

van den Ouweland, A. M., van der Est, M. N., Wesby-van Swaay, E., et al. DNA diagnosis of Prader-Willi and Angelman syndromes with the probe PW71 (D15S63). *Hum Genet* 95:562–567, 1995.

Vu, T. H. and Hoffman, A. R. Imprinting of the Angelman syndrome gene, UBE3A, is restricted to brain [letter]. *Nat Genet* 17:12–13, 1997.

Williams, C. A., Angelman, H., Clayton-Smith, J., et al. Angelman syndrome: consensus for diagnostic criteria. Angelman Syndrome Foundation. *Am J Med Genet* 56:237–238, 1995.

Williams, C. A., Gray, B. A., Hendrickson, J. E., Stone, J. W. and Cantu, E. S. Incidence of 15q deletions in the Angelman syndrome: a survey of twelve affected persons. *Am J Med Genet* 32:339–345, 1989.

Williams, C. A., Zori, R. T., Hendrickson, J., et al. Angelman syndrome. *Curr Probl Pediatr* 25:216–231, 1995.

Williams, C. A., Zori, R. T., Stone, J. W., Gray, B. A., Cantu, E. S. and Ostrer, H. Maternal origin of 15q11-q13 deletions in Angelman syndrome suggests a role for genomic imprinting [see comments]. *Am J Med Genet* 35:350–353, 1990.

Yamamoto, Y., Huibregtse, J. M. and Howley, P. M. The human E6-AP gene (UBE3A) encodes three potential protein isoforms generated by differential splicing. *Genomics* 41:263–266, 1997.

Zackowski, J. L., Nicholls, R. D., Gray, B. A., et al. Cytogenetic and molecular analysis in Angelman syndrome [see comments]. *Am J Med Genet* 46:7–11, 1993.

Zori, R. T., Hendrickson, J., Woolven, S., Whidden, E. M., Gray, B. and Williams, C. A. Angelman syndrome: clinical profile. *J Child Neurol* 7:270–280, 1992.

Chapter 8

The Beckwith-Wiedemann Syndrome (BWS)

The Beckwith-Wiedemann syndrome (OMIM 130650) involves another important imprinted segment of the human genome, on the short arm of chromosome 11. The etiopathogenic mechanism of the syndrome has been intensively investigated in the last several years, but our molecular understanding is still incomplete. The current molecular pathophysiology of BWS has been reviewed recently (Maher and Reik, 2000).

In the BWS and in related oncologic problems associated with chromosomal or epigenetic alterations of 11p, one major added factor of genomic imbalance is probably an overdose of some gene products that are normally hemizygously expressed. Therefore, a deregulation and overproduction of transcripts, normally committed to monoallelic expression, along with a suppression or restriction of activity from oppositely imprinted loci seem to contribute to the phenotypes of BWS.

I. THE CLINICAL PICTURE

Features and characteristics of the BWS have been thoroughly discussed in a paper presenting a remarkable familial observation of the condition (Piussan et al., 1980). The original descriptions go back to the early 1960s (Wiedemann, 1964; Beckwith, 1963). This Exomphalos-Macroglossia-Gigantism syndrome, with a reported incidence of approximately 1 in 13,500 live births (Junien, 1992; Thorburn et al., 1970) (probably an underestimate because milder cases may be undiagnosed), is characterized by numerous growth abnormalities, including visceromegaly and other occasional features, such as hemihypertrophy, earlobe creases and pits, adrenocortical cytomegaly, neonatal hypoglycemia, and predisposition to several childhood embryonal malignancies such as Wilms' tumor (nephroblastoma), adrenocortical carcinoma, hepatoblastoma, rhabdomyosarcoma, pancreatic tumor, and neuroblastoma. The omphalocele varies in size and may be replaced by a simple hernia; it exists in a majority of the cases and may be complicated by liver, bowel, and diaphragmatic malformations. Macroglossia, evident from birth, is variable, from moderate to severe, eventually causing respiratory difficulties and may require surgery although gradual regression is the rule. The degree of macrosomia is also quite variable, usually presented as neonatal gigantism, excessive weight, accelerated growth and advanced bone age, body hemihypertrophy with increased muscle mass and subcutaneous lipodystrophy. The extent of macrosomia is difficult to assess in stillborn and in premature birth of ill-defined gestational age. The organomegaly is clinically manifested as hepatomegaly, splenomegaly, enlarged kidneys, and eventual hypertrophy of the clitoris.

Other inconsistent signs include microcephaly, facial nevus flammeus, at times extending from the forehead to the nose, eyelids, even the upper lip and fading within some years. The earlobe crease, present in some familial cases, is lacking in others. Variable other malformations have been observed in different organs: skeletal, cardiac, genital (bifid uterus, undescended testes, hypospadias) and urinary (urethral strictures, hydronephrosis, polycystic kidneys, horse-shoe kidneys).

Neonatal hypoglycemia is of major concern in over half the cases and may be difficult to manage; it usually subsides by the second trimester of life. It is best explained by Langerhans islet hyperplasia and excess of insulin secretion, well-documented in some cases. In other cases, there is reactional hypoglycemia, in response to glucose loading or progressive prediabetic status. An excellent clinical description of BWS may also be found at (http://www.geneclinics.org/profiles/bws/).

The neonatal prognosis relies on the degree of hypoglycemia and a liability to infections. Infants with BWS show an approximately 20% mortality rate. The prognosis after the infantile period depends on the development of secondary malignant growth, including nephroblastomas, adrenal carcinomas, brain astrocytomas, ganglioneuromas, heart hamartomas, rhabdomyosarcomas, small bowel carcinoid tumors, and gonadoblastomas. The size of internal organs is increased, and on microscopic examination, there is diffuse cellular hyperplasia. Pancreatic islets are numerous and hypertrophic, showing acinar hyperplasia. The adrenal cortex is overdeveloped, with a striking degree of cytomegaly. Kidney glomeruli are numerous and dense with persistent glomerulogenesis (Piussan et al., 1980).

II. BWS WITH PATERNAL UPD11

There are several molecular abnormalities associated with BWS and their frequencies are shown in Figure 1. Approximately 10–20% of patients with BWS have a paternal UPD for a segment of chromosome 11p15. The paternal UPD in the majority of cases is limited to a short-arm segment, apparently resulting from

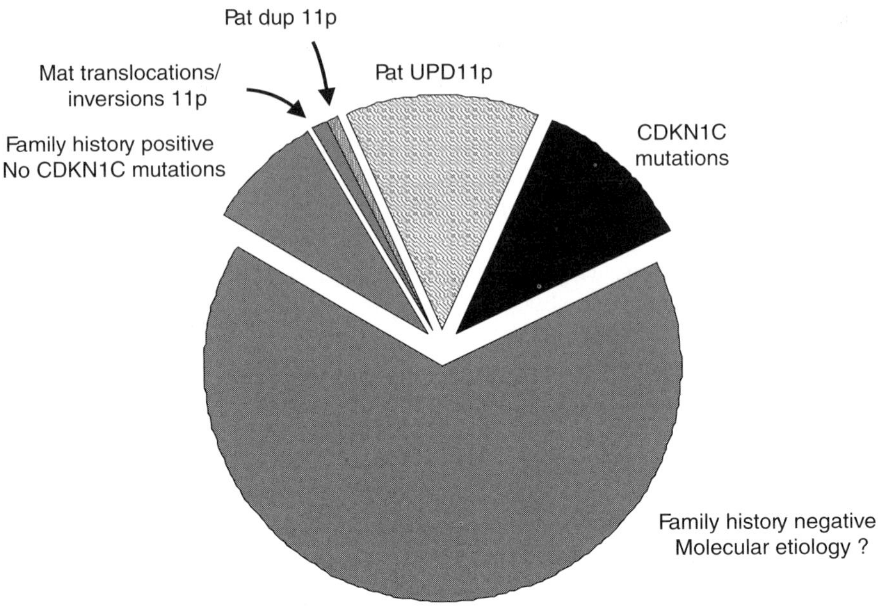

Figure 1 *Frequency of the different clinical, cytogenetic, and molecular subcategories of BWS.*

interchromatid somatic recombination. There follows clonal development of a cell population with a paternal isodisomic area on a portion of chromosome 11, while the remaining chromosome 11 is biparental. This mechanism has been discussed in Chapter 3 and Figure 7 there.

There are only three cases reported with paternal UPD for the entire length of chromosome 11 (see Chapter 4), suggesting a meiotic error in these cases.

Henry et al. (1991) first demonstrated paternal UPD in 3/8 informative sporadic cases of BWS. Subsequently, investigators in the same laboratory undertook a systemic study of 17 more cases for a total of 25 patients, to further document the nature and extent of the aberration in the etiology and natural history of BWS (Henry et al., 1993). This study revealed: (i) Five of the 25 cases had paternal UPD in the short-arm area of chromosome 11. (ii) The smallest region of proven disomy using informative markers always included the INS and IGF2 loci. (iii) Isodisomy was unambiguously proven in four cases for markers at loci TH, INS, D11S12, HBBP1, PTH, D11S774, HBB, IGF2, CALCA, and HRAS. A fifth case was uninformative for these or other markers, but exhibited homozygosity for 16 contiguous markers in the region, extending from HRAS to D11S325, strongly suggesting isodisomy as did homozygosity for two markers (D11S12, HBBP1), for which the father was heterozygous. (iv) The extent of the isodisomy could be determined by recognizing alleles unquestionably maternal in origin. This was observed, e.g., at the CALCA locus in one case, carrying an informative allele for which the mother was homozygous and different from the father. This observation, establishes segmental UPD and the likely mechanism of a postzygotic mitotic event. (v) Somatic mosaicism for cells with paternal UPD11p and biparental inheritance of 11p was shown in four patients by direct examination or after overexposure and/or overloading the Southern blots, since a maternal allele was observed for some markers in the region of UPD, at loci such as D11S774, HBBP1, INS, HRAS, or CALCA. Contamination by maternal DNA was ruled out by the absence of the second maternal allele. Furthermore, densitometer scanning of the respective signal intensities confirmed the presence of two types of cells with respect to the segment made uniparental by recombination, i.e., cells with biparental and those with uniparental contribution for this area. In fact, proof of a mosaic pattern confirming the finding that paternal UPD11p is of somatic and clonal origin was achieved by more direct studies. Using micromanipulation, single blood lymphocytes from a BWS patient were isolated and subjected to whole genome amplification, after which locus-specific microsatellite marker analyses were performed to determine the chromosome 11 origin. Two populations were thus detected, one wholly biparental and one with 11p paternal isodisomy (from D11S922 to D11S904) and biparental disomy of 11p14-11qter (Bischoff et al., 1995). Mosaics of cells heterozygous and homozygous for paternal 11p markers have also been seen in normal tissues of Wilms' tumor patients (Chao et al., 1993).

That such mosaics exist is, in fact, of no surprise since in nine cases of monozygotic twinning, each pair was discordant for the expression of BWS (Lubinsky and Hall, 1991; Henry et al., 1993). This is perfectly compatible with the postzygotic origin of paternal UPD11p, which is only present in somatic tissues of one of the identical twins. The real puzzle is that such somatic recombination events should yield, as a counterpart, a second clonal cell line with a maternal UPD11p that has so far never been detected, indicating that cells with such a maternal isodisomy are likely to have a selective disadvantage and are therefore eliminated.

The initial work from the Junien laboratory (Henry et al., 1991, 1993; Junien, 1992) was conclusively confirmed by a study from another laboratory (Slatter et al., 1994). UPD for chromosome 11p15 was detected in 9 of 32 (28%) informative patients and, in all cases with UPD, the probands showed mosaicism for a paternal isodisomy and a normal cell line, confirming that UPD had arisen as a postzygotic event. The estimated proportion of disomic cells varied between 23 and 79%. In three informative cases, UPD extended to the WT1 locus that maps at 11p13, whereas three others were found biparental at this same locus.

Phenotypic Presentation of BWS with Mosaic Paternal Isodisomy UPD11p15

In this latter study (Slatter et al., 1994), comparison of the nine cases with paternal UPD11p to those 23 BWS without paternal UPD11p indicated that the UPD cases were more likely to have hemihypertrophy (6/9 versus 1/23) and less likely to have exomphalos (0/0 versus 13/23), whereas no differences were seen with respect to other features such as neoplasia, developmental delay, hypoglycemia, and nephromegaly. An initial comparison between UPD11p and non-UPD11p BWS cases indicated that tumorigenesis was more frequent in UPD cases, but the collected number of observations was small (Henry et al., 1993). Mosaicism for normal and UPD11p cells might account for asymmetrical tissue growth and hemihypertrophy in BWS (Slatter et al., 1994).

III. BWS WITH MUTATIONS IN THE CDKN1C (p57KIP2) GENE

The chromosomal region 11p15.5 involved in BWS (see below) contains several genes including CDKN1C also called p57KIP2 (Figure 2). CDKN1C encodes a cyclin-dependent kinase inhibitor of the CIP family of cell-cycle regulators (Lee et al., 1995; Matsuoka et al., 1995). It negatively regulates cell proliferation and its overexpression can arrest cells in G1. This gene maps within the imprinted cluster on chromosome 11p15.5 (between loci NAP2 and L23). It is expressed in several fetal tissues, in lung and adult kidney from the maternally derived chromosome 11, and is almost inactive and methylated (low levels of expression) in the paternally derived chromosome 11 (Matsuoka et al., 1996; Hatada et al., 1996).

Several BWS patients have been identified with point mutations in the CDKN1C gene (Table 1 and the references contained there). Approximately 40% of familial cases contain heterozygous mutations in this gene and all these mutations result in loss of function of the CDKN1C protein. In all familial cases, the mutation was

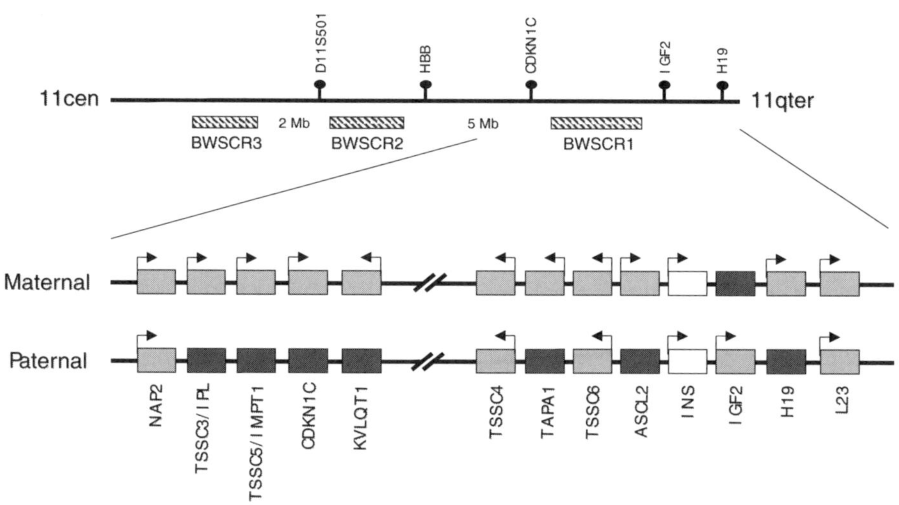

Figure 2 *Schematic representation of the imprinting cluster on chromosome 11p15.5 involved on BWS. The three breakpoint cluster regions (BWSCR) are indicated on the top panel. The bottom panel depicts selected genes between NAP2 and L23 and their imprinted status. Dark gray genes are not active in the parental chromosome shown. Light gray genes are active. The direction of transcription is also shown. The imprinting status of the insulin (INS) gene has not yet been elucidated in humans. This figure is modified from (Maher and Reik, 2000).*

TABLE 1 Mutations in the CDKN1C Gene in Patients with BWS

Mutation	BWS	Origin	References	Comments
Glu47Ter	Familial	Maternal	(Hatada et al., 1996; Bhuiyan et al., 1999)	Complete loss of cell-cycle inhibition and its nuclear localization
Pro70Leu		New mutation	(Lam et al., 1999)	
Ser247Ter		?	(Hatada et al., 1997)	
Ser282Ter	Familial	Maternal	(Lam et al., 1999)	
Arg316Trp		?	(Lam et al., 1999)	
Del1nt codon 34		?	(Lee et al., 1997)	
Del1nt codon 153		?	(Lee et al., 1997)	
Del9nt codon 61		?	(O'Keefe et al., 1997)	
Del16nt codon 167	Familial	Maternal	(Lam et al., 1999)	
Del2nt ins1nt codon 103	Familial	Maternal	(Hatada et al., 1997)	Patient sister had same mutation, mother had gigantism during infancy
Del1nt ins2nt codon 275	Familial	Maternal	(Hatada et al., 1996)	
IVS2 − 3C > G acceptor	Familial	Maternal	(Lam et al., 1999)	

inherited from the mother in whom this gene is active. The incidence of CDKN1C mutations among sporadic BWS cases is approximately 5–10%. Overall, the incidence of CDKN1C mutation in BWS patients is 10–15% since about 85% of cases have negative family history and 15% positive family history (Pettenati et al., 1986). A mouse with targeted disruption of the CDKN1C (Zhang et al., 1997) suffers from anterior abdominal wall defects, adrenal cortical cytomegaly, and renal medulla dysplasia; however, there is no prenatal overgrowth. Remarkably, there is high incidence of exomphalos (9 of 11 cases) in BWS with CDKN1C point inactivating mutations (Lam et al., 1999).

A total of 70 patients with classical BWS were investigated; 54 were sporadic with no evidence of UPD and 16 were familial from seven kindreds. Germ-line CDKN1C mutations were identified in five probands, 3/7 (43%) familial cases and 2/54 (4%) sporadic cases. There was no association between germ-line CDKN1C mutations and IGF2 or H19 epigenotype abnormalities (Lam et al., 1999). These findings show that germ-line CDKN1C mutations are a considerable cause of familial but not sporadic BWS, and suggest that CDKN1C mutations probably cause BWS independently of changes in IGF2/H19 imprinting.

It is also apparent that not all BWS patients with positive family history are associated with CDKN1C mutations, and that additional yet undiscovered mechanisms are involved in BWS. Recently, 32 patients with BWS were examined for mutations in the CDKN1C gene, including seven cases of familial BWS. Mutations were not detected in the coding region of the CDKN1C gene in any individual with BWS (Algar et al., 2000). In two patients, two G/A base substitutions at adjacent positions in the 5′UTR were detected and, as these substitutions were also found in normal controls, they are likely polymorphisms. Expression of CDKN1C in somatic tissues was examined in 18 of the 32 cases using semiquantitative RT-PCR.

CDKN1C expression was significantly reduced in the blood of three cases compared with controls. These results suggest that, although coding region mutations in the CDKN1C gene are rare in BWS, mutations disrupting CDKN1C expression may be found. Three of five informative patients exhibited biallelic CDKN1C expression in lymphocytes, cord blood, and kidney tissue, respectively. Biallelic expression was not associated with CDKN1C levels significantly different from those in controls. Patients who expressed CDKN1C biallelically, or who were low CDKN1C expressors, maintained monoallelic methylation in the differentially methylated region 2 (DMR2) of the IGF2 locus. One patient expressing CDKN1C biallelically maintained imprinted gene expression at the IGF2 locus. These results confirmed that the mechanisms associated with regulating CDKN1C expression and imprinting are separate from those regulating IGF2 imprinting.

IV. CHROMOSOMAL ABNORMALITIES AND BWS

A small proportion, around 1%, of BWS cases have chromosomal abnormalities involving 11p15. These abnormalities are either paternal 11p15 duplications or maternally derived 11p15 translocations or deletions.

11p Trisomy Duplication in BWS

These cases have been reviewed in recent years (Weksberg et al., 1993; Mannens et al., 1994), as they further delineated underlying genomic lesions causing BWS and associated tumors. From 1983–1993, 18 cases with duplications of 11p material had been reported. In 11 informative cases, all these were of paternal origin and the smallest duplication involved the area from 11p15 to pter. Similar duplications of 11p15 had also been found in five patients described earlier with the diagnosis of BWS; four of them were noted to occur on the paternally derived chromosome (Turleau and de Grouchy, 1985). Four of the BWS cases with duplications were restudied and confirmed (Mannens et al., 1994). Southern blotting using single-copy probes and FISH using cosmids for 11p allowed the localization of breakpoints at 11p14.1; 11p15.1-p14.3; 11p15.1-p15.3, and 11p15.3-p15.4. Clinically, these patients were typical for BWS, with increased birthweight or postnatal gigantism, macroglossia, earlobe grooves, omphalocele or hernia, and mild mental retardation.

Balanced Translocations or Inversions

These chromosomal abnormalities always involved the maternally derived chromosome 11. The breakpoints on 11p all mapped to a region 11p15.4-pter, as follows: inv(11)(p15.4q22.3); t(4;11)(p15.2;15.4); t(11;22)(p15.5;q12); t(9;11)(p11.2;p15.5); t(11;16)(p15.5;q12); and t(11;2)(p15.5;q13.1).

Of the above, five were personally analyzed by the authors (Mannens et al., 1994), and although macrosomia was less documented among these, most shared macroglossia, omphalocele or hernia, and earlobe grooves. These five patients and two more from the literature had inherited their balanced translocations or inversions from their healthy mothers.

The breakpoints of maternally inherited translocations and inversions map in three distinct breakpoint cluster regions BNSCR1, BNSCR2, and BNSCR3 on 11p15 (Figure 2). BNSCR1, the most common breakpoint cluster region, is approximately 450 kb and is at least 200 kb centromeric to IGF2, which is not disrupted, and telomeric to CDKN1C, which is also not disrupted. Breakpoints in this region interrupt the KVLQT1 gene (Lee et al., 1997). BWSCR2 and BWSCR3 are less frequent breakpoint clusters and map 5 Mb and 7 Mb centromeric to BWSCR1 (Redeker et al., 1994, 1995). BWSCR1 encompasses approximately 2 Mb.

The region of the imprinted cluster on chromosome 11p15.5 is shown in Figure 2 (adapted from (Maher and Reik, 2000). This cluster maps between NAP2 and L23 that do not show differential parentally specific expression. Genes TSSC4 and TSSC6 also do not show parental imprinting. Genes IPL/TSSC3, IMPT1/TSSC5, CDKN1C, KVLQT1, TAPA1, ASCL2, and H19 are expressed from the maternal chromosome and are silent from the paternal (Maher and Reik, 2000). IGF2 that encodes a fetal growth factor is silent from the maternal chromosome and active from the parental chromosome. The imprinting status of the INS gene has not yet been defined in humans.

V. IMPRINTED GENES IN THE BWS, 11p15.5 CRITICAL SEGMENT

IGF2 (and H19) in BWS

The IGF2 and H19 genes are closely linked on 11p15.5 and oppositely imprinted. As depicted in Figure 2, IGF2 is normally expressed from the paternal chromosome, whereas H19 is expressed from the maternal (Maher and Reik, 2000), suggesting a single mechanism for the reciprocal imprinting of these two genes. The paternally expressed IGF2 gene has been considered an important contributor to the overgrowth phenotypes of BWS. IGF2, insulinlike growth factor 2, encodes a fetal growth factor (Rappolee et al., 1992; Bartolomei and Tilghman, 1997); the locus lies about 100–200 kb upstream of HI9 and is expressed from only the paternal allele in most tissues. A specific paternal methylation of this monollelically active locus is observed within exon 9 in normal human tissues (Lalande, 1996; Reik et al., 1994; Schneid et al., 1993). There are several IGF2 isoforms due to transcripts originating from the alternative promoter use or alternate mRNA splicing (Vu and Hoffman, 1994; Ekstrom et al., 1995). The functional imprinting of these different products can be tissue-specific, vary throughout development, and even differ between individuals. In humans, alternative usage of the IGF2 four promoters may in the liver be responsible, at different developmental stage, for either monoallelic (P2-P4) or biallelic (P1) expression (Vu and Hoffman, 1994). In most humans and in mice embryonic tissues, the maternal allele is inactivated, except in the choroid plexus and leptomeninges, where IGF2 is expressed from both alleles (Lalande, 1996).

Overexpression of IGF2 in transgenic mice is associated with polyhydramnios, prenatal overgrowth, macroglossia, and organomegaly (Sun et al., 1997; Eggenschwiler et al., 1997). IGF2 is therefore considered a strong candidate for contributing to the phenotypes of BWS.

The H19 gene encodes a polyadenylated spliced messenger RNA without open reading frame and is assumed to act as an mRNA suppressing tumor growth in vitro (Hao et al., 1993). In humans and mice, only the maternal allele is expressed, and the same pattern of methylation is observed for the paternal (unexpressed) HI9 gene (Tremblay et al., 1995; Zhang et al., 1993). In normal mouse tissue, monoallelic transcription of the two loci, HI9 and IGF2, is observed in the same cell type, maternal for H19 and paternal for IGF2. A shared set of enhancers, located upstream of H19, may regulate the activity of promoters of both genes (Zemel et al., 1992; Bartolomei et al., 1993). When the maternal H19 allele "uses" the enhancers, the maternal IGF2 allele is inactive; on the paternal chromosome, the H19 allele is methylated and repressed, and the enhancers are left free to activate transcripts of the IGF2 allele [reviewed in (Lalande, 1996)]. A model of enhancer competition has been proposed to explain the reciprocal nature of HI9 and IGF2 imprinting (Banerjee and Smallwood, 1995). An alternative model has also been hypothesized to explain the observation in H19 knock-out mice that the transcriptional activity from the H19 promoter is not required for the imprinted silencing of the IGF2 gene (Schmidt et al., 1999). In this model, an epigenetic mark upstream of H19 serves as a boundary element on the unmethylated maternal chromosome. In fact, the imprinting of other genes in the cluster such as CDKN1C and KVLQT1 is independent of H19 (Caspary et al., 1998).

Loss of uniparental expression of IGF2 and appearance of biallelic expression from both parental chromosomes have been observed in a substantial number of sporadic BWS patients (Weksberg et al., 1993; Joyce et al., 1997; Lee et al., 1999). This transition from uniparental to biparental expression is often termed LOI, "loss of imprinting."

The methylation status on INS and IGF2, distal to the breakpoint of a BWS patient with a t(4;11)(p15.2;15.4)mat, was found to be diminished at the insulin locus relative to various controls (Mannens et al., 1994). At birth, both alleles were hypomethylated, but remarkably after the first three years only, the maternal allele remained as such. BWS patients have elevated levels of insulin and IGF2 so that hypomethylation of these genes could link their involvement in some features of the syndrome such as hypoglycemia, growth abnormalities, and neoplasia.

Abnormalities in imprinting have been proposed by a number of investigators to account for the diverse phenotypes of BWS. These hypotheses were based on (i) paternal origin of 11p duplications; (ii) maternal derivation of inversions and balanced translocations in affected individuals; (iii) maternal mode of transmission in familial cases without chromosome rearrangement (Moutou et al., 1992; Viljoen and Ramesar, 1992); (iv) homology of the loci involved, including IGF2 and H19 with mouse imprinted domains on chromosome 7 (Bartolomei et al., 1993); (v) methylation changes of these loci in health, development, and disease (Tremblay et al., 1995; Zhang et al., 1993; Schneid et al., 1993; Ohlsson et al., 1993).

Methylation Deregulation of IGF2/H19 in BWS

In patients with mosaic paternal segmental UPD11p, there is some demethylation at exon 9 of IGF2, but it is difficult to assess the H19 methylation since the cell population is not homogeneous.

There is a subset of non-UPD BWS with normal allelic methylation of both genes on peripheral blood DNA. Some of these carry mutations in the CDKN1C gene.

In another subset of the non-UPD BWS, there is an altered imprinting pattern in the IGF2/H19 domain. The IGF2 gene is biallelically expressed (IGF2 LOI) and there is silencing of the H19 from the maternal chromosome (Reik et al., 1995; Weksberg et al., 1993) (Figure 3). There is also a paternal type of methylation on both maternal and paternal chromosomes (i.e., hypermethylation of H19), suggesting an abnormality in the imprinting control region termed BWSIC1 (Reik et al., 1995). In these patients, the methylation status in the maternal chromosome of another imprinted gene KVLQT1-AS (antisense KVLQT1; see below) is not changed (Smilinich et al., 1999).

In a non-UPD BWS patient with an inherited 11p15.5 inversion, an IGF2 LOI (biallelic expression of IGF2) was observed but with normal H19 monoallelic (maternal) expression (Brown et al., 1996). An abnormality in a second imprinting center, BWSIC2, was proposed (Reik et al., 1995) (Figure 3). In addition, there was a change in the methylation status or expression of the KVLQT1-AS transcript

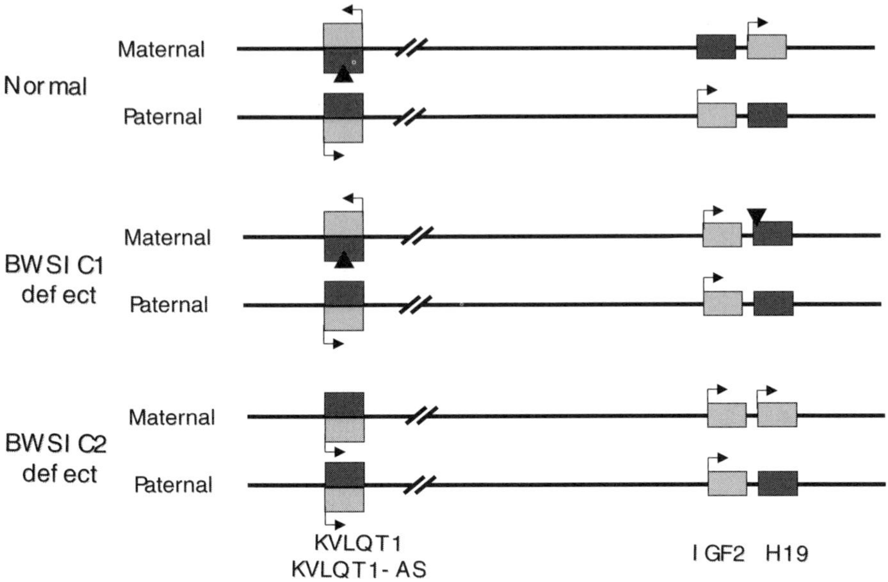

Figure 3 Schematic representation of the epigenetic changes in BWS patients with presumed imprinting center defects. Abnormalities in two imprinting centers are proposed. In each case, the gene activity from the maternal and paternal chromosome is shown. Dark gray boxes indicate inactive and light gray boxes the active copies of genes. The direction of transcription is indicated by arrows. The two overlapping transcripts KVLQT1 and KVLQT1-AS are shown as staggered boxes. Black triangles depict methylation. This figure is modified from (Maher and Reik, 2000).

(Smilinich et al., 1999). The nature of the defects in the BWSIC1 and BWSIC2 is unknown. The function of these yet hypothetical imprinting centers is related to the regulation of the IGF2 and H19 imprinting for BWSIC1, and to the regulation of imprinting of IGF2 independent of H19 and the antisense KVLQT1-AS transcript for BWSIC2. It is of interest to note that mice with a targeted disruption of the H19 gene mimic the epigenetic effects of BWSIC1 defects (Schmidt et al., 1999; Reik et al., 1995).

BWS, the KVLQT1 Gene, and the KVLQT1-AS (Antisense Transcript)

The gene for KVLQT1 (or KCNQ1) maps between the CDKN1C and IGF2 on 11p15.5 (Lee et al., 1997). When its imprinting was assessed, most fetal tissues showed monoallelic expression, the active allele being of maternal origin (Lee et al., 1997). However, some tissues displayed significant expression of both alleles, most markedly and constantly the heart. The 14 exons of KVLQT1 gene span about 300 kb, encode a potassium channel, and are usually disrupted by chromosomal rearrangements in the BWSCR1 region in patients with BWS, as well as by a balanced chromosomal translocation in an embryonal rhabdoid tumor. Mutations in

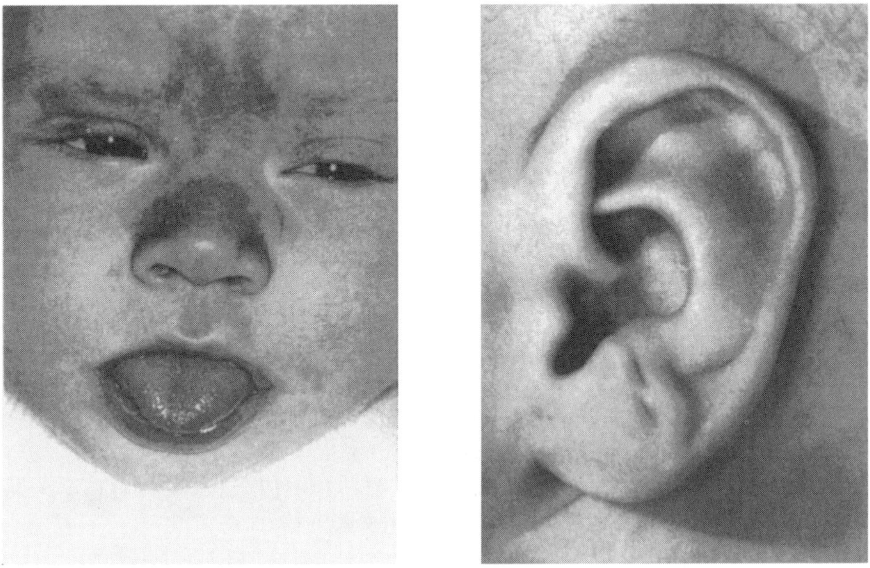

Figure 4 *Infants with BWS syndrome, showing macroglossia, telangiectatic nevi of the face, and linear ear creases (from Wiedmann and Kunze 1997. Reproduced with permission).*

KVLQT1 have been identified in patients with one form of the long QT syndrome (Wang et al., 1996) and one form of the Jervell and Lange-Nielsen syndrome (Neyroud et al., 1997). This gene encodes multiple transcripts and the function of each different transcript is not entirely clear. The role of the KVLQT1 gene in BWS is also not clear.

An KVLQT1 antisense overlapping transcript has also been described (named KVLQT1OT, or KVLQT1-AS or long QT intronic transcript 1, LIT1). KVLQT1-AS is expressed preferentially from the paternal allele and produced in most human tissues (Mitsuya et al., 1999; Smilinich et al., 1999; Lee et al., 1999). Methylation analysis revealed that an intronic CpG island was specifically methylated on the silent maternal allele and that 4 of 13 BWS patients showed complete loss of maternal methylation at the CpG island, suggesting that antisense regulation is involved in the development of human disease.

These results were confirmed by another study in which, in the majority of patients with BWS, KVLQT1-AS is abnormally expressed from both the paternal and maternal alleles. Eight of 16 informative BWS patients (50%) showed biallelic expression, i.e., loss of imprinting (LOI) of KVLQT1-AS. Similarly, 21 of 36 (58%) BWS patients showed loss of maternal allele-specific methylation of a CpG island upstream of KVLQT1-AS (Lee et al., 1999). The LOI of KVLQT1-AS was not linked to LOI of IGF2, which was found in 2 of 10 (20%) BWS patients. Thus, LOI of KVLQT1-AS was the most common genetic alteration in BWS in this study (Lee et al., 1999).

VI. GENETIC COUNSELING OF BWS

Laboratory Tests

Based on the available information regarding the molecular etiology of BWS, the following laboratory test are recommended:

1. To rule out a chromosomal abnormality that involves chromosome 11p15.5, it is necessary to perform a high-resolution karyotype with special emphasis on the region of interest. The yield of such abnormalities is about 1% of BWS cases. If positive, parental karyotypes should also be performed.
2. About 10–15% of BWS cases have segmental mosaic paternal UPD11p15. This could be diagnosed using analysis of DNA polymorphic markers.
3. Search for mutations in the CDKN1C gene; approximately 5–10% of BWS patients have maternally inherited mutations in this gene.
4. Studies of the methylation status of 1GF2, H19, and KVLQT1-AS are optional and could be performed on a research basis.

Recurrence Risk

The counseling of couples with a child with BWS regarding the recurrence risk (RR) is complex and depends on the molecular basis of the syndrome. The large majority of cases (80–85%) are those with negative family histories and normal karyotypes. The RR is up to 50% if a CDKN1C mutation is present in a parent. In contrast, the RR is very low if paternal UPD11p is diagnosed. In non-UPD11, non-CDKN1C mutation families, the RR is unknown but probably low (<5%).

Approximately 10–15% of BWS have a positive family history and normal karyotype. The RR is up to 50% in families with mutations in the CDKN1C gene. Similarly, the RR is up to 50% in families without CDKN1C mutations.

A very small fraction (1%) of BWS patients have a chromosomal abnormality on 11p (paternally derived duplication or maternally derived translocation/inversion). Parental karyotypes are necessary to provide an estimate of the recurrence risk. The RR for a mother with balanced translocation is up to 50%, whereas for a father the RR is increased but not exactly known.

The detection of CDKN1C mutations and UPD11p could be offered for fetal diagnosis in appropriate cases.

REFERENCES

Algar, E., Brickell, S., Deeble, G., Amor, D. and Smith, P. Analysis of CDKN1C in Beckwith Wiedemann syndrome. *Hum Mutat* 15:497–508, 2000.

Banerjee, S. and Smallwood, A. A chromatin model of IGF2/H19 imprinting [letter]. *Nat Genet* 11:237–238, 1995.

Bartolomei, M. S. and Tilghman, S. M. Genomic imprinting in mammals. *Annu Rev Genet* 31:493–525, 1997.

Bartolomei, M. S., Webber, A. L., Brunkow, M. E. and Tilghman, S. M. Epigenetic mechanisms underlying the imprinting of the mouse H19 gene. *Genes Dev* 7:1663–1673, 1993.

Beckwith, J. B. Extreme cytomegaly of the adrenal fetal cortex, omphalocele, hyperplasia of the kidneys and pancreas and Leydig cell hyperplasia; another syndrome? (Western Society of Pediatric Research). 1963.

Bhuiyan, Z. A., Yatsuki, H., Sasaguri, T., et al. Functional analysis of the p57KIP2 gene mutation in Beckwith-Wiedemann syndrome. *Hum Genet* 104:205–210, 1999.

Bischoff, F. Z., Feldman, G. L., McCaskill, C., Subramanian, S., Hughes, M. R. and Shaffer, L. G. Single cell analysis demonstrating somatic mosaicism involving 11p in a patient with paternal isodisomy and Beckwith-Wiedemann syndrome. *Hum Mol Genet* 4:395–399, 1995.

Brown, K. W., Villar, A. J., Bickmore, W., et al. Imprinting mutation in the Beckwith-Wiedemann syndrome leads to biallelic IGF2 expression through an H19-independent pathway. *Hum Mol Genet* 5:2027–2032, 1996.

Caspary, T., Cleary, M. A., Baker, C. C., Guan, X. J. and Tilghman, S. M. Multiple mechanisms regulate imprinting of the mouse distal chromosome 7 gene cluster. *Mol Cell Biol* 18:3466–3474, 1998.

Chao, L. Y., Huff, V., Tomlinson, G., Riccardi, V. M., Strong, L. C. and Saunders, G. F. Genetic mosaicism in normal tissues of Wilms' tumour patients. *Nat Genet* 3:127–131, 1993.

Eggenschwiler, J., Ludwig, T., Fisher, P., Leighton, P. A., Tilghman, S. M. and Efstratiadis, A. Mouse mutant embryos overexpressing IGF-II exhibit phenotypic features of the Beckwith-Wiedemann and Simpson-Golabi-Behmel syndromes. *Genes Dev* 11:3128–3142, 1997.

Ekstrom, T. J., Cui, H., Li, X. and Ohlsson, R. Promoter-specific IGF2 imprinting status and its plasticity during human liver development. *Development* 121:309–316, 1995.

Hao, Y., Crenshaw, T., Moulton, T., Newcomb, E. and Tycko, B. Tumour-suppressor activity of H19 RNA. *Nature* 365:764–767, 1993.

Hatada, I., Inazawa, J., Abe, T., et al. Genomic imprinting of human p57KIP2 and its reduced expression in Wilms' tumors. *Hum Mol Genet* 5:783–788, 1996.

Hatada, I., Nabetani, A., Morisaki, H., et al. New p57KIP2 mutations in Beckwith-Wiedemann syndrome. *Hum Genet* 100:681–683, 1997.

Hatada, I., Ohashi, H., Fukushima, Y., et al. An imprinted gene p57KIP2 is mutated in Beckwith-Wiedemann syndrome [see comments]. *Nat Genet* 14:171–173, 1996.

Henry, I., Bonaiti-Pellie, C., Chehensse, V., et al. Uniparental paternal disomy in a genetic cancer-predisposing syndrome [see comments]. *Nature* 351:665–667, 1991.

Henry, I., Puech, A., Riesewijk, A., et al. Somatic mosaicism for partial paternal isodisomy in Wiedemann-Beckwith syndrome: a post-fertilization event. *Eur J Hum Genet* 1:19–29, 1993.

Joyce, J. A., Lam, W. K., Catchpoole, D. J., et al. Imprinting of IGF2 and H19: lack of reciprocity in sporadic Beckwith-Wiedemann syndrome. *Hum Mol Genet* 6:1543–1548, 1997.

Junien, C. Beckwith-Wiedemann syndrome, tumourigenesis and imprinting [published erratum appears in Curr Opin Genet Dev 1992 Aug;2(4):651]. *Curr Opin Genet Dev* 2:431–438, 1992.

Lalande, M. Parental imprinting and human disease. *Annu Rev Genet* 30:173–195, 1996.

Lam, W. W., Hatada, I., Ohishi, S., et al. Analysis of germline CDKN1C (p57KIP2) mutations in familial and sporadic Beckwith-Wiedemann syndrome (BWS) provides a novel genotype phenotype correlation. *J Med Genet* 36:518–523, 1999.

Lee, M. H., Reynisdottir, I. and Massague, J. Cloning of p57KIP2, a cyclin-dependent kinase inhibitor with unique domain structure and tissue distribution. *Genes Dev* 9:639–649, 1995.

Lee, M. P., DeBaun, M., Randhawa, G., Reichard, B. A., Elledge, S. J. and Feinberg, A. P. Low frequency of p57KIP2 mutation in Beckwith-Wiedemann syndrome. *Am J Hum Genet* 61:304–309, 1997.

Lee, M. P., DeBaun, M. R., Mitsuya, K., et al. Loss of imprinting of a paternally expressed transcript, with antisense orientation to KVLQT1, occurs frequently in Beckwith-Wiedemann syndrome and is independent of insulin-like growth factor II imprinting. *Proc Natl Acad Sci USA* 96:5203–5208, 1999.

Lee, M. P., Hu, R. J., Johnson, L. A. and Feinberg, A. P. Human KVLQT1 gene shows tissue-specific imprinting and encompasses Beckwith-Wiedemann syndrome chromosomal rearrangements [see comments]. *Nat Genet* 15:181–185, 1997.

Lubinsky, M. S. and Hall, J. G. Genomic imprinting, monozygous twinning, and X inactivation [letter] [see comments]. *Lancet* 337:1288, 1991.

Maher, E. R. and Reik, W. Beckwith-Wiedemann syndrome: imprinting in clusters revisited [see comments]. *J Clin Invest* 105:247–252, 2000.

Mannens, M., Hoovers, J. M., Redeker, E., et al. Parental imprinting of human chromosome region 11p15.3-pter involved in the Beckwith-Wiedemann syndrome and various human neoplasia. *Eur J Hum Genet* 2:3–23, 1994.

Matsuoka, S., Edwards, M. C., Bai, C., et al. p57KIP2, a structurally distinct member of the p21CIP1 Cdk inhibitor family, is a candidate tumor suppressor gene. *Genes Dev* 9:650–662, 1995.

Matsuoka, S., Thompson, J. S., Edwards, M. C., et al. Imprinting of the gene encoding a human cyclin-dependent kinase inhibitor, p57KIP2, on chromosome 11p15. *Proc Natl Acad Sci USA* 93:3026–3030, 1996.

Mitsuya, K., Meguro, M., Lee, M. P., et al. LIT1, an imprinted antisense RNA in the human KvLQT1 locus identified by screening for differentially expressed transcripts using monochromosomal hybrids. *Hum Mol Genet* 8:1209–1217, 1999.

Moutou, C., Junien, C., Henry, I. and Bonaiti-Pellie, C. Beckwith-Wiedemann syndrome: a demonstration of the mechanisms responsible for the excess of transmitting females [see comments]. *J Med Genet* 29:217–220, 1992.

Neyroud, N., Tesson, F., Denjoy, I., et al. A novel mutation in the potassium channel gene KVLQT1 causes the Jervell and Lange-Nielsen cardioauditory syndrome [see comments]. *Nat Genet* 15:186–189, 1997.

O'Keefe, D., Dao, D., Zhao, L., et al. Coding mutations in p57KIP2 are present in some cases of Beckwith-Wiedemann syndrome but are rare or absent in Wilms tumors. *Am J Hum Genet* 61:295–303, 1997.

Ohlsson, R., Nystrom, A., Pfeifer-Ohlsson, S., et al. IGF2 is parentally imprinted during human embryogenesis and in the Beckwith-Wiedemann syndrome. *Nat Genet* 4:94–97, 1993.

Pettenati, M. J., Haines, J. L., Higgins, R. R., Wappner, R. S., Palmer, C. G. and Weaver, D. D. Wiedemann-Beckwith syndrome: presentation of clinical and cytogenetic data on 22 new cases and review of the literature. *Hum Genet* 74:143–154, 1986.

Piussan, C., Risbourg, B., Lenaerts, C., Delvallez, N., Gontier, M. F. and Vitse, M. Wiedemann and Beckwith syndrome: a new familial case. *J Genet Hum* 28:281–291, 1980.

Rappolee, D. A., Sturm, K. S., Behrendtsen, O., Schultz, G. A., Pedersen, R. A. and Werb, Z. Insulin-like growth factor II acts through an endogenous growth pathway regulated by imprinting in early mouse embryos. *Genes Dev* 6:939–952, 1992.

Redeker, E., Alders, M., Hoovers, J. M., Richard, C. W., Westerveld, A. and Mannens, M. Physical mapping of 3 candidate tumor suppressor genes relative to Beckwith-Wiedemann syndrome associated chromosomal breakpoints at 11p15.3. *Cytogenet Cell Genet* 68:222–225, 1995.

Redeker, E., Hoovers, J. M., Alders, M., et al. An integrated physical map of 210 markers assigned to the short arm of human chromosome 11. *Genomics* 21:538–550, 1994.

Reik, W., Brown, K. W., Schneid, H., Le Bouc, Y., Bickmore, W. and Maher, E. R. Imprinting mutations in the Beckwith-Wiedemann syndrome suggested by altered imprinting pattern in the IGF2-H19 domain. *Hum Mol Genet* 4:2379–2385, 1995.

Reik, W., Brown, K. W., Slatter, R. E., Sartori, P., Elliott, M. and Maher, E. R. Allelic methylation of H19 and IGF2 in the Beckwith-Wiedemann syndrome. *Hum Mol Genet* 3:1297–1301, 1994.

Schmidt, J. V., Levorse, J. M. and Tilghman, S. M. Enhancer competition between H19 and IGF2 does not mediate their imprinting. *Proc Natl Acad Sci USA* 96:9733–9738, 1999.

Schneid, H., Seurin, D., Vazquez, M. P., Gourmelen, M., Cabrol, S. and Le Bouc, Y. Parental allele specific methylation of the human insulin-like growth factor II gene and Beckwith-Wiedemann syndrome. *J Med Genet* 30:353–362, 1993.

Slatter, R. E., Elliott, M., Welham, K., et al. Mosaic uniparental disomy in Beckwith-Wiedemann syndrome. *J Med Genet* 31:749–753, 1994.

Smilinich, N. J., Day, C. D., Fitzpatrick, G. V., et al. A maternally methylated CpG island in KVLQT1 is associated with an antisense paternal transcript and loss of imprinting in Beckwith-Wiedemann syndrome. *Proc Natl Acad Sci USA* 96:8064–8069, 1999.

Sun, F. L., Dean, W. L., Kelsey, G., Allen, N. D. and Reik, W. Transactivation of IGF2 in a mouse model of Beckwith-Wiedemann syndrome [see comments]. *Nature* 389:809–815, 1997.

Thorburn, M. J., Wright, E. S., Miller, C. G. and Smith-Read, E. H. Exomphalos-macroglossia-gigantism syndrome in Jamaican infants. *Am J Dis Child* 119:316–321, 1970.

Tremblay, K. D., Saam, J. R., Ingram, R. S., Tilghman, S. M. and Bartolomei, M. S. A paternal-specific methylation imprint marks the alleles of the mouse H19 gene. *Nat Genet* 9:407–413, 1995.

Turleau, C. and de Grouchy, J. Beckwith-Wiedemann syndrome. Clinical comparison between patients with and without 11p15 trisomy. *Ann Genet* 28:93–96, 1985.

Viljoen, D. and Ramesar, R. Evidence for paternal imprinting in familial Beckwith-Wiedemann syndrome [see comments]. *J Med Genet* 29:221–225, 1992.

Vu, T. H. and Hoffman, A. R. Promoter-specific imprinting of the human insulin-like growth factor-II gene. *Nature* 371:714–717, 1994.

Wang, Q., Curran, M. E., Splawski, I., et al. Positional cloning of a novel potassium channel gene: KVLQT1 mutations cause cardiac arrhythmias. *Nat Genet* 12:17–23, 1996.

Weksberg, R., Shen, D. R., Fei, Y. L., Song, Q. L. and Squire, J. Disruption of insulin-like growth factor 2 imprinting in Beckwith-Wiedemann syndrome. *Nat Genet* 5:143–150, 1993.

Weksberg, R., Teshima, I., Williams, B. R., et al. Molecular characterization of cytogenetic alterations associated with the Beckwith-Wiedemann syndrome (BWS) phenotype refines the localization and suggests the gene for BWS is imprinted. *Hum Mol Genet* 2:549–556, 1993.

Wiedemann, H. R. Complex malformatif familial avec hernie ombilicale et macroglossie – un syndrome nouveau? *J Genet Hum* 13:223–232, 1964.

Wiedemann, H-R., and Kunze, J. *Clinical Syndromes*, third ed., Mosby-Wolfe (Times Mirror International Publisher Limited), London, 1997, pp. 148–151.

Zemel, S., Bartolomei, M. S. and Tilghman, S. M. Physical linkage of two mammalian imprinted genes, H19 and insulin-like growth factor 2. *Nat Genet* 2:61–65, 1992.

Zhang, P., Liegeois, N. J., Wong, C., et al. Altered cell differentiation and proliferation in mice lacking p57KIP2 indicates a role in Beckwith- Wiedemann syndrome. *Nature* 387:151–158, 1997.

Zhang, Y., Shields, T., Crenshaw, T., Hao, Y., Moulton, T. and Tycko, B. Imprinting of human H19: allele-specific CpG methylation, loss of the active allele in Wilms tumor, and potential for somatic allele switching. *Am J Hum Genet* 53:113–124, 1993.

Chapter 9

Genetic Counseling and Prenatal Diagnosis

Contributed by Célia D. DeLozier-Blanchet, Ph.D., Division of Medical Genetics, Geneva University Medical School.

This chapter has three sections, with overlapping goals, as follows:

(i) A guide to genetic testing and counseling in "new" syndromes associated with UPD (uniparental disomy), for chromosomes 2, 6, 7, 14, and 16
(ii) A specific look at UPD in relation to prenatal diagnosis, e.g., placental mosaicism, *de novo* and inherited structural chromosomal aberrations
(iii) Suggestions of when and where to search for UPD

Genetic counseling in the three principal UPD/imprinting syndromes was discussed in Chapter 6 (Prader-Willi syndrome or PWS), Chapter 7 (Angelman syndrome or AS), and Chapter 8 (Beckwith-Wiedemann syndrome or BWS). To understand the familial transmission of imprinting disorders, we recommend a model proposed earlier by (Engel, 1997).

I. GENETIC COUNSELING IN "NEW SYNDROMES"

Paternal UPD6 and Transient Neonatal Diabetes Mellitus

Transient neonatal diabetes mellitus (TNDM), the clinical features of which are reviewed in Chapter 5 (on old and new syndromes), is a rare pathology, with an incidence of 1/400,000–1/500,000 (Fosel, 1995; Shield et al., 1997). The unifying cause of TNDM appears to be a duplication of a paternally inherited gene or genes on chromosome 6q24.1-24.3 (Gardner et al., 1999). This could occur either through duplication of a segment of the paternal chromosome 6 (e.g., unequal meiotic crossing-over), or through paternal UPD6; the latter may account for 20% of cases of TNDM (Shield et al., 1997; Gardner et al., 1998).

The imprinted region associated with TNDM has been localized to a 300–400 kb segment of 6q24 and parent-of-origin-specific differences in methylation have been documented (Gardner et al., 2000). Some TNDM patients with neither chromosomal duplication nor UPD6 also demonstrate abnormal methylation patterns. If this finding is confirmed, it may lead to development of a diagnostic screening test using methylation differences, such as is commonly used in Prader-Willi and Angelman syndromes (ASHG/ACMG, 1996).

As always, genetic counseling can be optimal only if the specific chromosomal or molecular mechanism causing the disorder has been determined; high-resolution chromosomal analysis and detection/exclusion of UPD using DNA polymorphisms should be performed in all cases. The risk for siblings when TNDM is caused by paternal UPD6 is very little increased above the general population risk. TNDM secondary to a *de novo* chromosomal duplication could theoretically recur, but with a very low probability.

Most cases of TNDM arise sporadically (Christian et al., 1999). Autosomal inheritance of an imprinting mutation, expected by analogy with other imprinting disorders, has not yet been reported.

For offspring of an affected individual, the recurrence risk will again be very low in cases due to UPD6. However, the recurrence risk would be high (up to 50%) when TNDM is caused by regional duplication of chromosome 6q24 and transmitted by an affected male.

Prenatal testing for siblings of TNDM patients may not be warrented, given the low chance of recurrence and the availability of treatment. It would, however, be prudent to check for signs of neonatal diabetes in siblings. To our knowledge, no case of TNDM has been associated with trisomy 6 mosaicism or discrepancy on CVS or amniocentesis; chromosome 6 is, in fact, rarely implicated in feto-placental discrepancies (Hahnemann and Vejerslev, 1997B; Wolstenholme et al., 1994).

Maternal UPD7 and Russell-Silver Syndrome

The Russell-Silver syndrome (RSS), whose clinical features are reviewed in chapter 5, is an etiologically heterogeneous disorder, with less than 10% of cases being caused by maternal UPD7 (Table 1, Chapter 5). Maternal UPD7 is generally of the isodisomic type, indicating a mitotic duplication of one chromosome 7 (Kotzot et al., 1995; Eggerding et al., 1994; Eggermann et al., 1997). The finding of maternal UPD7 supports the hypothesis that one or several genes on chromosome 7 are involved in growth and development; three candidate genes (PEG1/MEST, gamma2-COP, and GRB10) have been identified on chromosome 7 (Miyoshi et al., 1998; Monk et al., 2000). Recently, 2 of 58 RSS patients were found to have

GRB10 mutations, both inherited from their mother (Yoshihashi et al., 2000). However, another recent report has shown that GRB10 is probably not the only gene on chromosome 7 that contributes to the Russell-Silver syndrome. A RSS patient had segmental UPD7 for a 35-Mb area of 7q31, but GRB10 was located within the region showing biparental inheritance (Hannula et al., 2001).

Most cases of RSS are sporadic (Tanner et al., 1975; Patton, 1988), but familial occurrence of some characteristics of the syndrome, particularly in the maternal family, has been reported several times (Duncan et al., 1990). Genetic counseling in RSS will be imprecise unless the etiological mechanism in a particular family can be determined; known causes and their probability of recurrence can be summarized as follows:

(i) In RSS due to a *de novo chromosomal 7 rearrangement*, the recurrence risk will be low for siblings, although not null since germinal mosaicism cannot be ruled out. If a chromosomal rearrangement is present in a parent (particularly the father), there would theoretically be a small risk of recurrence through the same type of meiotic rearrangement that occurred the first time. For potential offspring of a RSS patient, the probability of reproductive pathology—whether RSS or recombination aneusomy—will depend on the specific type of chromosomal rearrangement present.

(ii) Cases due to maternal UPD7, whether for the entire chromosome as in the great majority of cases, or segmental as recently described (Hannula et al., 2001), will have negligible recurrence risk for siblings or offspring.

(iii) In a minority of families, there is apparent monogenic transmission of RSS, which could be due to segregation of an undetected chromosomal rearrangement or to imprinting mutations. The maximum recurrence risks in these families correspond to those for autosomal dominant (or recessive) transmission, risks that might be modified according to the sex of the transmitting parent.

Prenatal diagnosis for subsequent pregnancies should be offered in familial chromosome 7 rearrangements, given the risk for segmental aneuploidy through meiotic recombination or the predisposition to UPD7. Exclusion of UPD7 should also be offered when prenatal diagnosis for other indications reveals mosaicism for chromosome 7, since this could be indicative of an initially trisomic conceptus that underwent trisomy rescue.

Maternal and Paternal UPD14

Maternal UPD14 has been reported in a total of 19 cases, and its clinical features are indexed in Chapter 5 (on new and old syndromes, see Table 4 there). The mechanisms that may lead to its production, discussed in more detail in Chapter 4, are given here for our consideration of genetic counseling:

(i) *Translocation t(13;14) and UPD14*. Eight cases have been described, of which five were *de novo* and three maternally inherited. The recurrence risk for siblings is low in both situations, since it would necessitate germinal mosaicism in the *de novo* cases and a second instance of associated UPD14 in the familial cases. The latter appears to occur infrequently, as shown in two series. In the first, a retrospective study of 64 individuals with a Robertsonian translocation, UPD was detected only once (James et al., 1994). In the second, a prospective study of translocations detected prenatally, one case of UPD was found among 168 fetuses/infants studied (Berend et al., 2000). Thus, the risk of UPD due to missegregation of a familial translocation is apparently less than 1%.

(ii) *Isochromosome i(14q) or homologous translocation t(14q;14q) and UPD14*. There are reports of six *de novo* cases in the literature. No recurrence risk would be expected, excluding (hypothetical) germinal mosaicism in a parent.

(iii) *UPD14 associated with confined placental mosaicism and apparent trisomy rescue*. There are three cases described to date. The recurrence risk to siblings should not be different than that of the general population of similar parental ages.

For patients who have UPD14 themselves, the probability of having an affected child should not be increased over that of the general population. In translocation cases, the risk will depend on the specific chromosomal anomaly, varying from <1% (see above) to a nearly 100% risk of trisomy or monosomy 14 in the case of an i(14q) or homologous t(14q;14q) translocation. Ironically, for such individuals to have a euploid child, UPD through gamete complementation must occur: A germ cell carrying the t(14q;14q) must encounter a complementary gamete nullisomic for chromosome 14. The phenotype of such a "miracle child" would also depend on the sex of the transmitting parent! Familial transmission of a homologous Robertsonian translocation has not been described for chromosome 14, but has been for chromosomes 13 and 22, that apparently do not contain imprinted genes (Slater et al., 1994; Palmer et al., 1980). These cases have been detected in families studied for multiple spontaneous abortions. Prenatal diagnosis in those rare individuals with homologous translocations should certainly include both chromosomal analysis and, if the fetus is euploid, analysis of DNA polymorphisms to exclude UPD.

Paternal UPD14 has been reported four times (Chapters 4 and 6), each time associated with translocations involving chromosome 14. The recurrence risk in cases with parental translocations is sufficiently high to warrant UPD research, once prenatal diagnosis has excluded aneuploidy. For potential siblings of *de novo* t(14q;14q) translocation cases (two reported), the risk for recurrence of UPD is unknown but probably very low; for an affected individual with a homologous translocation, the situation is the same as for maternal UPD14.

Prenatal diagnosis should be offered in all cases with parental chromosomal translocations, both for detection of a translocation-induced aneuploidy, and in the case of a fetus with apparently normal chromosomes to exclude UPD14.

Chromosomes 2 and 16

As was discussed in Chapter 5, there is some evidence of clinically recognizable syndromes associated with UPD2 and UPD16. Most cases have been diagnosed through mosaicism on prenatal diagnosis (particularly CVS), or because of trisomy 2 or trisomy 16 detected in the placenta, studied because of growth retardation. These cases are thus the result of meiotic nondisjunction with trisomy rescue; phenotypic anomalies may result from (undetected) fetal mosaicism for the chromosome in question.

The recurrence risk will be theoretically low for both siblings and offspring of patients with UPD2 or UPD16, unless one of the parents has an undetected (germinal or low-percentage somatic) mosaicism for the trisomy.

II. PRENATAL DIAGNOSIS AND UPD

The possibility of uniparental disomy and its potential effects needs to be considered in a number of situations in prenatal diagnosis:

(i) Chromosomal mosaicism detected on chorionic villus sampling or on amniocentesis
(ii) *De novo* or inherited translocations, inversions, and chromosomal markers

Chromosomal Mosaicism and Trisomy Rescue

The mechanisms by which UPD occurs are reviewed in Chapter 3. The production of *holochromosomal* UPD is the result of three principal mechanisms: gamete complementation, duplication of a monosomy, and loss of a chromosome following a trisomic conception (trisomy rescue). The latter mechanism appears to be responsible for the majority of cases of UPD. This is particularly true for those cases detected through prenatal diagnosis, where a relationship between increased maternal age and UPD has been documented (Ledbetter and Engel, 1995; Engel, 1993; Kalousek et al., 1991; Ginsberg et al., 2000). Trisomy rescue with the concomitant risk of UPD may be suspected because of the coexistence of normal and trisomic cell lines in chorionic villi (Engel, 1993; Robinson et al., 1997; Kalousek et al., 1991).

Chromosomal mosaicism (Level III, defined as a chromosomal anomaly present in some cells from at least two independent cell cultures) (Hsu et al., 1992) detected on *amniocentesis* is generally indicative of a fetal (fetoplacental) origin. Additional cytogenetic studies (second sample, umbilical cord blood sample), along with ultrasound examination may help resolve counseling issues concerning the extent of fetal mosaicism. In addition, depending on the chromosome implicated, the possibility of UPD needs to be investigated. For example, a low percentage of trisomy 15 cells, in a case of normal pregnancy evolution and ultrasound screen, should be tested for UPD15 since its presence would indicate Angelman or Prader-Willi syndrome in the fetus.

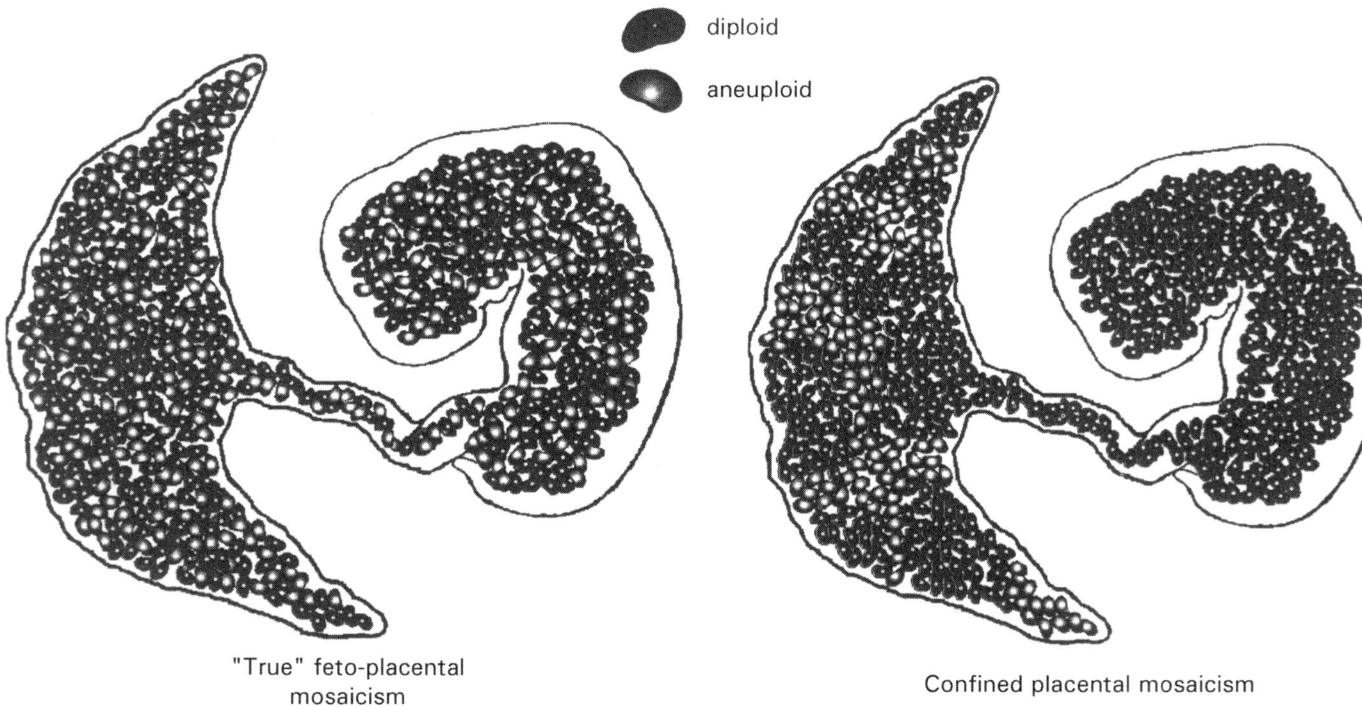

Figure 1 *Schematic representation of true fetoplacental mosaicism (a) as compared to confined placental mosaicism (CPM) (b).*

When chromosomal mosaisicm or a discrepant chromosomal result is observed in the study of *chorionic villus samples*, the abnormal cells may be limited to the placenta, a situation known as *confined placental mosaicism* (CPM, Figure 1) (Kalousek and Dill, 1983; Johnson et al., 1990; Hahnemann and Vejerslev, 1997B). The risk of associated uniparental disomy is limited to chromosomal errors of meiotic origin, followed by trisomy rescue. On the other hand, CPM is more often the result of a mitotic (postzygotic) error occurring in placental tissues with no associated risk of UPD.

The following factors point to meiotic nondisjunction in the etiology of trisomic cells, thus putting the fetus at increased risk for both residual fetal mosaicism and UPD (Hahnemann and Vejerslev, 1997A and B; DeLozier-Blanchet et al., 1995; Robinson et al., 1997):

(i) *The proportion of abnormal cells*. The higher the percentage of chromosomally abnormal cells, the more likely a meiotic origin of the anomaly (Kalousek et al., 1991; Hahnemann and Vejerslev, 1997B).

(ii) *The type of mosaicism* (Figure 2). Type III mosaicism (abnormal cell line present in both cytotrophoblast and cultured mesodermal core cells (long-term culture) is the combination most indicative of a meiotic error. However, mosaicism of level I (short-term culture) or II (aneuploidy cultured cells from mesodermal core) is still associated with a meiotic origin in some cases (Leschot et al., 1989; Hahnemann and Vejerslev, 1997B).

(iii) *The chromosome implicated*. Some autosomal trisomies, e.g., 3 and 7, are relatively often involved in CVS mosaicism, but rarely confirmed in the

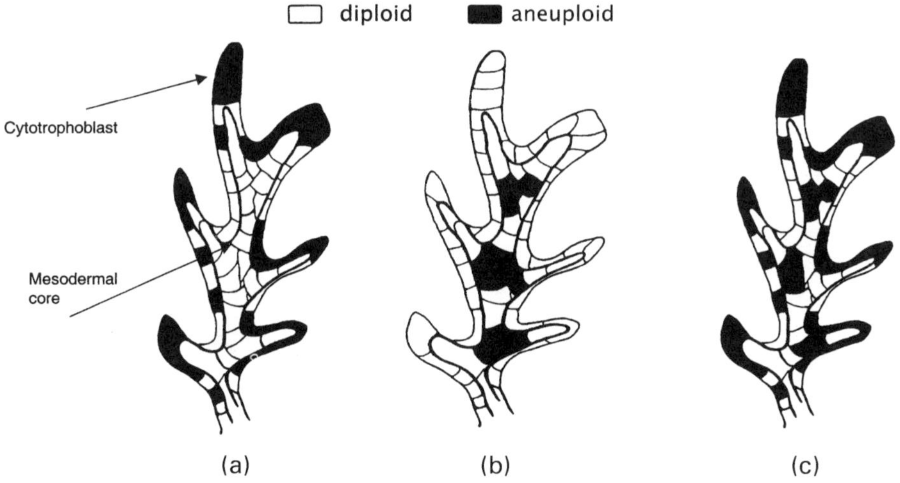

Figure 2 *Representation of the three types of chromosomal mosaicism observed on CVS mosaicism: In type I, (a) the anomaly is seen only in cytotrophoblast; in type II, (b) only in cultured cells derived from the mesodermal core of the villi; and in type III, (c) in cells from both embryologic origins.*

fetus. Others, such as trisomy 15 and 16, are more likely to be confirmed on amniocentesis and fetal biopsies, and associated with an abnormal phenotype. Still others, such as trisomies 2 and 22, are generally due to CPM but occasionally represent true mosaicism and thus need to be further investigated (Hahnemann and Vejerslev, 1997B; Leschot et al., 1989; Wolstenholme, 1996).

The European Collaborative Research on Mosaicism in Chorionic Villus Sampling (EUCROMIC 1986–1996) (Vejerslev and Mikkelsen, 1989; Hahnemann and Vejerslev, 1997A and B) investigated mosaicism or fetoplacental discrepancies in over 126,000 samples. Several other large series (Simoni et al., 1986; Wolstenholme et al., 1994; Kalousek et al., 1991) have also investigated clinical and cytogenetics correlations. The majority of these discrepancies (between the results of CVS and amniocentesis or fetal tissues) are apparently due to confined placental mosaicism, suggesting that the origin of this mosaicism is most often mitotic. The relationship between CPM and true fetoplacental mosaicism varies as a function of the particular chromosome (Table 1). It is more likely that trisomy 21 be confirmed, for example, than trisomy 13 or 18. Even for these three autosomes, the proportion of fetal mosaicism that is confirmed after mosaic results on CVS is less than half according to the largest study specifically addressing this question (Hahnemann and Vejerslev, 1997B). A second example of an autosomal trisomy needing extensive investigation is trisomy 15; about half of mosaic or discrepant results involving chromosome 15 have a meiotic origin, the others being apparently of a postconceptional nature [European Collaborative Research on Mosaicism in CVS (EUCROMIC), 1999; Robinson et al., 1997]. There is thus a need to further investigate

TABLE 1 CVS Mosaicism for Single Autosomal Trisomies[a]

Confirmed as True Fetal Chromosome	Never Confirmed as True Fetal Chromosome
8	2
	3
9	4[b]
	5[b]
12[b]	6[b]
	7
13	10[b]
	11[b]
15	14[b]
	17[b]
16	19[b]

[a] Based on the study of 769 autosomal trisomies with sufficient cytogenetic and/or clinical follow-up, as part of the EUCROMIC collaboration
[b] Less than 20 cases involving this chromosome were studied.
Source: Hahnemann and Vejerslev, 1997A and B and unpublished.

potential fetal mosaicism (by mesodermal core cell culture, amniocentesis, or cordocentesis) as well as to rule out UPD15 in the disomic cells from the fetus.

One of the goals of the EUCROMIC ancillary studies was to investigate the frequency and nature of UPD in cases of fetoplacental discrepancies after CVS. The retrospective study used DNA polymorphic analysis in children/fetuses and their parents to determine, in 105 conceptuses, the proportion of cases with UPD (DeLozier-Blanchet et al., 1995B). Seventeen of the 105 (16%) had UPD for chromosomes 2, 9, 11, 15, 16, and 22. For these chromosomes, a meiotic origin of the chromosomal error was thus confirmed. For chromosomes 15 and 16, the proportion of UPD cases approached the theoretical 1/3 to be expected if all cases were meiotic. No case of UPD was observed for chromosomes 7, 8, 13, 18, 21 or the sex chromosomes, suggesting that CPM for the latter chromosomes is most often of mitotic origin. A British study also concluded that trisomies for chromosomes 2, 3, 7, 8, and 9 are most often due to mitotic nondisjunction (Wolstenholme, 1996).

A similar proportion of UPD (17 of 91 cases, about 19%) was observed in placentas studied postabortion or at term for a variety of reproductive pathologies (intrauterine growth retardation being the major indication) (Robinson et al., 1997), as was found in the above-cited EUCROMIC study.

In summary, these data are particularly important for genetic counseling in prenatal diagnosis. When chromosomal analysis of CVS reveals "ambiguous" results, e.g., mosaicism or a nonmosaic chromosomal aberration unlikely to exist in fetal cells, it is necessary to confirm or refute the finding in cells of a different origin; this may be done on a long- term culture of mesodermal cells from chorionic villi, an amniocentesis, or a fetal blood sample. Monitoring of the pregnancy with ultrasound, as well as genetic counseling, is necessary (Leschot et al., 1989; Vejerslev and Mikkelsen, 1989). However, present experience would suggest that follow-up invasive tests are probably not necessary for some aneuploidies, such as trisomies 3 or 7, particularly when the proportion of abnormal cells is less than 20% (DeLozier-Blanchet et al., 1995A; Wolstenholme, 1996).

The situation of mosaicism/discrepancy on CVS for the autosomal trisomies 13, 18, and 21 is worth special consideration. Even for these "viable" trisomies, less than half of all cases will be confirmed in the fetus (Hahnemann and Vejerslev, 1997B). Amniocentesis, as well as ultrasound monitoring, is required in continuing pregnancies. However, UPD testing is not warranted, given that chromosomes 13 and 21 do not appear to contain imprinted genes, and no case of UPD18 has been reported to date (see Chapter 4).

For chromosome 15, however, UPD testing as well as amniocentesis for potential fetal mosaicism is indicated, since some 50% of cases of CVS mosaicism are meiotic [Robinson et al., 1996; European Collaborative Research on Mosaicism in CVS (EUCROMIC), 1999].

Suggested guidelines for UPD testing in association with prenatal diagnosis are presented in Table 2.

The finding of chromosomal mosaicism on CVS, even in the absence of fetal aneuploidy and/or UPD, may warrant that the pregnancy be considered high-risk. When CPM is widespread (thus probably of meiotic origin), reproductive patholo-

TABLE 2 Prenatal Diagnosis and UPD Testing
Suggestions for testing: *Analysis should be offered only for chromosomes with an imprinting-associated phenotype*

Significant Risk of UPD	Lower Risk of UPD
On Amniocentesis	*On Amniocentesis*
• Level III mosaicism	• Level II mosaicism for trisomies 15 and 16
• Structural anomaly (inherited or *de novo*) of chromosomes 2, 14, 15, 16, 22	• Structural anomaly involving chromosomes 7 or 11
On CVS	*On CVS*
• Type II and III mosaicism/discrepancies involving chromosomes 15, 16	• Any mosaicism involving chromosomes 2, 7, 11, 15, 16, 22
• Structural anomaly involving chromosomes 14, 15, 22	• Structural anomaly of chromosomes 6, 7, 11

gies including fetal demise and intrauterine growth retardation may occur (Johnson et al., 1990; Kennerknecht and Terinde, 1990; Kalousek et al., 1991). A EUCROMIC ancillary study found an increased incidence of low birthweight with CPM for chromosomes 8, 13, 16, and 22 (DeLozier-Blanchet et al., 1996).

Parental Translocations, Inversions and Chromosomal Markers

As discussed in Chapters 3 and 7, centric chromosomal fusions in association with UPD have been observed mainly for chromosomes 14 and 15. Whether *de novo* or inherited, apparently balanced heterologous translocations merit molecular investigation of potential UPD. The proportion of such conceptuses that actually do present UPD is, however, less than 1% (James et al., 1994; Berend et al., 2000) as presented earlier in this chapter. Prenatal testing for UPD of chromosomes with no known imprinting pathology, such as 13 and 21, should be discouraged, since the UPD apparently does not cause phenotypic abnormality (Ledbetter and Engel, 1995; Kotzot, 1999).

As for the unusual situation of *de novo homologous translocations*, a number of cases of UPD14 or UPD15 have been associated with such translocations. UPD should therefore be systematically considered in all prenatally tested cases with *de novo* centric fusion of homologues for imprinted chromosomes. Although the frequency of UPD in homologous chromosomal fusion is not known, it appears to be high in this class of patients, since one-half of these might be of meiotic origin, thus implying trisomy rescue to yield diploid genomes.

All homologous parental fusions should lead to monosomic or trisomic offspring. The finding of an offspring having inherited such a translocation is thus unexpected and implies gamete complementation or trisomy rescue for the chromosome involved in the fusion. Inherited homologous fusions, in offspring with balanced chromosomal complements (no associated trisomy), have been seen only for chromosomes 13 and 22 (Chapters 4 and 5).

Balanced Reciprocal Translocations

The risk of imprinting disorders is mostly related to translocations involving chromosomes 11, 14, and 15. As already noted in Chapter 3, they contribute principally to deletions—or duplications—of imprinted domains through meiotic recombination, rather than leading to nondisjunction and UPD. Here, of course, the sex of the transmitting parent will determine the phenotype. For instance, AS or PWS will be seen when a t(15;22) results in a 15q11-q13 deletion from either a mother (AS) or a father (PWS). Prenatal diagnosis should be offered to investigate both the cytogenetic and appropriate DNA polymorphisms of the fetus.

Peri and Paracentric Inversions

These may also be a predisposing factor in UPD for the chromosome involved in the rearrangement, although this must be even rarer than deletions or alterations potentially arising from meiotic recombination. Prenatal diagnosis could be offered to exclude recombination aneusomy and might include UPD analysis in cases where the inversion concerns chromosomes known (6, 7, 11, 14, 15) or suspected (2, 16) to harbor imprinted genes.

Small Marker Chromosomes

These are found in UPD cases under two sets of circumstances, namely, as the tiny remnant of an incompletely rescued trisomy or as a fragment of a truncated monosomic chromosome entirely duplicated in the uniparental pair (James et al., 1995). Prenatal molecular analysis is essential when the marker chromosome may carry imprinted genes; this is particularly important for chromosome 15, as it is involved in cases of Prader-Willi and Angelman syndromes (see Chapters 3, 6 and 7). Nearly half of all extra structurally abnormal chromosomes are invdup15. The most common form of the marker has been a type I (small) invdup15 (Webb, 1994), as reviewed in Chapter 3. Markers chromosomes have been associated with UPD only for a few other chromosomes, such as 6 and 21 (Chapter 3).

Prenatal Diagnosis when a Previous Pregnancy Involved UPD

An assessment of the index case and parents is needed to check on the cytogenetic parameters reviewed above. In the case of a chromosomal rearrangement, particularly if inherited, the risk of recurrence may be difficult to calculate, but would appear sufficient to justify a full prenatal work-up of subsequent pregnancies. Usually, no parental cytogenetic markers will be apparent but an increased parental age may be noted as UPD more often arises from an aneuploid event, chiefly a trisomy secondarily turned into a disomy. Genetic counseling in UPD without specific chromosomal rearrangements should address the problem of a subsequent fetus with aneuploidy more so than the problem of a recurrence of UPD.

In summary, the exclusion of UPD can be recommended in prenatal diagnosis in the following situations:

(i) Detection of mosaicism for a numerical chromosomal anomaly on either amniocentesis or chorionic villus sampling
(ii) Presence of a structural chromosomal anomaly (translocation, inversion, marker), whether familial or *de novo*, whenever that particular chromosome is known to contain imprinted genes and is associated with an abnormal phenotype

UPD testing should not be performed for chromosomes where UPD has not been reported in association with a recognizable phenotype or structural/developmental anomaly, as no information on the prognosis could be provided.

The probability of UPD is low in most cases, but increases whenever:

(i) There is a high percentage of abnormal cells (or the mosaicism is of type III).
(ii) Chromosomes 14, 15, or 16 are involved.
(iii) An abnormal phenotype, with fetal malformation or gestational pathology, is present.
(iv) A risk factor for nondisjunction, including increased maternal age, is present.

Genetic counseling and psychological support are essential in such cases of prenatal diagnosis, with prognostic uncertainty and multiple laboratory tests.

III. THE SEARCH FOR UPD

As UPD could potentially explain the following situations, molecular investigation should be considered when:

(i) Two distinct disorders are present in the same individual. The first molecularly proven cases of UPD7 were of this type (Chapters 1 and 4) (Spence et al., 1988), e.g., the individuals had both cystic fibrosis through homozygosity for a CF mutation due to maternal UPD7, as well as very short stature (due to the absence of the paternal copy of imprinted genes).
(ii) An autosomal recessive disorder in which only one parent is a carrier (Chapters 1 and 4). Since the recurrence risk will be much lower if the recessive phenotype results from UPD than through biparental transmission of the mutant gene, testing for UPD might be done in certain cases homozygous for an identical mutation or haplotype.
(iii) When an individual has two copies of a unique parental chromosomal heteromorphism (Betz et al., 1974; Carpenter et al., 1982).

(iv) In recognized clinical syndromes of unknown origin, particularly where the disorder has been associated with various chromosomal rearrangements and/or transmitted preferentially by one sex (e.g., Russell-Silver syndrome). In such disorders, the UPD might be expected to be *holochromosomic* in some patients and *segmental* in others.

Recent series have looked for UPD in empirically predisposed groups. In one study, an excess of UPD was found in individuals with unclassified developmental defects whose mothers were over 35 at the time of their birth, versus those whose mothers were under 35 (Ginsburg et al., 2000). Similar frequency estimates could be done, e.g., for early spontaneous pregnancy loss or to embryos that do not develop after in vitro fertilization.

REFERENCES

ASHG/ACMG. Diagnostic testing for Prader-Willi and Angleman syndromes: Report of the ASHG/ACMG Test and Technology Transfer Committee. *Am J Hum Genet* 58:1085–1088, 1996.

Berend, S. A., Horwitz, J., McCaskill, C. and Shaffer, L. G. Identification of Uniparental Disomy Following Prenatal Detection of Robertsonian Translocations and Isochromosomes. *Am J Hum Genet* 66:1787–1793, 2000.

Betz, A., Turleau, C. and de Grouchy, J. Heterozygosity and homozygosity for a pericentric inversion of human chromosone 3. *Ann Genet* 17:79–80, 1974.

Carpenter, N. J., Say, B. and Barber, N. D. A homozygote for pericentric inversion of chromosome 4. *J Med Genet* 19:469–471, 1982.

Christian, S. L., Rich, B. H., Loebl, C., et al. Significance of genetic testing for paternal uniparental disomy of chromosome 6 in neonatal diabetes mellitus [see comments]. *J Pediatr* 134:42–46, 1999.

DeLozier-Blanchet, C. D., Francipane, L., Morris, M. A., Hoovers, J. M., Leschot, N. J. and Cox, J. N. Trisomy 3 mosaicism on CVS: case report with literature review and propositions for investigation and counseling. *Clin Genet* 48:308–312, 1995A.

DeLozier-Blanchet, C. D., Hahnemann, J. M. and Vejerslev, L. O. Uniparental disomy and feto-placental discrepencies on chorionic villus sampling. *Am J Hum Genet* 57:A263, 1995B. (Abstract)

DeLozier-Blanchet, C. D., Pellegrini, B., Hahnemann, J. M. and Vejerslev, L. O. Birth weight analysis and mosaic/discrepant results on chorionic villus sampling: the EUCROMIC. *Am J Hum Genet* 59:A319, 1996. (Abstract)

Duncan, P. A., Hall, J. G., Shapiro, L. R. and Vibert, B. K. Three-generation dominant transmission of the Silver-Russell syndrome. *Am J Med Genet* 35:245–250, 1990.

Eggerding, F. A., Schonberg, S. A., Chehab, F. F., Norton, M. E., Cox, V. A. and Epstein, C. J. Uniparental isodisomy for paternal 7p and maternal 7q in a child with growth retardation. *Am J Hum Genet* 55:253–265, 1994.

Eggermann, T., Wollmann, H. A., Kuner, R., et al. Molecular studies in 37 Silver-Russell syndrome patients: frequency and etiology of uniparental disomy. *Hum Genet* 100:415–419, 1997.

Engel, E. Uniparental disomy revisited: the first twelve years. *Am J Med Genet* 46:670–674, 1993.

Engel, E. Imprinting is also a mechanism for immediate or delayed hemizygous expression of several uniparental haplotypes selected from the genome of each sex. *Am J Med Genet* 70:95–98, 1997.

European Collaborative Research on Mosaicism in CVS (EUCROMIC). Trisomy 15 CPM: probable origins, pregnancy outcome and risk of fetal UPD. *Prenat Diagn* 19:29–35, 1999.

Fosel, S. Transient and permanent neonatal diabetes. *Eur J Pediatr* 154:944–948, 1995.

Gardner, R. J., Mackay, D. J., Mungall, A. J., et al. An imprinted locus associated with transient neonatal diabetes mellitus. *Hum Mol Genet* 9:589–596, 2000.

Gardner, R. J., Mungall, A. J., Dunham, I., et al. Localisation of a gene for transient neonatal diabetes mellitus to an 18.72 cR3000 (approximately 5.4 Mb) interval on chromosome 6q. *J Med Genet* 36:192–196, 1999.

Gardner, R. J., Robinson, D. O., Lamont, L., Shield, J. P. and Temple, I. K. Paternal uniparental disomy of chromosome 6 and transient neonatal diabetes mellitus. *Clin Genet* 54:522–525, 1998.

Ginsburg, C., Fokstuen, S. and Schinzel, A. The contribution of uniparental disomy to congenital development defects in children born to mothers at advanced childbearing age. *Am J Med Genet* 95:454–460, 2000.

Hahnemann, J. M. and Vejerslev, L. O. European collaborative research on mosaicism in CVS (EUCROMIC)—fetal and extrafetal cell lineages in 192 gestations with CVS mosaicism involving single autosomal trisomy. *Am J Med Genet* 70:179–187, 1997A.

Hahnemann, J. M. and Vejerslev, L. O. Accuracy of cytogenetic findings on chorionic villus sampling (CVS)—diagnostic consequences of CVS mosaicism and non-mosaic discrepancy in centres contributing to EUCROMIC 1986–1992. *Prenat Diagn* 17:801–820, 1997B.

Hahnemann, J. M. and Vejerslev, L. O. True foetal mosaicism diagnosed on CVS. Data for EUCROMIC 1986–1995. *Medizin Genet* 9(suppl):56, 1997. (Abstract)

Hannula, K., Lipsanen-Nyman, M., Kontiokari, T. and Kere, J. A Narrow Segment of Maternal Uniparental Disomy of Chromosome 7q31-qter in Silver- Russell Syndrome Delimits a Candidate Gene Region. *Am J Hum Genet* 68:247–253, 2001.

Hsu, L. Y., Kaffe, S., Jenkins, E. C., et al. Proposed guidelines for diagnosis of chromosome mosaicism in amniocytes based on data derived from chromosome mosaicism and pseudomosaicism studies. *Prenat Diagn* 12:555–573, 1992.

James, R. S., Temple, I. K., Dennis, N. R. and Crolla, J. A. A search for uniparental disomy in carriers of supernumerary marker chromosomes. *Eur J Hum Genet* 3:21–26, 1995.

James, R. S., Temple, I. K., Patch, C., Thompson, E. M., Hassold, T. and Jacobs, P. A. A systematic search for uniparental disomy in carriers of chromosome translocations. *Eur J Hum Genet* 2:83–95, 1994.

Johnson, A., Wapner, R. J., Davis, G. H. and Jackson, L. G. Mosaicism in chorionic villus sampling: an association with poor perinatal outcome. *Obstet Gynecol* 75:573–577, 1990.

Kalousek, D. K. and Dill, F. J. Chromosomal mosaicism confined to the placenta in human conceptions. *Science* 221:665–667, 1983.

Kalousek, D. K., Howard-Peebles, P. N., Olson, S. B., et al. Confirmation of CVS mosaicism in term placentae and high frequency of intrauterine growth retardation, association with confined placental mosaicism. *Prenat Diagn* 11:743–750, 1991.

Kennerknecht, I. and Terinde, R. Intrauterine growth retardation associated with chromosomal aneuploidy confined to the placenta. Three observations: triple trisomy 6,21,22; trisomy 16; and trisomy 18. *Prenat Diagn* 10:539–544, 1990.

Kotzot, D. Abnormal phenotypes in uniparental disomy (UPD): fundamental aspects and a critical review with bibliography of UPD other than 15. *Am J Med Genet* 82:265–274, 1999.

Kotzot, D., Schmitt, S., Bernasconi, F., et al. Uniparental disomy 7 in Silver-Russell syndrome and primordial growth retardation. *Hum Mol Genet* 4:583–587, 1995.

Ledbetter, D. H. and Engel, E. Uniparental disomy in humans: development of an imprinting map and its implications for prenatal diagnosis. *Hum Mol Genet* 4 Spec No:1757–1764, 1995.

Leschot, N. J., Wolf, H., Van Prooijen-Knegt, A. C., et al. Cytogenetic findings in 1250 chorionic villus samples obtained in the first trimester with clinical follow-up of the first 1000 pregnancies. *Br J Obstet Gynaecol* 96:663–670, 1989.

Miyoshi, N., Kuroiwa, Y., Kohda, T., et al. Identification of the Meg1/Grb10 imprinted gene on mouse proximal chromosome 11, a candidate for the Silver-Russell syndrome gene. *Proc Natl Acad Sci USA* 95:1102–1107, 1998.

Monk, D., Wakeling, E. L., Proud, V., et al. Duplication of 7p11.2-p13, including GRB10, in Silver-Russell syndrome. *Am J Hum Genet* 66:36–46, 2000.

Palmer, C. G., Schwartz, S. and Hodes, M. E. Transmission of a balanced homologous t(22q;22q) translocation from mother to normal daughter. *Clin Genet* 17:418–422, 1980.

Patton, M. A. Russell-Silver syndrome. *J Med Genet* 25:557–560, 1988.

Robinson, W. P., Barrett, I. J., Bernard, L., et al. Meiotic origin of trisomy in confined placental mosaicism is correlated with presence of fetal uniparental disomy, high levels of trisomy in trophoblast, and increased risk of fetal intrauterine growth restriction. *Am J Hum Genet* 60:917–927, 1997.

Robinson, W. P., Langlois, S., Schuffenhauer, S., et al. Cytogenetic and age-dependent risk factors associated with uniparental disomy 15. *Prenat Diagn* 16:837–844, 1996.

Shield, J. P., Gardner, R. J., Wadsworth, E. J., et al. Aetiopathology and genetic basis of neonatal diabetes. *Arch Dis Child Fetal Neonatal Ed* 76:F39–F42, 1997.

Simoni, G., Gimelli, G., Cuoco, C., et al. First trimester fetal karyotyping: one thousand diagnoses. *Hum Genet* 72:203–209, 1986.

Slater, H., Shaw, J. H., Dawson, G., Bankier, A. and Forrest, S. M. Maternal uniparental disomy of chromosome 13 in a phenotypically normal child [see comments]. *J Med Genet* 31:644–646, 1994.

Spence, J. E., Perciaccante, R. G., Greig, G. M., et al. Uniparental disomy as a mechanism for human genetic disease. *Am J Hum Genet* 42:217–226, 1988.

Tanner, J. M., Lejarraga, H. and Cameron, N. The natural history of the Silver-Russell syndrome: a longitudinal study of thirty-nine cases. *Pediatr Res* 9:611–623, 1975.

Vejerslev, L. O. and Mikkelsen, M. The European collaborative study on mosaicism in chorionic villus sampling: data from 1986 to 1987. *Prenat Diagn* 9:575–588, 1989.

Webb, T. Inv dup(15) supernumerary marker chromosomes. *J Med Genet* 31:585–594, 1994.

Wolstenholme, J. Confined placental mosaicism for trisomies 2, 3, 7, 8, 9, 16, and 22: their incidence, likely origins, and mechanisms for cell lineage compartmentalization. *Prenat Diagn* 16:511–524, 1996.

Wolstenholme, J., Rooney, D. E. and Davison, E. V. Confined placental mosaicism, IUGR, and adverse pregnancy outcome: a controlled retrospective U.K. collaborative survey. *Prenat Diagn* 14:345–361, 1994.

Yoshihashi, H., Maeyama, K., Kosaki, R., et al. Imprinting of human GRB10 and its mutations in two patients with Russell-Silver syndrome. *Am J Hum Genet* 67:476–482, 2000.

Chapter 10

Genomic Imprinting in the Mouse

Contributed by Robert Lyle, Ph.D., Division of Medical Genetics, University of Geneva Medical School.

Mammals are diploid for all autosomal genes and the majority of genes are assumed to be expressed from both alleles within the cell. However, a number of different mechanisms exist for ensuring that some genes are only expressed monoallelically. For example, X inactivation in female mammals involves silencing a large number of genes on the X chromosome (Lyon, 1999), olfactory cells express a single receptor subtype from the thousands of genes possible (Chess et al., 1994), and certain genes involved in the development of the immune system are expressed from only one allele (Holländer et al., 1998; Nutt et al., 1999; Rivière et al., 1998).

In addition, a subset of genes is expressed from only one allele in a parent-specific manner. Such genes are subject to the phenomenon of genomic imprinting. The study of genomic imprinting in mice and humans has been complementary through a number of different observations and experimental approaches. From the manipulation of early mouse embryos to the observation of parental effects in human genetic disease, work on both species has contributed to an understanding of imprinting and its clinical effects. This chapter will focus on some of the data that led to the identification of imprinting in both mice and humans and discuss how the study of imprinted genes in both species continues to provide insights into the molecular basis of imprinting.

NONEQUIVALENCE OF PARENTAL GENOMES IN MICE AND HUMANS

The manipulation of mouse embryos *in vitro* to produce diploid embryos with only maternal or paternal genomes provided the first clear evidence for the phenomenon of gametic imprinting. Initially, the production of parthenogenetic embryos, in a number of vertebrate species, indicated that normal development could occur with only a maternally derived diploid genome. However, the demonstration that mammalian parthenogenotes are not viable suggested this was not true for all species (Kaufman et al., 1977). A number of reasons were postulated for this failure to complete development: homozygosity for recessive lethal alleles, lack of an extragenetic contribution from the fertilizing sperm, or nonequivalence of the parental genomes (Graham, 1974).

Experiments to distinguish between these alternatives used nuclear transplantation to create embryos with different parental constitutions (Figure 1). Neither parthenogenotes (Kaufman et al., 1977) nor diploid biparental gynogenetic (Surani and Barton, 1983) or androgenetic (McGrath and Solter, 1984a) embryos develop to term. Both androgenetic and gynogenetic embryos can develop to the blastocyst stage, but die shortly after implantation. The fact that these embryos were derived from inbred mice rules out the possibility that the lethality was due to homozygosity for recessive mutations. Since the only difference between parthenogenotes and gynogenotes is that gynogenotes have been activated by contact with sperm, the demonstration that gynogenetic embryos develop to the same stage as parthenogenotes suggested that it was not a cytoplasmic contribution from sperm that led to the developmental failure (Surani and Barton, 1983). This was confirmed by experiments in which eggs were injected with pronuclei from fertilized eggs: Only those eggs that obtained a male pronucleus developed to term; those that obtained a female pronucleus did not. Thus, the cytoplasm of activated eggs is capable of supporting development (Surani et al., 1984). Nuclear transplantation experiments with pronuclei derived from T^{hp} mice (a maternal effect mutation on mouse chromosome 17) (Johnson, 1974) also confirmed that the phenotypic effect was nuclear and not cytoplasmic. A $T^{hp/+}$ pronucleus injected into an anucleate egg derived from wild-type mice did not produce viable offspring, whereas the reciprocal nuclear transfer was viable (McGrath and Solter, 1984b).

The conclusion from these experiments was that the parental genomes are not equivalent in the information they contribute to the embryo. From this it follows that one or both parental genomes must be marked in some way to modify their genetic information (McGrath and Solter, 1984a; Surani et al., 1984). This mark that confers parental "memory" has become known as the imprint. The imprint has been shown to remain on the parental chromosomes after cell division: Transplantation of nuclei from haploid early preimplantation embryos (two to eight cells) back into fertilized eggs, from which one pronucleus had been removed, develop to term, only if the donor nucleus was derived from the opposite sex of the remaining pronucleus. This demonstrated that the imprinting mark is heritable (Surani et al., 1986). During early development, gynogenotes and androgenotes are generally much smaller than normal littermates (Surani and Barton, 1983). In addition, the parental genomes do not contribute equally to the tissues of the developing embryo, apparently fulfilling complementary functions. Androgenotes have relatively poorly developed embryonic tissue, but well-developed extraembryonic membranes and trophoblast (Barton et al., 1984), gynogenotes have a small but relatively normal embryo,

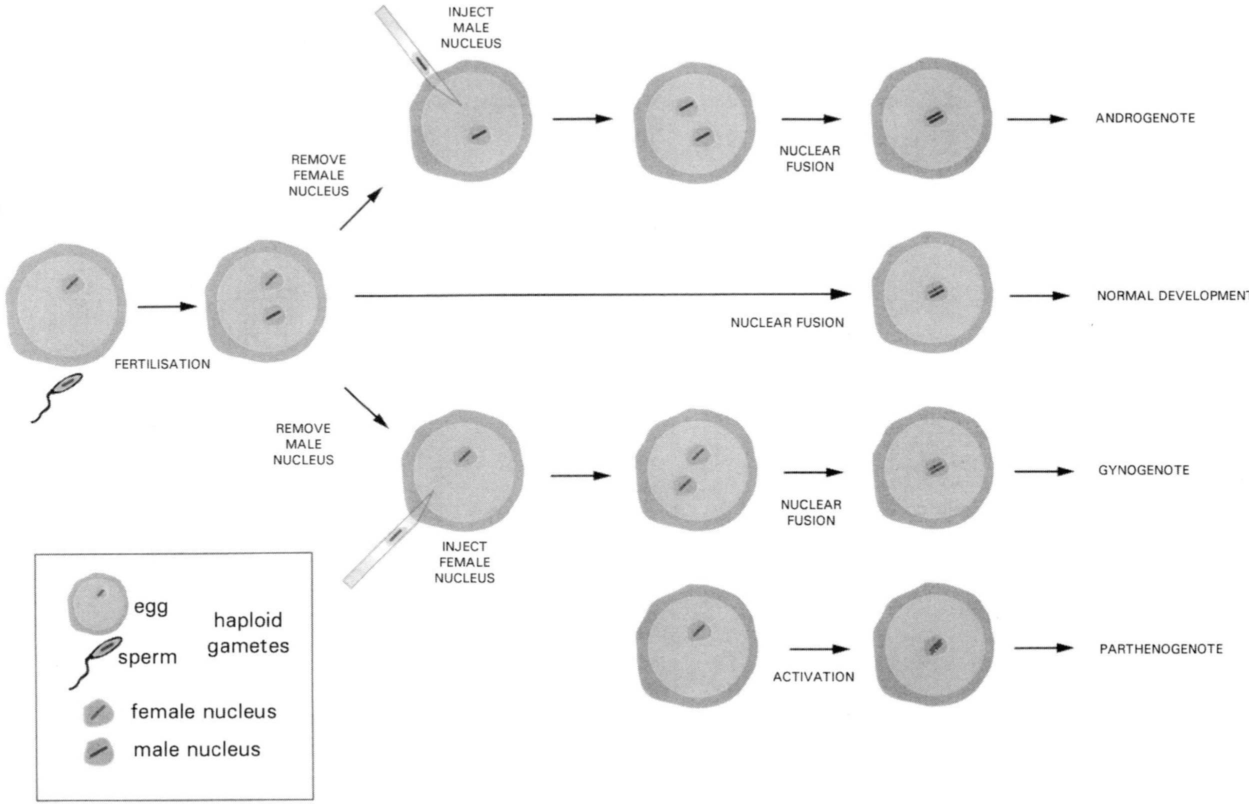

Figure 1 *Manipulation of early embryos to produce androgenotes, gynogenotes, and parthogenotes. The ability to distinguish male and female pronuclei in a fertilized egg means that they can be removed before pronuclear fusion. Subsequent injection of different pronuclei allows generation of diploid embryos with genomes derived from only paternal or maternal contributions.*

whereas the extraembryonic tissues are not well developed (Barton et al., 1985; Surani and Barton, 1983; Surani et al., 1984). The paternal genome thus contributes more to extraembryonic tissues and the maternal genome to postimplantation development of the embryo. This conclusion was tested, and confirmed, by constructing chimaeras between androgenetic and parthenogenetic cells: Androgenetic cells preferentially contribute to the extraembryonic tissues and gynogenetic cells to the embryo proper (Mann et al., 1990; Surani et al., 1987). Further analysis of chimaeras has shown that parthenogenetic, androgenetic, and gynogenetic cells also contribute unequally to the embryonic organs (Barton et al., 1991; Hardy and Handyside, 1996).

Naturally occurring equivalents of androgenetic and parthenogenetic mouse embryos exist in humans. Hydatidiform moles, like androgenotes, are diploid for the paternal genome and arise through fertilization of an anucleate oocyte (Jacobs et al., 1980). These moles are similar to androgenetic embryos in that they show hypertrophy of extraembryonic structures and lack embryonic tissue. Ovarian teratomas are benign embryonal tumors that develop with only a diploid maternal genome and are thus parthenogenetic (Surti et al., 1990). Similar to mouse parthenogenotes, ovarian teratomas have embryonic tissues but do not contain extraembryonic structures.

A MINOR PORTION OF THE MAMMALIAN GENOME IS IMPRINTED

More detailed analyses of the regions of the genome that contribute to parent-dependent phenotypic effects have been made possible by the use of translocation strains of mice that show a high rate of nondisjunction. Offspring derived from these mice often show uniparental disomy (UPD) for specific chromosomes; they are diploid but a region of the genome is of maternal or paternal origin only (Cattanach and Kirk, 1985; Figure 2). Selective breeding in mice has allowed the analysis of most of the genome (Cattanach and Beechey, 1990), and naturally occurring cases of UPD have allowed similar, although less definitive, analyses in humans (Ledbetter and Engel, 1995). The phenotypes of UPD mice range from embryonic lethal to postnatal growth effects (Cattanach and Beechey, 1990). For some chromosomal regions, the opposite, complementing phenotypes are produced dependent on the parent-of-origin of the UPD, reminiscent of the results from androgenotes and parthenogenotes. For example, for proximal chromosome 11, paternal UPD results in growth enhancement and maternal UPD in a growth decrease, suggestive of a paternally expressed growth enhancer or a maternally expressed negative regulator of growth mapping to this region. Such studies have shown that the effects observed in experiments with parthenogenetic, gynogenetic, and androgenetic embryos are attributable to a relatively small subset of the genome [reviewed in (Cattanach and Beechey, 1990)]. Large chromosomal regions have been identified that appear to show no imprinting effects (Beechey and Cattanach, 1996; Cattanach et al., 1993). So far, 15 regions of the mouse genome have been identified that display parental-specific effects [see (http://www.mgu.har.mrc.ac.uk/imprinting/imptables.html.

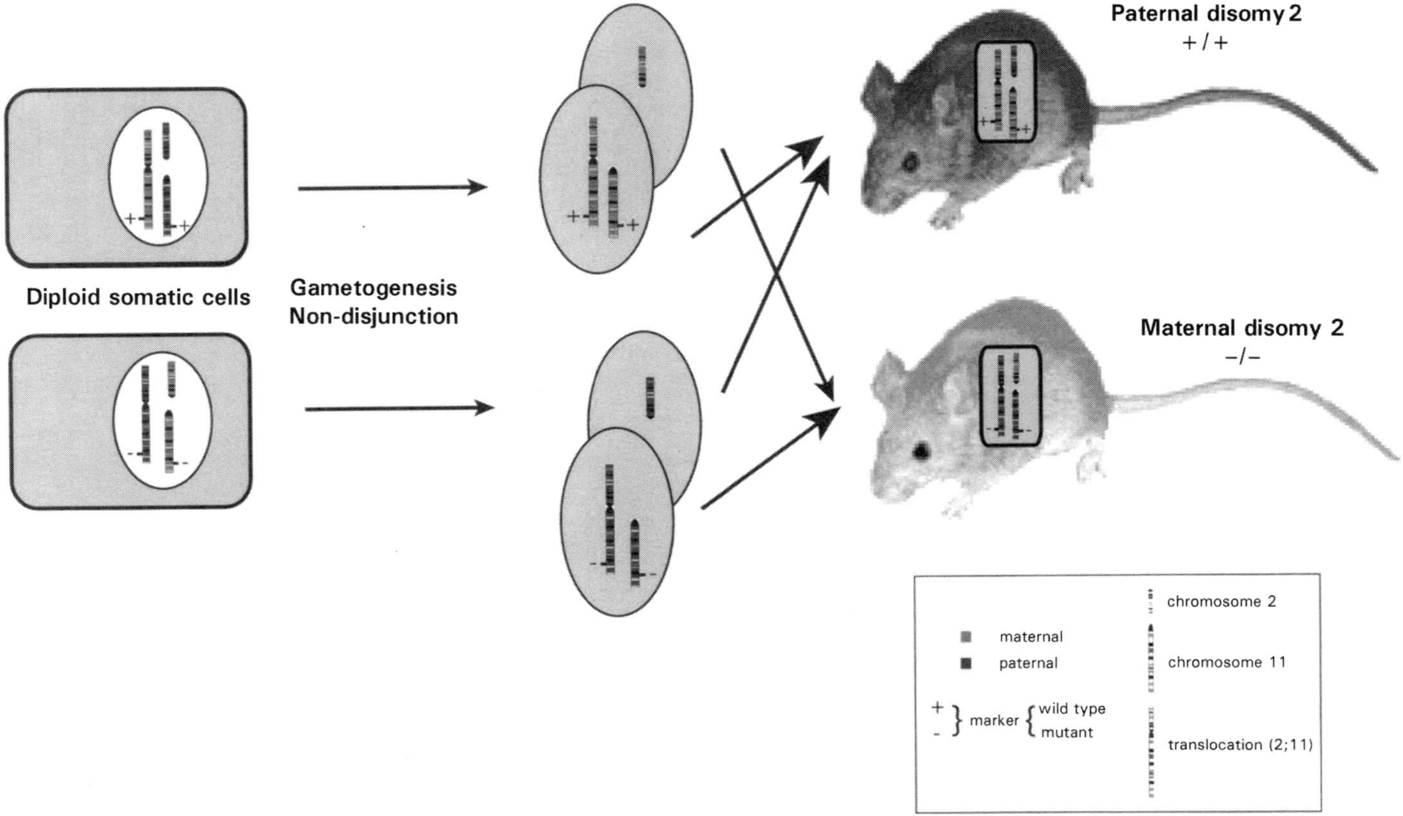

Figure 2 Mice with UPD for specific chromosomal regions can be generated using strains that show a high degree of nondisjunction. Marker genes (indicated by − and + and coat color) can be used to identify the UPD progeny. Only progeny showing UPD are indicated in the figure.

impregs) for an updated list]. Presumably, such screens will miss certain regions associated with mild phenotypes, or those modified by genetic background. In humans, UPD has been associated with a number of diseases that have been used to construct a similar imprinting map (Ledbetter and Engel, 1995; see Chapter 5). Such a map, although not as detailed as the mouse map, also reveals that not all chromosomes show parental origin effects, and that the identified chromosomes have homology of synteny with the regions identified in the mouse. Interestingly, three out of the four regions defined by Ledbetter and Engel (1995) as showing certain imprinting effects (on chromosomes 7, 11, and 14) are associated with a growth abnormality phenotype, as are the corresponding regions in mice (Figure 3).

In conclusion, the nuclear transplantation and genetic experiments demonstrate that maternal and paternal genomes make different contributions to the developing embryo, and indicate that a subset of genes within the genome display parent-of-origin effects.

IDENTIFYING IMPRINTED GENES

The first imprinted gene to be identified was the maternally expressed mouse insulinlike growth factor type-2 receptor (Igf2r), which was isolated by positional cloning as the gene responsible for the T maternal effect (Tme) phenotype, a naturally occurring mouse mutant known to display parental origin effects (Barlow et al., 1991; Lau et al., 1994; Wang et al., 1994). This was quickly followed by the paternally expressed Igf2, which was shown to be imprinted by examining the phenotypes of progeny from reciprocal crosses heterozygous for a targeted mutation (DeChiara et al., 1991). The abnormal phenotype was observed only when the mutant gene was inherited from the father, demonstrating that Igf2 is expressed only from the paternal allele. Thus, the first real evidence for the existence of imprinted genes came from studies of mice. Since these discoveries almost a decade ago, a number of approaches have been taken to identify more imprinted genes in both mice and humans. The most successful method so far has involved genetic mapping data. Imprinted genes have been identified both by testing genes that map close to known imprinted genes, e.g., the mouse H19 gene (Bartolomei et al., 1991), and by identifying and testing novel genes within imprinted regions, e.g., within the PWS/AS region in humans (Lee and Wevrick, 2000).

In addition to mapping strategies, systematic screens have been made to identify imprinted genes. Two strategies involving detecting methylation differences between alleles of an imprinted locus have been used: restriction landmark genome screening (RLGS; Hatada et al., 1993) and methylation-sensitive representational difference analysis (RDA; Kelsey et al., 1999). Both these strategies rely on the use of restriction enzymes that have the same recognition sequence, but differ in their sensitivity to CpG methylation to identify regions of the genome that have one methylated and one unmethylated allele. Potentially expressed sequences associated with these differentially methylated regions (DMRs) then have to be identified and

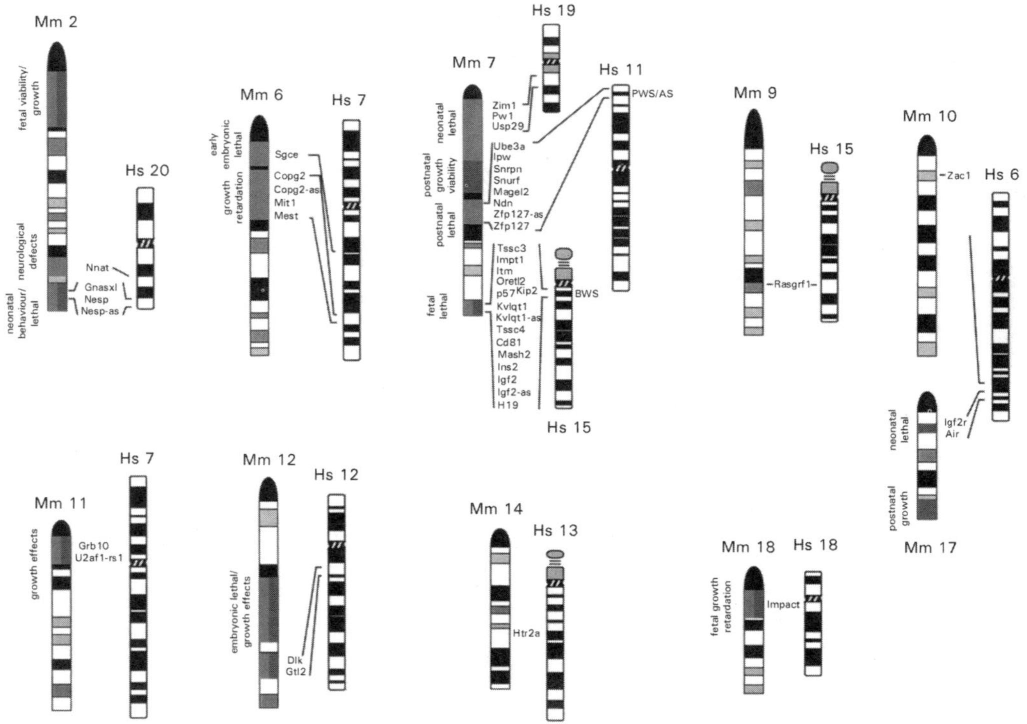

Figure 3 *Phenotypes of mouse UPD strains [adapted from Beechey, 2000; http://www.mgu.har.mrc.a-.mgu.har.mrc.ac.uk/imprinting/imptables.html#impregs)]. Also shown are imprinted genes identified in mice and humans. Mm, mouse chromosome; Hs, human chromosome; PWS, Prader-Willi syndrome; AS, Angelman syndrome; BWS, Beckwith-Wiedeman syndrome. Dark gray and light gray bars on mouse chromosomes indicate paternal and maternal UPD effect, respectively. Genes in dark gray and light gray are paternally or maternally expressed, respectively.*

tested for monoallelic expression. Other approaches have involved direct identification of differentially expressed genes (Kaneko-Ishino et al., 1995) or proteins (Bowden et al., 1996) from normal and parthenogenetic embryos. With the increasing availability of sequence information, bioinformatic approaches have also been used to identify potential imprinted genes (Wylie et al., 2000).

At least 49 imprinted genes have now been reported, and imprinting seems to be very well conserved between mice and humans, with only a few of the studied genes not imprinted in both species (Table 1, Figure 2). It is difficult to estimate the total number of imprinted genes present in mice and humans but, based on mouse mutants (either natural or engineered knock-outs) that display parental effects, 100–200 have been estimated (Barlow, 1995). Thus, if this estimate is accurate, a significant proportion of the total number of imprinted genes has already been identified.

Imprinted genes are often assumed to be monoallelically expressed in all tissues at all times, but this is not true for all imprinted genes, and perhaps not for any. Generally, expression studies analyse only a few tissues or developmental stages and thus do not give a complete picture of the patterns of expression of imprinted genes. This information is obviously important, as illustrated by two examples, kvLQT1 and UBAE3A. Mutations in KVLQT1, a potassium channel, cause long QT syndrome (LQT; a cardiac arrhythmia) and Jervell and Lange-Nielsen cardioauditory syndrome (JLN) (Neyroud et al., 1997). Neither of these syndromes show evidence of imprinting effects. Interestingly, KVLQT1 maps within the BWS region and is imprinted *except* in heart (Lee, 1997). In contrast, UBE3A was originally excluded as a candidate for Angelman syndrome (AS) because it was shown to be biallelically expressed in lymphocytes and fibroblasts (Nakao et al., 1994). However, subsequently mutations were found in UBE3A in AS patients (Kishino et al., 1997). This apparent paradox was resolved when it was shown that UBE3A is expressed only from the maternal allele in the brain (Vu and Hoffman, 1997). Thus in both these cases detailed knowledge of the tissue-specific imprinted expression was necessary to evaluate their involvement in disease.

The high degree of conservation of imprinted genes between mice and humans suggests that a comparison of sequence data can be used to identify conserved elements important for imprinting. Onyango et al. compared over 1 Mb of sequence from the human 11p15 region involved in Beckwith-Wiedeman syndrome (BWS) with the orthologous mouse chromosome 7 region and showed overall structural conservation of the region in terms of genes and CpG islands, but also identifed potentially novel regulatory elements (Onyango et al., 2000). However, differences are important too. For example, recent work on the gene IMPACT, which is imprinted in mice but not humans, indicates that it is the presence of a differentially methylated intronic CpG island which is necessary for imprinting (Okamura et al., 2000). This observation is strikingly similar to the results obtained in a mouse model of imprinting at the Igf2r locus, where transgenes were used to show that an intronic CpG island was necessary to imprint Igf2r (Wutz et al., 1997).

Most imprinted genes in humans and mice map to regions previously identified as displaying UPD effects (Figure 3). However, some imprinted genes map outside

TABLE 1 Imprinted Genes Identified in Mice and Humans
(The genes are grouped together based on map postion in the mouse to emphasize the clustered nature of imprinted genes in both mice and humans)

	Expression		Chromosome			
Gene	Mouse	Human	Mouse	Human	Description	Reference
Wt1[a]	BAE	mat/pat/BAE	prox. 2	11p13	Zinc finger, tumor suppressor	Jinno et al. (1995)
Wt1as	ND	pat	prox. 2	11p13		Malik et al. (2000)
Nnat	pat	ND	dist. 2	20q11-12	Neuronatin	Kagitani et al. (1997)
Gnas	mat	mat	dist. 2	20q13	Guanine nucleotide binding protein (G protein), alpha-stimulating activity polypeptide 1	Hayward et al. (1998a); Hayward et al. (1998b); Wroe et al. (2000); Yu et al. (1998)
Gnasxi	pat	pat	dist. 2	20q13	GNAS extralarge	Peters et al. (1999); Wroe et al. (2000); Yu et al. (1998)
Nesp	mat	mat	dist. 2	20q13	Neuroendocrine secretory protein	Peters et al. (1999); Wroe et al. (2000); Yu et al. (1998)
Nespas	pat	pat	dist. 2	20q13	Noncoding RNA	Hayward and Bonthron (2000); Wroe et al. (2000)
Arh1	ND	pat	4	1p31	Ras-related protein	Yu et al. (1999)
Sgce	pat	pat	6 cen.	7q21-22	Sarcoglycan epsilon	Piras et al. (2000)
Peg1/Mest	pat	pat	prox. 6	7q32	Hydrolase	Kaneko-Ishino et al. (1995); Nishita et al. (1996)
Copg2[a]	mat	pat	prox. 6	7q32	Coatomer protein complex subunit Gamma	Lee et al. (2000); Blagitko et al. (1999) Yamasaki et al. (2000)
Copg2as[a]	pat	ND	prox. 6	7q32	Noncoding RNA	Lee et al. (2000)
Mit1/lb9	pat	pat	prox. 6	7q32	—	Lee et al. (2000)
Zim1	mat	ND	prox. 7	19q13	Zinc finger transcription factor	Kim et al. (1999)
Peg3/Pw1	pat	ND	prox. 7	19q13	Zinc finger transcription factor	Kuroiwa et al. (1996)
Usp29	pat	ND	prox. 7	19q13	Ubiquitin-specific processing protease 29	Kim et al. (2000)
Snrpn	pat	pat	cen. 7	15q12	snRNP-associated polypeptide, splicing factor	Cattanach et al. (1992); Glenn et al. (1993b); Leff et al. (1992); Reed and Leff (1994)
Magel2	pat	pat	cen. 7	15q12	MAGe-like 2	Boccaccio et al. (1999)

(*continued*)

TABLE 1 (*continued*)

Gene	Expression		Chromosome		Description	Reference
	Mouse	Human	Mouse	Human		
Ndn	pat	pat	cen. 7	15q12	Unknown	Jay et al. (1997)
Zfp127	pat	pat	cen. 7	15q12	Zinc finger transcription factor	Jong et al. 1993; Saitoh et al. (1996)
Zfp127as	pat	pat	cen. 7	15q12	Noncoding RNA	
Ipw	pat	pat	cen. 7	15q12	Noncoding RNA	Wevrick and Francke (1997); Wevrick et al. (1994)
Par1	ND	pat	cen. 7	15q12	Noncoding RNA	Sutcliffe et al. (1994)
Par5	ND	pat	cen. 7	15q12	Noncoding RNA	Sutcliffe et al. (1994)
Ube3a	mat	mat	cen. 7	15q12	Ubiquitin conjugating enzyme E3A	Albrecht et al. (1997)
H19	mat	mat	dist. 7	11p15	Noncoding RNA	Bartolomei et al. (1991); Rainier et al. (1993)
Igf2	pat	pat	dist. 7	11p15	Insulinlike growth factor 2, mitogenic growth factor	DeChiara et al. (1991); Giannoukakis et al. (1993); Ohlsson et al. (1993)
Igf2as	pat	pat	dist. 7	11p15	Noncoding RNA	Okutsu et al. (2000)
Ins2	pat	BAE	dist. 7	11p15	Insulin	Giddings et al. (1994)
Mash2	mat	mat	dist. 7	11p15	bHLH transcription factor	Alders et al. (1997); Guillemot et al. (1995)
Kvlqt1	mat	mat	dist. 7	11p15	Potassium channel involved in long QT syndrome	Lee (1997)
Kvlqt1-as	pat	pat	dist. 7	11p15	Noncoding RNA	Lee et al. (1999b)
$p57^{KIP2}$	mat	mat	dist. 7	11p15	CDK inhibitor	Hatada et al. (1996); Matsuoka et al. (1996)
Impt1	mat	mat	dist. 7	11p15	Solute carrier family 22 (organic cation transporter), member 1-like	Dao et al. (1999)
Ipl/Tssc3	mat	mat	dist. 7	11p15	Tumor-suppressing subtransferable candidate 3	Schwlenbacher et al. (1998)
Tssc4	mat	BAE?	dist. 7	11p15	—	Lee et al. (1999a)

Rasgrf1	pat	pat	9	15q24	Ras protein-specific guanine nucleotide-releasing factor 1	Plass et al. (1996)
Zac1	pat	pat	10	6q24-25	Zinc finger transcription factor, putative tumor suppressor	Piras et al. (2000)
Hyma1	ND	pat	10	6q24-25	—	Arima et al. (2000)
Meg1/Grb10	mat	mat	prox. 11	7p12	Growth factor receptor-bound protein 10	Miyoshi et al. (1998); Yoshihashi et al. (2000)
U2af1-rs1	pat	BAE	prox. 11	5q23-31	U2AF auxiliary factor small subunit, splicing factor	Hatada et al. (1995); Kitagawa et al. (1995); Pearsall et al. (1996)
Dlk	pat	pat	dist. 12	14q32	Homologue of Drosophila deltalike protein	Schmidt et al. (2000); Wylie et al. (2000)
Gtl2	mat	mat	dist. 12	14q32	Unknown function	Miyoshi et al. (2000); Schmidt et al. (2000); Wylie et al. (2000)
Htr2a	mat	mat	dist. 14	13q14-21	Serotonin receptor 2A	Kato et al. (1996)
Igf2r	mat	mat	prox. 17	6q25.3	Lysosomal transport receptor	Barlow et al. (1991); Smrzka et al. (1995)
Air (Igf2ras)	pat	ND	prox. 17	6q25.3	Noncoding RNA	Lyle et al. (2000); Wutz et al. (1997)
Impact	pat	BAE	prox. 18	18q11	Unknown	Hagiwara et al. (1998); Okamura et al. (2000)
Ins1[a]	pat	—	19	No homologue	Insulin	Giddings et al. (1994)
Xist	pat/ran	pat/ran	X	X	Induces X inactivation	Goto et al. (1998); Kay et al. (1994)

pat = paternal expression; mat = maternal expression; BAE = biallelic expression.

[a] The data concerning the imprinting status of these genes are conflicting.

these regions and genomewide screens for imprinted genes may be expected to find more. The identification and analysis of imprinted gene function enable us to ask if individual genes are responsible for the phenotypes of UPD cases. For example, Igf2r$^{-/-}$ mice and the Tme mutant mice have a very similar phenotype, strongly arguing that this gene is responsible for the phenotypes observed for paternal UPD or maternal deletion of proximal mouse chromosome 17 (Barlow et al., 1991; Wang et al., 1994). Further examples are discussed later in the context of mouse models of Prader-Willi syndrome.

As more imprinted genes have been identified, it has become increasingly clear that they are clustered within the genome, which suggests that there may be a functional basis for grouping imprinted genes. In mice, deletion of the H19 gene or its enhancers results in disruption of imprinting of two upstream genes, Igf2 and Ins2 (Leighton et al. 1995a, 1995b). In humans, a postulated imprinting center (IC) has been suggested to control the imprinting of a number of genes within the Prader-Willi syndrome region (Buiting et al., 1995; Sutcliffe et al., 1994) (see later discussion). In addition to these effects on the expression of genes, imprinted domains also show parental differences in replication timing (Kitsberg et al., 1993; Simon et al., 1999) and meiotic recombination (Paldi et al., 1995), and imprinted regions may also show the homologous association at certain stages of the cell cycle (LaSalle and Lalande, 1995). It is not known how these domain effects, which affect imprinted and nonimprinted genes within the regions, relate mechanistically to imprinting. However, these data taken together do suggest the existence of functional imprinted domains in both mice and humans.

THE IMPRINTING MECHANISM

The ability to express one allele in a parent-specific manner in a diploid somatic cell implies that the alleles can be distinguished throughout development. Such a distinguishing mark has been termed the imprint. It is important to realize that imprinted gene expression can be considered in a number of separate stages: first, establishment of the imprint; second, maintenance of the imprint; third, reading of this imprint; fourth, imprint erasure (Figure 4).

The imprint must be established while the two parental genomes are separate, and this is most likely to occur during gametogenesis (Figure 4*a*). It may be predicted that certain sequences are imprinted in the male germ line and others imprinted in the female germ line, but such sequences remain elusive and what distinguishes a maternal from a paternal imprint remains unknown. The imprint must be stable and copied through DNA replication in somatic cells to maintain imprinted expression in daughter cells (Figure 4*b*). Reading the imprint by the transcriptional machinery (Figure 4*c*) can be considered a separate step since, e.g., in Igf2r (Stöger et al., 1993), the imprint is present during preimplantation development but imprinted gene expression is not established until after implantation. And finally, since imprinting for a particular sequence is switched on passage through the germ line of the

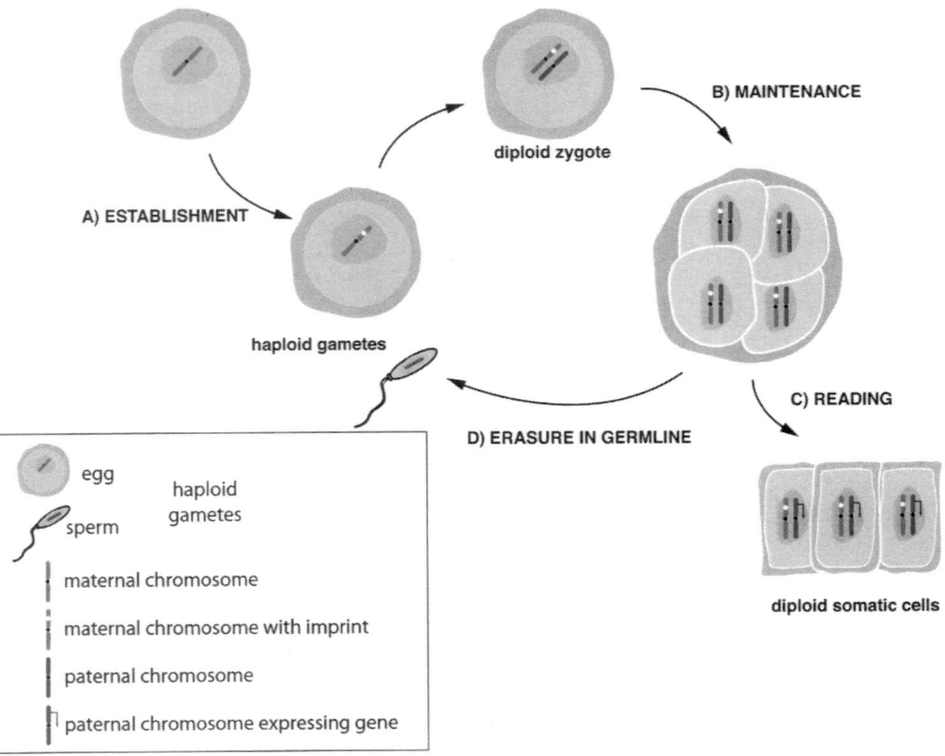

Figure 4 *Different stages in the imprinting mechanism. The example shown considers a maternally imprinted, paternally expressed gene. (A) Establishment of the imprint on the maternal chromosome during oogenesis. (B) Maintenance of the imprint throughout early development and in somatic cells. (C) Reading the imprint by the various mechanisms involved in imprinting, resulting in monoallelic expression. (D) Erasure of the imprint from the maternal chromosome during spermatogenesis.*

opposite sex, the imprint must be erased and then reestablished (Figure 4*d*). These criteria suggest that the imprint is maintained as an epigenetic mark on the DNA.

The best candidate so far identified for the imprint is methylation of cytosine within CpG dinucleotides [for a review of CpG methylation, see (Cross and Bird, 1995)]. The first indication of the importance of methylation in imprinting came with the observation that imprinted genes contain regions of allele-specific methylation [e.g., see (Stöger et al., 1993)]. Indeed, all imprinted genes examined so far have such regions, showing that methylation is used to distinguish the two alleles. However, not all these differentially methylated regions (DMRs) can be considered the imprint in the sense that methylation is present in the gametes and maintained throughout development. Only for three genes, Igf2r (Stöger et al., 1993), H19 (Bartolomei et al., 1993) and Snrpn (Shemer et al., 1997), have such candidate imprints been identified. Other DMRs are methylation-free during gametogenesis and early development, only acquiring methylation later, probably after imprinting gene expression has been established. Thus, DMRs may be primary imprints whose function is to determine the pattern of imprinted gene expression and secondary imprints that function to maintain this pattern during development.

The importance of methylation to imprinting was underlined by analyzing mice that had a targeted disruption of the maintenance methyltransferase gene (Dnmt1; Li et al., 1993); Dnmt1$^{-/-}$ mice fail to maintain monoallelic expression of imprinted genes. Although these experiments established that methylation can meet the criteria for the imprint and is required to maintain imprinted expression, it is known that methylation is completely removed from these allele-specific methylated regions in germ-line passage through the opposite sex, indicating that mechanisms other than methylation are involved in establishing the imprint.

Although we are far from a complete understanding of the imprinting mechanism, and it is still not clear if all imprinted genes are regulated in the same way, some common themes are present. As mentioned above, all imprinted genes have regions of differential methylation. In addition, imprinted genes are always in groups of two or more and there is evidence for imprinting control regions that regulate the imprinted expression of one or more genes (e.g., SNRPN and LIT1). How the imprint is read to produce monoallelic expression is being studied in many genes and some examples are discussed below to illustrate the apparent themes in imprinted gene expression (Figure 5).

Methylation is an attractive candidate for the imprint since a great deal of work has shown that methylation represses transcription (Bird and Wolffe, 1999), thus providing an obvious direct link between the imprint itself and the outcome of reading the imprint, which is transcription. Interestingly, in Dnmt1$^{-/-}$ mice the H19 gene becomes biallelically expressed and the Igf2r gene is completely silenced (Li et al., 1993). This implies that in some cases, methylation is repressing transcription, in other cases activating transcription. This apparent paradox can be understood if we consider that a methylated imprint may either repress transcription by modifying a *cis*-activating activator (e.g., a promoter or enhancer) or activate transcription by modifying a *cis*-acting repressor (e.g., a repressor-binding site; Figure 5).

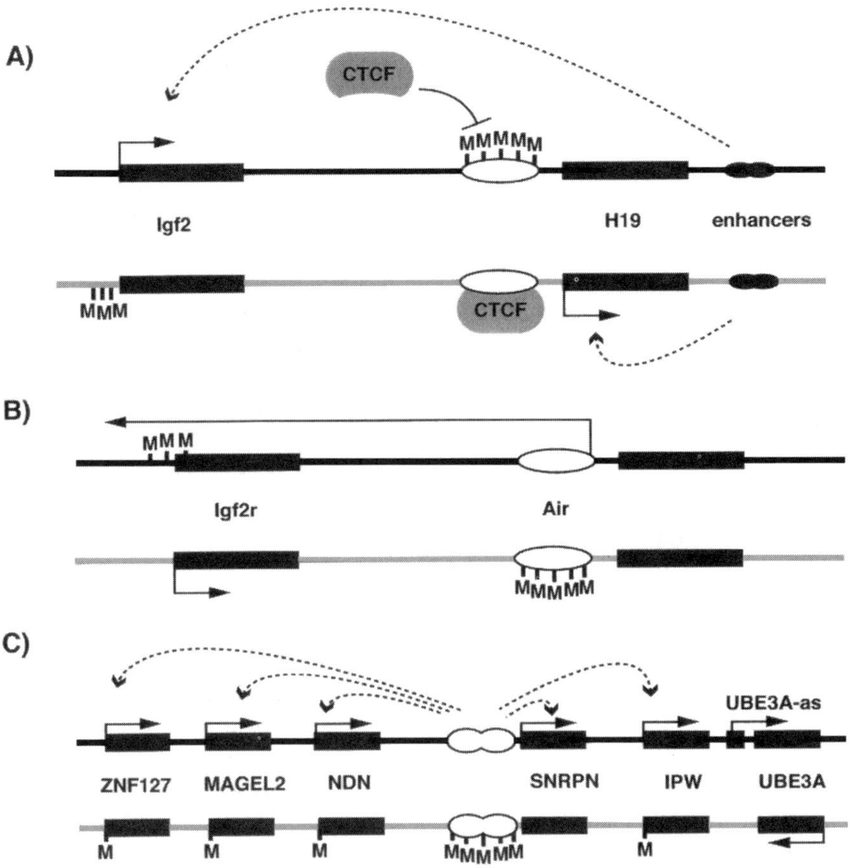

Figure 5 *Examples of different mechanisms involved in imprinted gene expression. (A) At the Igf2/H19 locus, paternal methylation at the differentially methylated region (DMR) 5′ to the H19 gene prevents the repressor CTCF binding; this, in turn, allows the enhancers to interact with the Igf2 promoter. Paternal H19 expression is presumed to be repressed because of the methylation at its promoter. On the maternal chromosome, CTCF can interact with the unmethylated DMR to set up a boundary of expression so that the enhancers can only interact with H19. Subsequent methylation of the Igf2 promoter helps to maintain this pattern of expression. (B) At Igf2r, an intronic DMR is known to be essential to repress Igf2r expression on the paternal chromosome. This DMR is also the promoter for an antisense RNA (Air), raising the possibilty of repression via an antisense RNA mechanism. On the maternal chromosome, this DMR shows imprinted methylation, Air is not expressed and Igf2r is expressed. (C) Imprinted expression of genes at the PWS region is controlled by an IC that is responsible for paternal gene expression throughout the region. How this long-range effect is achieved is not understood. Expression of an antisense gene UBE3A-AS, under control of the IC, may regulate the Angelman syndrome gene (UBE3A) in a manner similar to Igf2r/Air. Black oblongs represent genes; white ovals represent imprinting control regions (ICs)/imprinted differentially methylated regions (DMRs). Dark gray and light gray lines represent paternal and maternal chromosomes, respectively. CTCF, repressor; M, methylation; arrows with solid lines represent gene expression; arrrows with dotted lines indicate interactions between imprinting control regions (B and C) or enhancers (A) with genes. See text for further explanation.*

This second mechanism is, in fact, suggested by recent work on H19 and Igf2 (Bell and Felsenfeld, 2000; Hark et al., 2000). Previous work showed that there is a region of paternal-specific methylation between the maternally expressed H19 and paternally expressed Igf2, and that both genes require access to enhancers downstream of H19 (Bartolomei et al., 1993) for expression. The reciprocal imprinting of this gene pair was originally suggested to be controlled by enhancer competition: the region of differential methylation silenced the promoter for H19, and thus Igf2 could gain access to the enhancers and was expressed (Figure 5*a*). However, it now appears that the differentially methylated region is a binding site for the repressor CTCF (Bell and Felsenfeld, 2000; Hark et al., 2000). Thus when methylated, CTCF cannot bind and Igf2 is expressed. The presence of potential CTCF binding sites at another imprinting gene pair (DLK1/GTL2; Wylie et al., 2000) raises the exciting possibility that boundary elements may be a common feature regulating imprinted gene expression.

Another well-studied imprinted gene pair that appears similar to Igf2/H19 is Igf2r/Air (Figure 5*b*). Igf2r and Air are reciprocally imprinted and both have differentially methylated CpG island promoters (Lyle et al., 2000; Stöger et al., 1993). Region 2 is methylated throughout development and is the promoter for the noncoding RNA, Air. Deletion of region 2 in mice results in biallelelic Igf2r expression (Wutz et al., 1997), indicating that region 2/Air functions as a *cis* repressor of Igf2r expression, possibly via an antisense RNA mechanism (Lyle et al., 2000). The presence of many noncoding, antisense RNAs at other imprinted loci (Table 1) indicates that this is a common mechanism. For example, at the UBE3A locus, the paternally expressed antisense RNA (UBE3A-as) may repress paternal UBE3A expression (Rougeulle et al., 1998) in a similar way to Air/Igf2r. However, it remains to be shown whether these RNAs are the cause of the imprinting or are simply a secondary consequence of, e.g., local changes in chromatin structure.

Certain regions thus appear to control the imprinted expression of genes. This is illustrated very clearly in the PWS/AS region (as discussed in Chapter 6). Approximately 5% of PWS and AS patients have imprinting mutations, revealed as abnormal methylation and gene expression throughout the region (Glenn et al., 1993a). Around half these patients have small deletions in the first exon of SNRPN in the case of PWS (Buiting et al., 1995; Sutcliffe et al., 1994) and deletion upstream of the first exon in AS (Buiting et al., 1995; Dittrich et al., 1996; Saitoh et al., 1996) (Figure 5*c*). These patients have a normal biparental content in these region, but only the maternal epigenotype in PWS and only the paternal epigenotype in AS. Thus, there is a bipartite imprinting control (IC) center in the 5′ region of SNRPN controlling the epigenetic status, and thus gene expression, over a 2-Mb region (Figure 5*c*).

In summary, imprinted expression of certain genes seems to be regulated by small imprinting control (IC) regions. What remains intriguing is the variety of mechanisms (boundary elements, antisense RNA, methylation) involved at these ICs. Is there a unifying mechanism underlying the expression of all imprinted genes?

USING MICE TO MODEL IMPRINTING

The clustering of imprinted genes and the existence of coordinate regulation may make it difficult to interpret genotype-phenotype correlations in UPD and imprinting syndromes. Only by analyzing the individual phenotypic effects of imprinted genes, either by targeted mutation in mice or identification of mutations in humans, can we begin to understand the complex phenotypes seen in UPD. Working with mice has the advantage that specific genetic changes can be introduced into the germ line. These can be broadly split into two categories: first, targeted mutations of single genes or control regions; second, deletions of much larger chromosomal regions containing many genes. The usefulness of these approaches is discussed below with reference to dissecting the phenotype of the Prader-Willi syndrome.

PWS has a complex phenotype with developmental, behavioral and mental aspects (see Chapter 6 for a more detailed discussion). Unlike for AS, where approximately 20% of cases are caused by mutation of a single gene (UBE3A; Kishino et al., 1997; Matsuura et al., 1997), no single gene defects have been found. This implies that PWS is a contiguous gene syndrome caused by alteration in the expression of a number of genes. Although patients may exist with mutations in single genes from the PWS region, they would not be recognized as PWS. Thus in order to dissect the contribution of single or multiple genes to the phenotype, and to gain insights into the imprinting mechanism in this region, the mouse has been used as a model system (Table 2).

The first mouse model of PWS (Cattanach et al., 1992) used translocation mouse strains to produce mice with UPD for the central region of mouse chromosome 7, the region homologous to human 15q11-13. Maternal UPD for this region is early postnatal lethal in these mice, clearly indicating an imprinting effect in this region. However, while the authors suggest that these mat UPD mice have a possibly reduced suckling activity reminiscent of the failure to thrive of PWS patients, the usefulness of this model is limited by the large number of genes that map within the translocated segment.

Three other studies have examined large deletions within this region to evaluate the contribution of multiple genes to PWS. A radiation-induced deletion (P^{30PUb}) at the mouse p locus includes genes from p to Ipw, but does not show an imprinted early lethal phenotype (Johnson et al., 1995). Genes within this region are therefore unlikely to contribute, at least grossly, to the PWS phenotype. A transgene insertion into mouse chromosome 7C resulted in a large deletion encompassing Zfp127 to Herc2, thus including all genes within the PWS/AS region (Gabriel et al., 1999). Mice inheriting this deletion paternally had a much reduced growth rate and died within 1 week of birth. Using Cre/lox technology, Tsai et al. (1999) generated mice with a deletion from Snrpn to Ube3a. Paternal inheritance of this deletion resulted in severe growth retardation, hypotonia, poor feeding, decreased movement, and approximately 80% of pups died before weaning. Together, these data would suggest that a gene or genes lying between Snrpn and Ipw is responsible for the early lethality and feeding defect and at least three imprinted transcripts lie within this region (Lee and Wevrick, 2000). In humans, genes within this area may therefore be

TABLE 2 Mouse Models Created to Evaluate Contributions of Genes within PWS Region to the Phenotype

Type of Mutation	Genetic Region	Phenotype	Reference
Single gene targeted disruption	Snrpn	No obvious phenotype	Yang et al. (1998)
	Snurf (Snrpn exon2)	No obvious phenotype	Tsai et al. (1999b)
	Ipw	No obvious phenotype	Goss et al. (unpublished)[a]
	Znp127	No obvious phenotype	Carey et al. (unpublished)[a]
	Ndn[b]	Strain- dependent failure to thrive, respiratory defects; partial postnatal lethality on paternal transmission	Gérard et al. (1999); Tsai et al. (1999a)
Deletion of IC	IC and Snrpn	Neonatal lethal, impaired feeding	Yang et al. (1998)
	IC	Failure to switch imprint	Bielinska et al. (2000)
Construction of UPD strains	Central chromosome 7	Maternal UPD is postnatal lethal	Cattanach et al. (1992)
Deletion of multiple genes	Deletion of p-Ipw	No obvious phenotype	Johnson et al. (1995)
	Deletion of Zfp127-Herc2	Early postnatal lethal	Gabriel et al. (1999)
	Deletion of Snrpn-Ube3a	Hypotonia, growth retardation, partial lethality	Tsai et al. (1999b)

[a] Unpublished data cited in (Nicholls, 1999).
[b] The data for Ndn are conflicting; see text for details.
IC = imprinting control center.

responsible for the hypotonia and poor nursing seen in PWS. The fact that neither of these mouse models displays the hypogonadism seen in PWS could be due to developmental differences between mice and humans or the fact that genes outside the regions studied are involved.

Since data from PWS patients suggested evidence of an imprinting control center located 5′ to SNRPN, two groups have targeted this IC region and the Snrpn gene itself in mice to test its function. Yang et al. (1998) created two mice strains, one with a targeted mutation of Snrpn, and the other with a deletion containing Snrpn and the putative IC. Mice with deletion of Snrpn alone showed no obvious phenotypic defects, indicating that the loss of SmN (the product of the SNRPN gene) is not responsible for any aspects of the PWS phenotype. In contrast, paternal inheritance of a deletion covering the IC and Snrpn is neonatally lethal. Mutant mice were approximately 20% smaller than wild-type littermates and showed evidence of reduced feeding activity. Thus, mice with a deletion of the IC and Snrpn mimic at least one aspect of PWS, but they do not exhibit the hypotonia or gonadal hypoplasia common in PWS. At the molecular level, they do not express the imprinted genes Ipw, Ndn, and Zfp127 when the IC deletion is inherited paternally, indicating a failure to switch the imprint in the male germ line, and showing conservation of the IC and imprinting mechanism between mice and humans.

The studies described above were all aimed at creating models where the expression of multiple genes was affected. However, it is equally important to assess the contribution of single genes to aspects of the PWS phenotype. To date, targeted mutagenesis has been reported for three genes, Snrpn, Snurf, and Ndn (Table 2). The SNRPN coding region has an unusual bicistronic structure with the potential to code for two proteins, SNURF and SmN (Gray et al., 1999). Loss of either Snrpn (Yang et al., 1998) or Snurf (Tsai et al., 1999b) has no obvious phenotypic effect. In addition, unpublished data for Zfp127 and Ipw have been cited (Nicholls, 1999), indicating that they do not make a significant contribution to PWS.

Two conflicting reports of targeted deletion of the necdin (Ndn) have been published (Gérard et al., 1999; Tsai et al., 1999a). First, Tsai et al. found no abnormal phenotype in $Ndn^{-/-}$ mice. In contrast, Gérard et al. reported hypotonia, postnatal resipratory distress and lethality. Interestingly, the phenotypes observed by Gérard et al. were found to be dependent on the background mouse strain, perhaps explaining the lack of a phenotype observed by Tsai et al. Mice that survived the early critical period showed normal postnatal development, indicating that Ndn does not play a role in PWS beyond this early period. Thus, Ndn is the sole example of a gene responsible for part of the PWS phenotype, and this is strain-dependant.

Mouse models of PWS with mutations affecting a number of genes have reproduced some of the associated phenotypes. Single-gene mutations have proved less successful. However, given that PWS is almost certainly caused by multiple genes, this may not be surprising, and breeding mice with double or triple mutations will hopefully provide us with more appropriate models to understand this complex syndrome.

One other advantage of the mouse that should not be overlooked is the ability to carry out detailed expression studies. For example, the gene that is responsible for

Angelman syndrome (UBE3A; Fang et al., 1999; Malzac et al., 1998) was shown to be expressed and imprinted in the human brain (Rougeulle et al., 1998; Vu and Hoffman, 1997). However, *in situ* hybridization studies in the mouse revealed that Ube3a is only imprinted in hippocampal neurons, Purkinje cells, and olfactory mitral cells (Albrecht et al., 1997), thus giving a clearer picture of the neurological cells involved in the complex Angelman syndrome phenotype.

REFERENCES

Albrecht, U., Sutcliffe, J. S., Cattanach, B. M., Beechey, C. V., Armstrong, D., Eichele, G. and Beaudet, A. L. Imprinted expression of the murine Angelman syndrome gene, Ube3a, in hippocampal and Purkinje neurons. *Nature Genet* 17:75–78, 1997.

Alders, M., Hodges, M., Hadjantonakis, A.-M., Postmus, J., van Wijk, I., Bliek, J., de Meulemeester, M., et al. The human *Achaete-Scute* homologue 2 (ASCL2, HASH2) maps to chromosome 11p15.5, close to IGF2 and is expressed in extravillus trophoblasts. *Hum Molec Genet* 6:859–867, 1997.

Arima, T., Drewell, R. A., Oshimura, M., Wake, N. and Surani, M. A. A novel imprinted gene, HYMAI, is located within an imprinted domain on human chromosome 6 containing ZAC. *Genomics* 67:248–255, 2000.

Barlow, D. P. Gametic imprinting in mammals. *Science* 270:1610–1613, 1995.

Barlow, D. P., Stöger, R., Herrmann, B. G., Saito, K. and Schweifer, N. The mouse insulin-like growth factor type 2 receptor is imprinted and closely linked to the Tme locus. *Nature* 349:84–87, 1991.

Bartolomei, M. S., Webber, A. L., Brunkow, M. E. and Tilghman, S. M. Epigenetic mechanisms underlying the imprinting of the mouse H19 gene. *Genes Dev* 7:1663–1673, 1993.

Bartolomei, M. S., Zemel, S. and Tilghman, S. M. Parental imprinting of the mouse H19 gene. *Nature* 351:153–155, 1991.

Barton, S. C., Adams, C. A., Norris, M. L. and Surani, M. A. H. Development of gynogenetic and parthenogenetic inner cell mass and trophectoderm tissues in reconstituted blastocysts in the mouse. *Journal of Embryology and Experimental Morphology* 90:267–285, 1985.

Barton, S. C., Ferguson-Smith, A. C., Fundele, F. and Surani, M. A. Influence of paternally imprinted genes on development. *Development* 113:679–688, 1991.

Barton, S. C., Surani, M. A. H. and Norris, M. L. Role of paternal and maternal genomes in mouse development. *Nature* 311:374–376, 1984.

Beechey, C. V. and Cattanach, B. M. Genetic imprinting map. *Mouse Genome* 94:96–99, 1996.

Bell, A. C. and Felsenfeld, G. Methylation of a CTCF-dependent boundary controls imprinted expression of the Igf2 gene. *Nature* 405:482–485, 2000.

Bielinska, B., Blaydes, S. M., Buiting, K., Yang, T., Krajewska-Walasek, M., Horsthemke, B. and Brannan, C. I. De novo deletions of SNRPN exon 1 in early human and mouse embryos result in a paternal to maternal imprint switch [see comments]. *Nature Genet* 25:74–78, 2000.

Bird, A. P. and Wolffe, A. P. Methylation-induced repression—belts, braces, and chromatin. *Cell* 99:451–454, 1999.

Blagitko, N., Schulz, U., Schinzel, A. A., Ropers, H. H. and Kalscheuer, V. M. gamma2-COP, a novel imprinted gene on chromosome 7q32, defines a new imprinting cluster in the human genome. *Hum Molec Genet* 8:2387–2396, 1999.

Boccaccio, I., Glatt-Deeley, H., Watrin, F., Roeckel, N., Lalande, M. and Muscatelli, F. The human MAGEL2 gene and its mouse homologue are paternally expressed and mapped to the Prader-Willi region. *Hum Molec Genet* 8:2497–2505, 1999.

Bowden, L., Klose, J. and Reik, W. Analysis of parent-specific gene expression in early mouse embryos and embryonic stem cells using high- resolution two-dimensional electrophoresis of proteins. *International Journal of Developmental Biology* 40:499–506, 1996.

Buiting, K., Saitoh, S., Gross, S., Dittrich, B., Schwartz, S., Nicholls, R. D. and Horsthemke, B. Inherited microdeletions in the Angelman and Prader-Willi syndromes define an imprinting center on human chromosome 15. *Nature Genet* 9:395–400, 1995.

Cattanach, B. M., Barr, J. A., Evans, E. P., Burtenshaw, M., Beechey, C. V., Leff, S. E., Brannan, C. I., et al. A candidate mouse model for Prader-Willi syndrome which shows an absence of Snrpn expression. *Nature Genet* 2:270–274, 1992.

Cattanach, B. M., Beechey, C. V. Autosomal and X-chromosome imprinting. *Development Supp* 63–72, 1990.

Cattanach, B. M., Burtenshaw, M. D., Rasberry, C. and Evans, E. P. Large deletions and other gross forms of chromosome imbalance compatible with viability and fertility in the mouse. *Nature Genet* 3:56–61, 1993.

Cattanach, B. M. and Kirk, M. Differential activity of maternally and paternally derived chromosome regions in mice. *Nature* 315:496–498, 1985.

Chess, A., Simon, I., Cedar, H. and Axel, R. Allelic inactivation regulates olfactory receptor gene-expression. *Cell* 78:823–834, 1994.

Cross, S. H. and Bird, A. P. CpG islands and genes. *Curr Opin Genet Dev* 5:309–314, 1995.

Dao, D., Walsh, C. P., Yuan, L., Gorelov, D., Feng, L., Hensle, T., Nisen, P., et al. Multipoint analysis of human chromosome 11p15/mouse distal chromosome 7: inclusion of H19/IGF2 in the minimal WT2 region, gene specificity of H19 silencing in Wilms' tumorigenesis and methylation hyper-dependence of H19 imprinting. *Hum Molec Genet* 8:1337–1352, 1999.

DeChiara, T. M., Robertson, E. J. and Efstratiadis, A. Parental imprinting of the mouse insulin-like growth factor II gene. *Cell* 64:849–859, 1991.

Dittrich, B., Buiting, K., Korn, B., Rickard, S., Buxton, J., Saitoh, S., Nicholls, R. D., et al. Imprint switching on human chromosome 15 may involve alternative transcripts of the SNRPN gene. *Nature Genet* 14:163–170, 1996.

Fang, P., Lev-Lehman, E., Tsai, T. F., Matsuura, T., Benton, C. S., Sutcliffe, J. S., Christian, S. L., et al. The spectrum of mutations in UBE3A causing Angelman syndrome. *Hum Molec Genet* 8:129–135, 1999.

Gabriel, J. M., Merchant, M., Ohta, T., Ji, Y., Caldwell, R. G., Ramsey, M. J., Tucker, J. D., et al. A transgene insertion creating a heritable chromosome deletion mouse model of Prader-Willi and angelman syndromes. *Proc Natl Acad Sci USA* 96:9258–9263, 1999.

Gérard, M., Hernandez, L., Wevrick, R. and Stewart, C. L. Disruption of the mouse necdin gene results in early post-natal lethality. *Nature Genet* 23:199–202, 1999.

Giannoukakis, N., Deal, C., Paquette, J., Goodyer, C. G. and Polychronakos, C. Parental genomic imprinting of the human IGF2 gene. *Nature Genet* 4:98–101, 1993.

Giddings, S. J., King, C. D., Harman, K. W., Flood, J. F. and Carnaghi, L. R. Allele-specific inactivation of insulin 1 and 2, in the mouse yolk sac, indicates imprinting. *Nature Genet* 6:310–313, 1994.

Glenn, C. C., Nicholls, R. D., Robinson, W. P., Saitoh, S., Niikawa, N., Schinzel, A., Horsthemke, B., et al. Modification of 15q11-q13 DNA methylation imprints in unique Angelman and Prader-Willi patients. *Hum Molec Genet* 2:1377–1382, 1993a.

Glenn, C. C., Porter, K. A., Jong, M. T. C., Nicholls, R. D. and Driscoll, D. J. Functional imprinting and epigenetic modification of the human SNRPN gene. *Hum Molec Genet* 2:2001–2005, 1993b.

Goto, T., Christians, E. and Monk, M. Expression of an Xist promoter-luciferase construct during spermatogenesis and in preimplantation embryos: regulation by DNA methylation. *Mol Reprod Dev* 49:356–367, 1998.

Graham, C. F. The production of parthenogenetic mammalian embryos and their use in biological research. *Biol Rev* 49:399–422, 1974.

Gray, T. A., Saitoh, S. and Nicholls, R. D. An imprinted, mammalian bicistronic transcript encodes two independent proteins. *Proc Natl Acad Sci USA* 96:5616–5621, 1999.

Guillemot, F., Caspary, T., Tilghman, S. M., Copeland, N. G., Gilbert, D. J., Jenkins, N. A., Anderson, D. J., et al. Genomic imprinting of Mash2, a mouse gene required for trophoblast development. *Nature Genet* 9:235–242, 1995.

Hagiwara, Y., Hirai, M., Nishimiyama, K., Kanazawa, I., Ueda, T., Sakaki, Y. and Ito, T. Screening for imprinted genes by allelic message display: identification of a paternally expressed gene impact on mouse chromosome 18. *Proceedings of the National Academy of Science USA* 94:9249–9254, 1998.

Hardy, K. and Handyside, A. H. Metabolism and cell allocation during parthenogenetic preimplantation mouse development. *Molecular Reproduction and Development* 43:313–322, 1996.

Hark, A. T., Schoenherr, C. J., Katz, D. J., Ingram, R. S., Levorse, J. M. and Tilghman, S. M. CTCF mediates methylation-sensitive enhancer-blocking activity at the H19/Igf2 locus. *Nature* 405:486–489, 2000.

Hatada, I., Inazawa, J., Abe, T., Nakayama, M., Kaneko, Y., Jinno, Y., Niikawa, N., et al. Genomic imprinting of human p57KIP2 and its reduced expression in Wilms' tumors. *Hum Molec Genet* 5:783–788, 1996.

Hatada, I., Kitagawa, K., Yamaoka, T., Wang, X. D., Arai, Y., Hashido, K., Ohishi, S., et al. Allele-specific methylation and expression of an imprinted U2af1-rs1 (Sp2) gene. *Nucleic Acids Res* 23:36–41, 1995.

Hatada, I., Sugama, T. and Mukai, T. A new imprinted gene cloned by a methylation-sensitive genome scanning method. *Nucleic Acids Res* 21:5577–5582, 1993.

Hayward, B. E. and Bonthron, D. T. An imprinted antisense transcript at the human GNAS1 locus. *Hum Molec Genet* 9:835–841, 2000.

Hayward, B. E., Kamiya, M., Strain, L., Moran, V., Campbell, R., Hayashizaki, Y. and Bonthron, D. T. The human GNAS1 gene is imprinted and encodes distinct paternally and biallelically expressed G proteins. *Proc Natl Acad Sci USA* 95:10038–10043, 1998a.

Hayward, B. E., Moran, V., Strain, L. and Bonthron, D. T. Bidirectional imprinting of a single gene: GNAS1 encodes maternally, paternally, and biallelically derived proteins. *Proc Natl Acad Sci USA* 95:15475–15480, 1998b.

Holländer, G. A., Zuklys, S., Morel, C., Mizoguchi, E., Mobisson, K., Simpson, S., Terhorst C., et al. Monoallelic expression of the interleukin-2 locus. *Science* 279:2118–2121, 1998.

Jacobs, P. A., Wilson, C. M., Sprenkle, J. A., Rosensheim, N. B. and Migeon, B. R. Mechanism of origin of complete hydatidiform moles. *Nature* 286:714–716, 1980.

Jay, P., Rougeulle, C., Massacrier, A., Moncla, A., Mattei, M. G., Malzac, P., Roeckel, N., et al. The human necdin gene, NDN, is maternally imprinted and located in the Prader-Willi syndrome chromosomal region. *Nature Genet* 17:357–361, 1997.

Jinno, Y., Ikeda, Y., Yun, K., Maw, M., Masuzaki, H., Fukuda, H., Inuzuka, K., et al. Establishment of functional imprinting of the H19 gene in human developing placentae. *Nature Genet* 10:318–324, 1995.

Johnson, D. K., Stubbs, L. J., Culiat, C. T., Montgomery, C. S., Russell, L. B. and Rinchik, E. M. Molecular analysis of 36 mutations at the mouse pink- eyed dilution (p) locus. *Genetics* 141:1563–1571, 1995.

Johnson, D. R. Hairpin-Tail: a case of post-reductional gene action in the mouse egg? *Genetics* 76:795–805, 1974.

Jong, M. T. C., Carey, A. H., Stewart, C. L., Rinchik, E. M., Glenn, C. C., Driscoll, D. J. and Nicholls, R. D. The Znf127 gene encodes a novel C_3HC_4 zinc finger protein and its expression is regulated by genomic imprinting. *Am J Hum Genet* 53:697, 1993.

Kagitani, F., Kuroiwa, Y., Wakana, S., Shiroishi, T., Miyoshi, N., Kobayashi, S., Nishida, M., et al. Peg5/Neuronatin is an imprinted gene located on sub-distal chromosome 2 in the mouse. *Nucleic Acids Res* 25:3428–3432, 1997.

Kaneko-Ishino, T., Kuroiwa, Y., Miyoshi, N., Kohda, T., Suzuki, R., Yokoyama, M., Viville, S., et al. Pegl/Mest imprinted gene on chromosome 6 identified by cDNA subtraction hybridization. *Nature Genet* 11:52–59, 1995.

Kato, M. V., Shimizu, T., Nagayoshi, M., Kaneko, A., Sasaki, M. S. and Ikawa, Y. Genomic imprinting of the human serotonin-receptor (HTR2) gene involved in development of retinoblastoma. *Am J Hum Genet* 59:1084–1090, 1996.

Kaufman, M. H., Barton, S. C. and Surani, M. A. Normal postimplantation development of mouse parthenogenetic embryos to the forelimb bud stage. *Nature* 265:53–55, 1997.

Kay, G. F., Barton, S. C., Surani, M. A. and Rastan, S. Imprinting and X chromosome counting mechanisms determine Xist expression in early mouse development. *Cell* 77:639–650, 1994.

Kelsey, G., Bodle, D., Miller, H. J., Beechey, C. V., Coombes, C., Peters, J. and Williamson, C. M. Identification of imprinted loci by methylation-sensitive representational difference analysis: application to mouse distal chromosome 2. *Genomics* 62:129–138, 1999.

Kim, J., Lu, X. and Stubbs, L. Zim1, a maternally expressed mouse Kruppel-type zinc-finger gene located in proximal chromosome 7. *Hum Molec Genet* 8:847–854, 1999.

Kim, J., Noskov, V. N., Lu, X., Bergmann, A., Ren, X., Warth, T., Richardson, P., et al. Discovery of a novel, paternally expressed ubiquitin-specific processing protease gene through comparative analysis of an imprinted region of mouse chromosome 7 and human chromosome 19q13.4. *Genome Res* 10:1138–1147, 2000.

Kishino, T., Lalanded, M. and Wagstaff, J. UBE3A/E6-AP mutations cause Angelman syndrome. *Nature Genet* 15:70–73, 1997.

Kitagawa, K., Wang, X., Hatada, I., Yamaoka, T., Nojima, H., Inazawa, J., Abe, T., et al. Isolation and mapping of human homologs of an imprinted mouse gene U2af1-rs1. *Genomics* 30:257–263, 1995.

Kitsberg, D., Selig, S., Brandeis, M., Simon, I., Keshet, I., Driscoll, D. J., Nicholls, R. D., et al. Allele-specific replication timing of imprinted gene regions. *Nature* 364:459–463, 1993.

Kuroiwa, Y., Kaneko-Ishino, T., Kagitani, F., Kohda, T., Li, L.-L., Tada, M., Suzuki, R., et al. Peg3 imprinted gene on proximal chromosome 7 encodes for zinc finger protein. *Nature Genet* 12:186–190, 1996.

LaSalle, J. M. and Lalande, M. Domain organization of allele-specific replication within the GABRB3 gene-cluster requires a biparental 15q11-13 contribution. *Nature Genet* 9:386–394, 1995.

Lau, M. M. H., Stewart, C. E. H., Liu, Z. Y., Bhatt, H., Rotwein, P. and Stewart, C. L. Loss of the imprinted Igf2/cation-independent mannose 6-phosphate receptor results in fetal overgrowth and perinatal lethality. *Genes Dev* 8:2953–2963, 1994.

Ledbetter, D. H. and Engel, E. Uniparental disomy in humans—development of an imprinting map and its implications for prenatal diagnosis. *Hum Molec Genet* 4:1757–1764, 1995.

Lee, M. P. Human KVLQT1 gene shows tissue-specific imprinting and encompasses the Beckwith-Wiedemann syndrome chromosomal rearrangements. *Nature Genet* 15:181–185, 1997.

Lee, M. P., Brandenburg, S., Landes, G. M., Adams, M., Miller, G. and Feinberg, A. P. Two novel genes in the center of the 11p15 imprinted domain escape genomic imprinting. *Hum Molec Genet* 8:683–690, 1999a.

Lee, M. P., DeBaun, M. R., Mitsuya, K., Galonek, H. L., Brandenburg, S., Oshimura, M. and Feinberg, A. P. Loss of imprinting of a paternally expressed transcript, with antisense orientation to KVLQT1, occurs frequently in Beckwith-Wiedemann syndrome and is independent of insulin-like growth factor II imprinting. *Proc Natl Acad Sci USA* 96:5203–5208, 1999b.

Lee, S. and Wevrick, R. Identification of novel imprinted transcripts in the Prader-Willi syndrome and Angelman syndrome deletion region: further evidence for regional imprinting control. *Am J Hum Genet* 66:848–858, 2000.

Lee, Y. J., Park, C. W., Hahn, Y., Park, J., Lee, J., Yun, J. H., Hyun, B., et al. Mit1/Lb9 and Copg2, new members of mouse imprinted genes closely linked to Peg1/Mest(1). *FEBS Lett* 472:230–234, 2000.

Leff, S. E., Brannan, C. I., Reed, M. L., Ozcelik, T., Francke, U., Copeland, N. G. and Jenkins, N. A. Maternal imprinting of the mouse Snrpn gene and conserved linkage homology with the human Prader-Willi syndrome region. *Nature Genet* 2:259–264, 1992.

Leighton, P. A., Ingram, R. S., Eggenschwiler, J., Efstratiadis, A. and Tilghman, S. M. Disruption of imprinting caused by deletion of the H19 gene region in mice. *Nature* 375:34–39, 1995a.

Leighton, P. A., Saam, J. R., Ingram, R. S., Stewart, C. L. and Tilghman, S. M. An enhancer deletion affects both H19 and Igf2 expression. *Genes Dev* 9:2079–2089, 1995b.

Li, E., Beard, C. and Jaenisch, R. Role for DNA methylation in genomic imprinting. *Nature* 366:362–365, 1993.

Lyle, R., Watanabe, D., te Vruchte, D., Lerchner, W., Smrzka, O. W., Wutz, A., Schageman, J., et al. The imprinted antisense RNA at the Igf2r locus overlaps but does not imprint Mas1. *Nature Genet* 25:19–21, 2000.

Lyon, M. F. X-chromosome inactivation. *Curr Biol* 9:R235–R237, 1999.

Malik, K., Salpekar, A., Hancock, A., Moorwood, K., Jackson, S., Charles, A. and Brown, K. W. Identification of differential methylation of the WT1 antisense regulatory region and relaxation of imprinting in Wilms' tumor. *Cancer Res* 60:2356–2360, 2000.

Malzac, P., Webber, H., Moncla, A., Graham, J. M., Kukolich, M., Williams, C., Pagon, R. A., et al. Mutation analysis of UBE3A in Angelman syndrome patients. *Am J Hum Genet* 62:1353–1360, 1998.

Mann, J. R., Gadi, I., Harbison, M. L., Abbondanzo, S. J., Stewart and C. L. Androgenetic mouse embryonic stem-cells are pluripotent and cause skeletal defects in chimeras—implications for genetic imprinting. *Cell* 62:251–260, 1990.

Matsuoka, S., Thompson, J. S., Edwards, M. C., Barletta, J. M., Grundy, P., Kalikin, L. M., Harper, J. W. et al. Imprinting of the gene encoding a human cyclin-dependent kinase inhibitor, p57KIP2, on chromosome 11p15. *Proceedings of the National Academy of Science USA* 93:3026–3030, 1996.

Matsuura, T., Sutcliffe, J. S., Fang, P., Galjaard, R.-J., Jiang, Y.-h., Benton, C. S., Rommens, J. M., et al. *De novo* truncation mutations in E6-AP ubiquitin-protein ligase gene (UBE3A) in Angelman syndrome. *Nature Genet* 15:74–77, 1997.

McGrath, J. and Solter, D. Completion of mouse embryogenesis requires both the maternal and paternal genomes. *Cell* 37:179–183, 1984a.

McGrath, J. and Solter, D. Maternal Thp lethality in the mouse is nuclear, not cytoplasmic, defect. *Nature* 308:550–551, 1984b.

Miyoshi, N., Kuroiwa, Y., Kohda, T., Shitara, H., Yonekawa, H., Kawabe, T., Hasegawa, H., et al. Identification of the Meg1/Grb10 imprinted gene on mouse proximal chromosome 11, a candidate for the Silver-Russell syndrome gene. *Proc Natl Acad Sci USA* 95:1102–1107, 1998.

Miyoshi, N., Wagatsuma, H., Wakana, S., Shiroishi, T., Nomura, M., Aisaka, K., Kohda, T., et al. Identification of an imprinted gene, Meg3/Gtl2 and its human homologue MEG3, first mapped on mouse distal chromosome 12 and human chromosome 14q. *Genes Cells* 5:211–220, 2000.

Nakao, M., Sutcliffe, J. S., Durtschi, B., Mutirangura, A., Ledbetter, D. H. and Beaudet, A. L. Impriting analysis of three genes in the Prader- Willi/Angelman region: SNRPN, E-6 associated protein, and PAR-2 (D15S225E). *Hum Molec Genet* 3:309–315, 1994.

Neyroud, N., Tesson, F., Denjoy, I., Leibovici, M., Donger, C., Barhanin, J., Fauré, S., et al. A novel mutation in the potassium channel gene KVLQT1 causes the Jervell and Lange-Nielsen cardioauditory syndrome. *Nature Genet* 15:186–189, 1997.

Nicholls, R. D. Incriminating gene suspects, Prader-Willi style. *Nature Genet* 23:132–134, 1999.

Nishita, Y., Yoshida, I., Sado, T. and Takagi, N. Genomic imprinting and chromosomal location of the human MEST gene. *Genomics* 36:539–542, 1996.

Nutt, S. L., Vambrie, S., Steinlein, P., Kozmik, Z., Rolink, A., Weith, A. and Busslinger, M. Independent regulation of the two Pax5 alleles during B- cell development. *Nature Genet* 21:390–395, 1999.

Ohlsson, R., Nyström, A., Pfeifer-Ohlsson, S., Töhönen, V., Hedborg, F., Schofield, P., Flam, F., et al. IGF2 is parentally imprinted during human embryogenesis and in the Beckwith-Wiedemann syndrome. *Nature Genet* 4:94–97, 1993.

Okamura, K., Hagiwara-Takeuchi, Y., Li, T., Vu, T. H., Hirai, M., Hattori, M., Sakaki, Y., et al. Comparative genome analysis of the mouse imprinted gene impact and its nonimprinted human homolog IMPACT: toward the structural basis for species-specific imprinting [In Process Citation]. *Genome Res* 10:1878–1889, 2000.

Okutsu, T., Kuroiwa, Y., Kagitani, F., Kai, M., Aisaka, K., Tsutsumi, O., Kaneko, Y., et al. Expression and imprinting status of human PEG8/IGF2AS, a paternally expressed antisense transcript from the IGF2 locus, in Wilms' tumors. *J Biochem (Tokyo)* 127:475–483, 2000.

Onyango, P., Miller, W., Lehoczky, J., Leung, C. T., Birren, B., Wheelan, S., Dewar, K., et al. Sequence and comparative analysis of the mouse 1-megabase region orthologous to the human 11p15 imprinted domain [In Process Citation]. *Genome Res* 10:1697–1710, 2000.

Paldi, A., Gyapay, G. and Jami, J. Imprinted chromosomal regions of the human genome display sex-specific meiotic recombination frequencies. *Current Biology* 5:1030–1035, 1995.

Pearsall, R. S., Shibata, H., Brozowska, A., Yoshino, K., Okuda, K., deJong, P. J., Plass, C., et al. Absence of imprinting in U2AFBPL, a human homologue of the imprinted mouse gene U2afbp-rs. *Biochem Biophys Res Comm* 222:171–177, 1996.

Peters, J., Wroe, S. F., Wells, C. A., Miller, H. J., Bodle, D., Beechey, C. V., Williamson, C. M., et al. A cluster of oppositely imprinted transcripts at the Gnas locus in the distal imprinting region of mouse chromosome 2. *Proc Natl Acad Sci USA* 96:3830–3835, 1999.

Piras, G., El Kharroubi, A., Kozlov, S., Escalante-Alcalde, D., Hernandez, L., Copeland, N. G., Gilbert, D. J., et al. Zac1 (Lot1), a potential tumor suppressor gene, and the gene for epsilon-sarcoglycan are maternally imprinted genes: identification by a subtractive screen of novel uniparental fibroblast lines. *Mol Cell Biol* 20:3308–3315, 2000.

Plass, C., Shibata, H., Kalcheva, I., Mullins, L., Kotelevtseva, N., Mullins, J., Kato, R., et al. Identification of Grfl on mouse chromosome 9 as an imprinted gene by RLGS-M. *Nature Genet* 14:106–109, 1996.

Rainier, S., Johnson, L. A., Dobry, C. J., Ping, A. J., Grundy, P. E. and Feinberg, A. P. Relaxation of imprinted genes in human cancer. *Nature* 362:747–749, 1993.

Reed, M. L. and Leff, S. E. Maternal imprinting of human SNRPN, a gene deleted in Prader-Willi syndrome. *Nature Genet* 6:163–167, 1994.

Rivèire, I., Sunshine, M. J. and Littman, D. R. Regulation of IL-4 expression by activation of individual alleles. *Immunity* 9:217–228, 1998.

Rougeulle, C., Cardoso, C., Fontés, M., Colleaux, L. and Lalande, M. An imprinted antisense RNA overlaps UBE3A and a second maternally expressed transcript. *Nature Genet* 19:15–16, 1998.

Saitoh, S., Buiting, K., Rogan, P. K., Buxton, J. L., Driscoll, D. J., Arnemann, J., Konig, R., et al. Minimal definition of the imprinting center and fixation of a chromosome 15q11-q13 epigenotype by imprinting mutations. *Proceedings of the National Academy of Sciences USA* 93:7811–7815, 1996.

Schmidt, J. V., Matteson, P. G., Jones, B. K., Guan, X. J. and Tilghman, S. M. The Dlk1 and Gtl2 genes are linked and reciprocally imprinted. *Genes Dev* 14:1997–2002, 2000.

Schwienbacher, C., Sabbioni, S., Campi, M., Veronese, A., Bernardi, G., Menegatti, A., Hatada, I., et al. Transcriptional map of 170-kb region at chromosome 11p15.5: identification and mutational analysis of the BWR1A gene reveals the presence of mutations in tumor samples. *Proc Natl Acad Sci USA* 95:3873–3878, 1998.

Shemer, R., Birger, Y., Riggs, A. D. and Razin, A. Structure of the imprinted mouse Snrpn gene and establishment of its parental-specific methylation pattern. *Proceedings of the National Academy of Science USA* 94:10267–10272, 1997.

Simon, I., Tenzen, T., Reubinoff, B. E., Hillman, D., McCarrey, J. R. and Cedar, H. Asynchronous replication of imprinted genes is established in the gametes and maintained during development. *Nature* 401:929–932, 1999.

Smrzka, O. W., Fae, I., Stöger, R., Kurzbauer, R., Fischer, G. F., Henn, T., Weith, A., et al. Conservation of a maternal-specific methylation signal at the human IGF2R locus. *Hum Molec Genet* 4:1945–1952, 1995.

Stöger, R., Kubicka, P., Liu, C. G., Kafri, T., Razin, A., Cedar, H. and Barlow, D. P. Maternal-specific methylation of the imprinted mouse Igf2r locus identifies the expressed locus as carrying the imprinting signal. *Cell* 73:61–71, 1993.

Surani, M. A. H. and Barton, S. C. Development of gynogenetic eggs in the mouse—implications for parthenogenetic embryos. *Science* 222:1034–1036, 1983.

Surani, M. A. H., Barton, S. C. and Norris, M. L. Development of reconstituted mouse eggs suggests imprinting of the genome during gametogenesis. *Nature* 308:548–550, 1984.

Surani, M. A. H., Barton, S. C. and Norris, M. L. Nuclear transplantation in the mouse: heritable differences between parental genomes after activation of the embryonic genome. *Cell* 45:127–136, 1986.

Surani, M. A. H., Barton, S. C. and Norris, M. L. Influence of parental chromosomes on spatial specificity in androgenetic parthenogenetic chimeras in the mouse. *Nature* 326:395–397, 1987.

Surti, U., Hoffner, L., Chakravarti, A. and Ferrell, R. E. Genetics and biology of human ovarian teratomas1. Cytogenetic analysis and mechanism of origin. *Am J Hum Genet* 47:635–643, 1990.

Sutcliffe, J. S., Nakao, M., Christian, S., örstavik, K. H., Tommerup, N., Ledbetter, D. H. and Beaudet, A. L. Deletions of a differentially methylated CpG island at the SNRPN gene define a putative imprinting control region. *Nature Genet* 8:52–58, 1994.

Tsai, T. F., Armstrong, D. and Beaudet, A. L. Necdin-deficient mice do not show lethality or the obesity and infertility of Prader-Willi syndrome [letter]. *Nature Genet* 22:15–16, 1999a.

Tsai, T. F., Jiang, Y. H., Bressler, J., Armstrong, D. and Beaudet, A. L. Paternal deletion from Snrpn to Ube3a in the mouse causes hypotonia, growth retardation and partial lethality and provides evidence for a gene contributing to Prader-Willi syndrome. *Hum Molec Genet* 8:1357–1364, 1999b.

Vu, T. H. and Hoffman, A. R. Imprinting of the Angelman syndrome gene, UBE3A, is restricted to brain [letter]. *Nat Genet* 17:12–13, 1997.

Wang, Z. Q., Fun, M. R., Barlow, D. P. and Wagner, E. F. Regulation of embryonic growth and lysosomal targeting by the imprinted Igf2/Mpr gene. *Nature* 372:464–467, 1994.

Wevrick, R. and Francke, U. An imprinted mouse transcript homologous to the human imprinted in Prader-Willi syndrome (IPW) gene. *Hum Molec Genet* 6:325–332, 1997.

Wevrick, R., Kerns, J. A. and Francke, U. Identification of a novel paternally expressed gene in the Prader-Willi syndrome region. *Hum Molec Genet* 3:1877–1882, 1994.

Wroe, S. F., Kelsey, G., Skinner, J. A., Bodle, D., Ball, S. T., Beechey, C. V., Peters, J., et al. An imprinted transcript, antisense to Nesp, adds complexity to the cluster of imprinted genes at the mouse Gnas locus. *Proc Natl Acad Sci USA* 97:3342–3346, 2000.

Wutz, A., Smrzka, O. W., Schweifer, N., Schellander, K., Wagner, E. F. and Barlow, D. P. Imprinted expression of the Igf2r gene depends on an intronic CpG island. *Nature* 389:745–749, 1997.

Wylie, A. A., Murphy, S. K., Orton, T. C. and Jirtle, R. L. Novel imprinted DLK1/GTL2 domain on human chromosome 14 contains motifs that mimic those implicated in IGF2/H19 regulation [In Process Citation]. *Genome Res* 10:1711–1718, 2000.

Yamasaki, K., Hayashida, S., Miura, K., Masuzaki, H., Ishimaru, T., Niikawa, N. and Kishino, T. The novel gene, gamma2-COP (COPG2), in the 7q32 imprinted domain escapes genomic imprinting [In Process Citation]. *Genomics* 68:330–335, 2000.

Yang, T., Adamson, T. E., Resnick, J. L., Leff, S., Wevrick, R., Francke, U., Jenkins, N. A., et al. A mouse model for Prader-Willi syndrome imprinting- centre mutations. *Nature Genet* 19:25–31, 1998.

Yoshihashi, H., Maeyama, K., Kosaki, R., Ogata, T., Tsukahara, M., Goto, Y., Hata, J., et al. Imprinting of human GRB10 and its mutations in two patients with Russell-Silver syndrome. *Am J Hum Genet* 67:476–482, 2000.

Yu, S., Yu, D., Lee, E., Eckhaus, M., Lee, R., Corria, Z., Accili, D., et al. Variable and tissue-specific hormone resistance in heterotrimeric Gs protein alpha-subunit (Gsalpha) knockout mice is due to tissue-specific imprinting of the gsalpha gene. *Proc Natl Acad Sci USA*95:8715–8720, 1998.

Yu, Y., Xu, F., Peng, H., Fang, X., Zhao, S., Li, Y., Cuevas, B., et al. NOEY2 (ARHI), an imprinted putative tumor suppressor gene in ovarian and breast carcinomas. *Proc Natl Acad Sci USA* 96:214–219, 1999.

Chapter 11

Epilogue of an Unfinished Story

The important thing is not to stop questioning.

Albert Einstein (1879–1955)

The discovery of the fascinating phenomena of uniparental disomy (UPD) and genomic imprinting generated a plethora of important biological questions that need to be addressed in the next few years. The answers to these questions will enhance our understanding of the importance of parental-specific gene expression, its contribution to normal development, and association with disease phenotypes.

Clearly, the UPD and genomic imprinting is an unfolding and thus unfinished story and much more fun in understanding both lies ahead in the years to come. The list of yet unanswered questions is long and some of these questions are presented below. This list is by no means exhaustive and only represents the biased views of the authors.

UPD and Human Disorders

- How many genes (and which ones) are imprinted from the maternal and the paternal genome?
- How many additional disorders are due to UPD or imprinted gene dysfunction?
- What is the role of segmental UPD in somatic cells? What is the spectrum of human disorders due to segmental UPD?
- What is the involvement of UPD in embryonic lethality?
- What is the involvement of UPD in male or female sterility (subfertility)?

The Genetic Control of UPD and Imprinting

- How exactly is parental-specific gene expression generated in the germ cells and maintained in the somatic cells?
- Are there *cis*-genomic sequences that distinguish/mark an imprinted gene or an imprinted genomic region?
- What is the biochemical basis of gene silencing in the imprinted chromosomal regions?
- How is the imprinting mechanism regulated? What are the different levels of control?
- Which are the proteins involved in the recognition of imprinted genes and regulate their expression?

Treatment of UPD/Imprinting-Related Phenotypes

- Can we treat abnormalities of imprinting? Could we imagine to exogenously regulate the paternal or maternal expression of a given gene?

Evolutionary Significance

- What is the evolutionary significance of parental imprinting?

The completion of the initial phase of the sequence of the entire human genome (International Human Genome Sequencing Consortium, 2001; Venter and et al., 2001) and the identification of all genes will greatly enhance the knowledge on UPD and imprinting. It is anticipated that we will soon identify not only the complete set of genes with parental-specific gene expression, but also the sequences necessary for the regulation of imprinting. The discovery of large numbers of DNA polymorphisms of the human genome, both SNPs (single nucleotide polymorphisms) (The International SNP Map Working Group, 2001; Mullikin et al., 2000; Deutsch et al., 2001) and SSRs (short-sequence repeats) (NIH/CEPH Collaborative Mapping Group, 1992), that could distinguish chromosomal or protein alleles will enhance our chances to identify additional disorders with full or segmental UPD. Furthermore, the sequence of the mouse chromosomes and the identification of murine homologues of the human imprinted genes [see, e.g., (Onyango et al., 2000)] will provide more experimental opportunities for *in vivo* testing of hypotheses in embryos that could not be done in humans. The advances in methods of comparison of global gene expression (Brown and Botstein, 1999; Velculescu et al., 1995) will also facilitate the study of imprinted genes in cells, tissues, and organs at different developmental stages in health and disease. Finally, the methods for global analysis of proteins (Pandey and Mann, 2000) would assist in the characterization of gene products of imprinted genes.

The mouse model will continue to be an excellent experimental organism to study UPD and imprinting (see Chapter 10) and many more transgenic and knock-out mice will be developed in the years to come that would undoubtedly elucidate many processes and answer many questions. But, what are the differences (in terms of imprinting) between us and the mouse? Could we study all the human pathologies using the mouse model or do we need to invest in human embryo research to understand aspects of human disorders that we cannot study in the mouse?

The sequence of the genomes of other laboratory model organisms, such as the fruitfly *Drosophila* (Adams et al., 2000) and the *C. Elegans* worm (The *C. Elegans* Sequencing Consortium, 1998), also provides an opportunity to study the function of the human homologues of the genes with parental-specific expression. An obvious question is that of the significance of imprinting in these organisms. If such phenomenon exists in these model organisms, then the laboratory investigation of many imprinting-related questions may be done in these species.

Below we enumerate some future topics of research of clinical significance.

ARE THERE LETHAL UPDs?

It is likely that some of the UPDs so far undetected are embryonic lethal and therefore selected against. Thirteen years after the description of the first clinical case of UPD, none has been found for human chromosomes 12, 18, and 19 (see Chapter 4). In other instances, only one of the two parental types has been observed (i.e., only the maternal or paternal UPD for a given chromosome); this also raises the issue of viability as a function of the parental source of UPD. To mention but one example, the somatic segmental UPD11p15.5 was only observed for the paternally derived chromosome 11; the maternal segmental UPD11p15.5 should have been theoretically equally observed. The fact that this is not the case argues for either the lethality of the phenotype, or selective disadvantage of the somatic cells or no phenotype at all.

SEGMENTAL UPDs ARE PERHAPS COMMON AND PATHOGENETICALLY IMPORTANT

The UPD for a segment of a chromosome is likely to occur as a result of the rare mitotic reciprocal somatic crossing-over. Except from the segmental clonal UPD11p mentioned above, such UPDs have been reported for chromosome 14 (Martin et al., 1999) and for either 6p (Lopez-Gutierrez et al., 1998) or part of 6q (Das et al., 2000). The identification of additional examples is not trivial, considering the limited size of the chromosome segments involved. Although the molecular structure may predispose some chromosomal areas to rearrangements (Ji et al., 2000), one would predict that segmental UPD resulting from homologous chromatid pairing and exchanges could be more frequent than currently appreciated.

The role of such somatic segmental UPDs may also become important as one step in the sequential events of genetic changes involved in cancer as already noted in some cases (White et al., 1996; Rousseau-Merck et al., 1999). Loss of heterozygosity is usually the hallmark of such cases.

FREQUENCY OF UPD

The frequency of UPD is currently unknown and may remain unknown for some time. Biases toward the assessment of this frequency include (at opposite ends of the spectrum) nonviability for some UPDs and lack of phenotypic expression for others. One of the authors (Engel, 1998) estimated a frequency of about 1/33,000 births for the cases of meiotic origin UPD (of chromosomes 15 and 6) and of somatic origin (UPD11p, in WBS). This estimated frequency only includes 4 of the 33 types of UPD currently identified, so that one would guess a UPD of some type may occur in the range of 1 in 3000–5000 viable births. More studies including large series of newborn babies, spontaneously aborted fetuses, pregnancies of mothers of advanced age (Ginsburg et al., 2000), and targeted phenotypic groups (i.e., people with handicaps, cognitive impairment, cancers) may provide a more accurate estimate of UPD frequency in health and disease.

ELUCIDATION OF MECHANISMS INVOLVED IN UPD AND ITS "CORRECTION"

It seems reasonable to infer that most cases of UPD for entire chromosomes result from two unrelated coincidental errors in meiosis or mitosis. It would be of interest to study potential predispositions for such events due to the nature of chromosomal centromeres, location and frequency of recombination events, or the proteins involved in chromosomal segregation and recombination. In addition, the mechanism of the events for correction of chromosome aneuploidies may reveal fundamental phenomena of gene copy counting and gene product imbalance.

ADDITIONAL DIAGNOSTIC TESTS

The elucidation of the molecular bases of UPDs is likely to provide more diagnostic tests in patients with unrecognized syndromes. For example, the discovery of the imprinted genes on chromosomes 6q, 7, or 14 associated with emerging phenotypes will provide more diagnostic options and tools. The abundance of DNA polymorphisms and the automation of their detection are likely to enhance our ability to recognize segmental and full UPD for any chromosomal region.

NEW UPD SYNDROMES

In recent years, some new syndromes emerged as a result of the detection of UPD (see Chapter 5). Other conditions (such as paternal or maternal UPD14, maternal UPD2 and maternal UPD16) are in the process of being better characterized. Undoubtedly, many surprises and exciting molecular etiologies lie ahead in the elucidation of causes of hereditary or congenital conditions.

IMPLICATION OF UPD IN RECESSIVE DISORDERS

The isodisomy UPD uncovers mutant alleles for recessive disorders (see Table 1, Chapter 4 for a list). In the few instances where recessive disorders have been screened serially to document the parental source of the mutation, a frequency of single mutant allele duplications by UPD (so-called reduction to homozygosity) has been obtained in around one case out of 50 [1/55 in cystic fibrosis (Voss et al., 1989); 2/54 in cartilage hair hypoplasia (Sulisalo et al., 1997); and 1/61 for junctional epidermolysis bullosa (Pulkkinen et al., 1997)]. The genetic counseling of these cases is certainly different from that of the usual recessive inheritance. We anticipate that the chromosomal regions of isodisomy UPD would be extensively studied for homozygosity of polymorphic alleles of all the genes of the region for associations of these alleles with normal or abnormal phenotypic characteristics.

THE EVOLUTIONARY ROLE OF IMPRINTING

Various theories have been proposed for the evolutionary significance of genomic imprinting. A discussion of the evolutionary aspects of imprinting is beyond the scope of this book. The reader is referred to the original literature for a more thorough discussion. Briefly, the following hypotheses have been proposed:

- A parental competition between paternal and maternal genes to balance the developing potential of an offspring (Moore and Haig, 1991)
- A host-defense mechanism akin to methylation of "foreign" DNA methylation (Barlow, 1993)
- A suppressive regulation of chromosomal aneuploidy to control malignant clonal developments in gonadal and somatic tissues (Varmuza and Mann, 1994; Thomas, 1995)
- A preferential gene expression pattern for the conservation of some traits inherited from one parent only (Engel, 1997)

It is apparent that all of the above are only theories and hypotheses. Clearer understanding will require more detailed knowledge of the molecular events of imprinting and its regulation in different species.

We hope that some of the readers of this book will become so fascinated with the parental-specific contribution to gene expression that they, in turn, devote part of their professional life to contributions to this field and advance our knowledge and understanding of this phenomenon. To these people, and those who care for patients with the UPD-imprinting-related disorders, we dedicate this book.

REFERENCES

Adams, M. D., Celniker, S. E., Holt, R. A., et al. The genome sequence of Drosophila melanogaster. *Science* 287:2185–2195, 2000.

Barlow, D. P. Methylation and imprinting: from host defense to gene regulation? *Science* 260:309–310, 1993.

Brown, P. O. and Botstein, D. Exploring the new world of the genome with DNA microarrays. *Nat Genet* 21:33–37, 1999.

Das, S., Lese, C. M., Song, M., et al. Partial Paternal Uniparental Disomy of Chromosome 6 in an Infant with Neonatal Diabetes, Macroglossia, and Craniofacial Abnormalities. *Am J Hum Genet* 67:2000.

Deutsch, S., Iseli, C., Bucher, P., Antonarakis, S. E. and Scott, H. S. A cSNP Map and Database for Human Chromosome 21. *Genome Res* 11:300–307, 2001.

Engel, E. Imprinting is also a mechanism for immediate or delayed hemizygous expression of several uniparental haplotypes selected from the genome of each sex. *Am J Med Genet* 70:95–98, 1997.

Engel, E. Uniparental disomies in unselected populations. *Am J Hum Genet* 63:962–966, 1998.

Ginsburg, C., Fokstuen, S. and Schinzel, A. The contribution of uniparental disomy to congenital development defects in children born to mothers at advanced childbearing age. *Am J Med Genet* 95:454–460, 2000.

International Human Genome Sequencing Consortium. Initial sequencing and analysis of the human genome. *Nature* 409:860–921, 2001.

Ji, Y., Eichler, E. E., Schwartz, S. and Nicholls, R. D. Structure of chromosomal duplicons and their role in mediating human genomic disorders. *Genome Res* 10:597–610, 2000.

Lopez-Gutierrez, A. U., Riba, L., Ordonez-Sanchez, M. L., Ramirez- Jimenez, S., Cerrillo-Hinojosa, M. and Tusie-Luna, M. T. Uniparental disomy for chromosome 6 results in steroid 21-hydroxylase deficiency: evidence of different genetic mechanisms involved in the production of the disease. *J Med Genet* 35:1014–1019, 1998.

Martin, R. A., Sabol, D. W. and Rogan, P. K. Maternal uniparental disomy of chromosome 14 confined to an interstitial segment (14q23-14q24.2) [In Process Citation]. *J Med Genet* 36:633–636, 1999.

Moore, T. and Haig, D. Genomic imprinting in mammalian development: a parental tug-of-war. *Trends Genet* 7:45–49, 1991.

Mullikin, J. C., Hunt, S. E., Cole, C. G., et al. An SNP map of human chromosome 22. *Nature* 407:516–520, 2000.

NIH/CEPH Collaborative Mapping Group. A comprehensive genetic linkage map of the human genome. *Science* 258:67–86, 1992.

Onyango, P., Miller, W., Lehoczky, J., et al. Sequence and comparative analysis of the mouse 1-megabase region orthologous to the human 11p15 imprinted domain. *Genome Res* 10:1697–1710, 2000.

Pandey, A. and Mann, M. Proteomics to study genes and genomes. *Nature* 405:837–846, 2000.

Pulkkinen, L., Bullrich, F., Czarnecki, P., Weiss, L. and Uitto, J. Maternal uniparental disomy of chromosome 1 with reduction to homozygosity of the LAMB3 locus in a patient with Herlitz junctional epidermolysis bullosa. *Am J Hum Genet* 61:611–619, 1997.

Rousseau-Merck, M. F., Versteege, I., Legrand, I., et al. hSNF5/INI1 inactivation is mainly associated with homozygous deletions and mitotic recombinations in rhabdoid tumors. *Cancer Res* 59:3152–3156, 1999.

Sulisalo, T., Makitie, O., Sistonen, P., et al. Uniparental disomy in cartilage-hair hypoplasia. *Eur J Hum Genet* 5:35–42, 1997.

The C.elegans Sequencing Consortium. Genome sequence of the nematode C. elegans: a platform for investigating biology. *Science* 282:2012–2018, 1998.

The International SNP Map Working Group. A map of human genome sequence variation containing 1.42 million single nucleotide polymorphisms. *Nature* 409:928–933, 2001.

Thomas, J. H. Genomic imprinting proposed as a surveillance mechanism for chromosome loss. *Proc Natl Acad Sci USA* 92:480–482, 1995.

Varmuza, S. and Mann, M. Genomic imprinting—defusing the ovarian time bomb [see comments]. *Trends Genet* 10:118–123, 1994.

Velculescu, V. E., Zhang, L., Vogelstein, B. and Kinzler, K. W. Serial analysis of gene expression. *Science* 270:484–487, 1995.

Venter, J. C. and et al. The Sequence of the Human Genome. *Science* 291:1304–1351, 2001.

Voss, R., Ben-Simon, E., Avital, A., et al. Isodisomy of chromosome 7 in a patient with cystic fibrosis: could uniparental disomy be common in humans? *Am J Hum Genet* 45:373–380, 1989.

White, V. A., McNeil, B. K., Thiberville, L. and Horsman, D. E. Acquired homozygosity (isodisomy) of chromosome 3 during clonal evolution of a uveal melanoma: association with morphologic heterogeneity. *Genes Chromosomes Cancer* 15:138–143, 1996.

Index